THE PHYSICS OF CARBON NANOTUBE DEVICES

MICRO & NANO TECHNOLOGIES

Series Editor: Jeremy Ramsden
Professor of Nanotechnology
Microsystems and Nanotechnology Centre, Department of Materials
Cranfield University, United Kingdom

The aim of this book series is to disseminate the latest developments in small scale technologies with a particular emphasis on accessible and practical content. These books will appeal to engineers from industry, academia and government sectors.

For more information about the book series and new book proposals please contact the Publisher, Dr. Nigel Hollingworth, at nhollingworth@williamandrew.com.

http://www.williamandrew.com/MNT

THE PHYSICS OF CARBON NANOTUBE DEVICES

François Léonard

Sandia National Laboratories, Livermore, California

William Andrew

Norwich, NY, USA

ISBN: 978-0-8155-1573-9

Library of Congress Cataloging-in-Publication Data

Leonard, Francois, 1972-
The physics of carbon nanotube devices / Francois Leonard.
p. cm. -- (Micro & nano technologies)
ISBN 978-0-8155-1573-9
1. Electronic apparatus and appliances--Materials. 2. Nanotubes--Electric properties. 3. Nanotubes--Analysls. 4. Microphysics. 5. Fullerenes--Structure. I. Title.
TK7871.15.C35L46 2008
620'.5--dc22
2008026529

Printed and bound by CPI Group (UK) Ltd, Croydon, CR0 4YY
Transferred to Digital Print 2012

10 9 8 7 6 5 4 3 2 1

Published by:
William Andrew Inc.
13 Eaton Avenue
Norwich, NY 13815
1-800-932-7045
www.williamandrew.com

NOTICE

Contents

Series Editor's Preface

There is no doubt that carbon nanotubes (CNT) are one of the more spectacular icons of the nanotechnology revolution. Be as it may that Neanderthal man synthesized them in the smoky depths of his caves, clearly CNT only became a scientific phenomenon after they had been minutely characterized. This book constitutes an admirable compendium of the work done since they were launched onto the world scientific scene.[1] Furthermore, it also includes a very careful and thorough treatment of the fundamental science underlying the phenomenology of carbon nanotubes. The field is still very new and research is burgeoning. Clearly, to make significant advances oneself one needs to have a firm grasp of the fundamentals, and be well aware of what has already been done. The careful reader of this book will acquire both that grasp and the awareness. The book is also a most valuable resource for all those interested in the technological applications of carbon nanotubes, many of which have doubtless not even been thought of as yet. The study of this book should be amply repaid by the gain of clear perceptions of the limitations and, more importantly, of the vast potential of this extremely important sector of nanotechnology.

Jeremy Ramsden
Cranfield University, United Kingdom
April 2008

[1] Readers may wish to refer to the article by Bojan Boskovic, Nanotechnology Perceptions 3 (2007) 141–158, for a meticulous account of the history of CNT.

Preface

Due to their low dimensionality, nanostructures such as quantum dots, nanowires and carbon nanotubes possess unique properties that make them promising candidates for future technology applications. However, to truly harness the potential of nanostructures, it is essential to develop a fundamental understanding of the basic physics that governs their behavior in devices. This is especially true for carbon nanotubes, where, as will be discussed in this book, research has shown that the concepts learned from bulk device physics do not simply carry over to nanotube devices, leading to unusual device operation. For example, the properties of bulk metal/semiconductor contacts are usually dominated by Fermi level pinning; in contrast, the quasi-one-dimensional structure of nanotubes leads to a much weaker effect of Fermi level pinning, allowing for tailoring of contacts by metal selection. Similarly, while strain effects in conventional silicon devices have been associated with mobility enhancements, strain in carbon nanotubes takes an entirely new perspective, with strain-induced bandgap and conductivity changes.

Carbon nanotubes also present a unique opportunity as one of the few systems where atomistic-based modeling may reach the experimental device size, thus in principle allowing the experimental validation of computational approaches and computational device design. While similar approaches are in development for nanoscale silicon devices, the different properties of carbon nanotubes require an entirely separate field of research.

The field of carbon nanotube devices is one that is rapidly evolving. Thus, even as this book is being written, new discoveries are enhancing our understanding of carbon nanotubes. Therefore, this book presents a snapshot of the status of the field at the time of writing, and the reader is encouraged to follow up on selected topics through the ongoing discussions in the scientific literature. The same can be said about the range of topics covered in this book: many interesting areas of carbon nanotube applications are not covered, and in fact, there is a bias towards electronic devices. The author's own personal interests are to blame. Additional topics that may interest the reader include the synthesis of carbon nanotubes, assembly of nanotube devices, the use of carbon nanotubes in composites, thermal properties, and high-frequency applications.

This book presents recent experimental and theoretical work that has highlighted the new physics of carbon nanotube devices. The intent is not to only discuss the fundamental physical aspects of carbon nanotubes or to give an overview of carbon nanotube devices, but to present the physics behind carbon nanotube devices. The book is intended for applied scientists, engineers or technical managers looking for a basic understanding of the properties of carbon nanotubes that affect device applications; it should also be useful to early graduate level students in universities. The book begins in Chapter 1 with an introduction to the atomic and electronic structure of carbon nanotubes, establishing the basic concepts that will be used in later chapters. Chapter 2 discusses the properties of metallic carbon nanotubes for carrying electronic current, where potential applications include interconnects. Concepts of the intrinsic conductance, capacitance and inductance are introduced as well as a discussion of the role of phonon and defect scattering. Chapter 3 addresses the important issue of contacts between metals and carbon nanotubes, focusing on the role of Fermi-level pinning, properties of contacts in carbon nanotube transistors and development of ultrathin oxides for contact insulation. The discussion in Chapter 4 focuses on electronic devices with semiconducting carbon nanotubes as the active elements. The simplest such device, the *p–n* junction is extensively discussed as an example to illustrate the differences between carbon nanotubes and traditional devices. This discussion is followed by an exposé on metal/semiconductor rectifiers where both the metal and the semiconductor are carbon nanotubes. A significant part of this chapter is devoted to carbon nanotube transistors, and to the recent scientific progress aimed at understanding their basic modes of operation. Chapter 5 examines progress in nanoelectromechanical devices such as actuators and resonators, and strain effects on the electronic structure and conductance. The subject of field emission is addressed in Chapter 6, and the emerging area of optoelectronics with carbon nanotubes is discussed in Chapter 7, reviewing the aspects of photoconductivity and electroluminescence. The book concludes with a chapter on chemical and biological sensors using carbon nanotubes.

I would like to acknowledge M. P. Anantram, without whom this book would not have been possible. The book evolved from a review article on the physics of carbon nanotubes that Anant and I co-authored, and his fingerprints can be found in many sections of this book. I wish that he could have joined me in this endeavor and I hope that he will not be disappointed with the manuscript. I also thank Sandia National Laboratories and Art Pontau, for giving me the time and resources to produce this book. In addition, several of my colleagues at Sandia have provided valuable insight, and this preface would not be complete without mentioning them: Diego Kienle, Xinjian Zhou, Alec Talin, Norman Bartelt, Kevin McCarthy, and Alf Morales. I am also indebted to Stefan Heinze, Vasili Perebeinos and Catalin Spataru for providing figures. Finally, to Marie-Josée, from my heart to yours, thank you for being there.

François Léonard
Livermore, CA
April 2008

1 Introduction

Carbon nanotubes are high aspect ratio hollow cylinders with diameters ranging from one to tens of nanometers, and with lengths up to centimeters. As the name implies, carbon nanotubes are composed entirely of carbon, and represent one of many structures that carbon adopts in the solid state. Other forms of solid carbon include for example diamond, graphite, and buckyballs. These many different forms arise because of the ability of carbon to form hybridized orbitals and achieve relatively stable structures with different bonding configurations. Carbon nanotubes exist because of sp^2 hybridization, the same orbital structure that leads to graphite. In this chapter, we discuss the atomic and electronic structure of carbon nanotubes, and establish the basic nanotube properties that will be utilized in the following chapters on nanotube devices.

1.1 Structure of Carbon Nanotubes

To understand the atomic structure of carbon nanotubes, one can imagine taking graphite, as shown in Fig. 1.1, and removing one of the two-dimensional planes, which is called a graphene sheet; a single graphene sheet is shown in Fig. 1.2 (a). A carbon nanotube can be viewed as a strip of graphene (strip in Fig. 1.2) that is rolled-up to form a closed cylinder. The basis vectors $\vec{a}_1 = a(\sqrt{3}, 0)$ and $\vec{a}_2 = a(\sqrt{3}/2, 3/2)$ generate the graphene lattice, where $a = 0.142$ nm is the carbon–carbon bond length. The two atoms marked A and B in the figure are the two atoms in the unit cell of graphene. In cutting the rectangular strip, one defines a "circumferential" vector $\vec{C} = n\vec{a}_1 + m\vec{a}_2$ corresponding to the edge of the graphene strip that will become the nanotube circumference. The nanotube radius is obtained from $\vec{C}$ as

$$R = C/2\pi = \left(\sqrt{3}/2\pi\right) a\sqrt{n^2 + m^2 + nm}. \tag{1.1}$$

There are two special cases shown in Fig. 1.2 that deserve special mention. First, when the circumferential vector lies purely along one of the two basis vectors, the carbon nanotube is said to be of "zigzag" type. The example in Fig. 1.2 (a) shows the generation of a (10,0) zigzag nanotube. Second, when the circumferential vector is along the direction exactly between the two basis vectors, $n = m$, and the carbon nanotube is said to be of "armchair" type. The example in Fig. 1.2 (b) shows the generation of a (5,5) armchair nanotube, whereas in Fig. 1.2 (c) a chiral nanotube is shown where the strip is generated by $m \neq n$.

In a planar graphene sheet, the bonds to the three nearest neighbors of a carbon atom, $\vec{r}_1 = a(0, 1)$, $\vec{r}_2 = a(\sqrt{3}/2, -1/2)$ and $\vec{r}_3 = a(-\sqrt{3}/2, -1/2)$

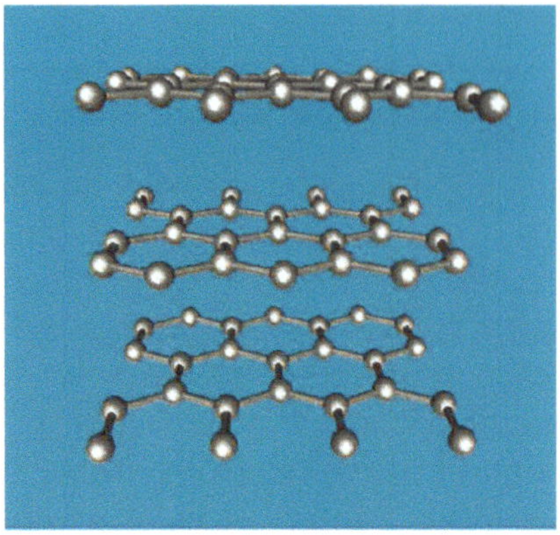

Figure 1.1 Illustration of the graphite structure, showing the parallel stacking of two-dimensional planes, called graphene sheets. Figure from Ref. [1].

(Fig. 1.2 (a)) are equivalent. Rolling up a graphene sheet however causes differences between the three bonds. In the case of zigzag nanotubes the bonds oriented at a nonzero angle to the axis of the cylinder are identical, but different from the axially oriented bonds which remain unaffected upon rolling up the graphene strip. For armchair nanotubes the bonds oriented at a nonzero angle to the circumference of the cylinder are identical, but different from the circumferentially oriented bonds. All three bonds are slightly different for other chiral nanotubes.

We discussed the single wall nanotube, which consists of a single layer of rolled-up graphene strip. Nanotubes, however are found in other closely related forms and shapes as shown in Fig. 1.3. Fig. 1.3 (b) shows a bundle of single wall nanotubes with the carbon nanotubes arranged in a triangular lattice. The individual tubes in the bundle are attracted to their nearest neighbors via van der Waals interactions, with typical distances between nanotubes being comparable to the interplanar distance of graphite which is 3.1 Å. The cross-section of an individual nanotube in a bundle is circular if the diameter is smaller than 15 Å and deforms to a hexagon as the diameter of the individual tubes increases [2]. A close allotrope of the single wall carbon nanotube is the multi wall carbon nanotube (MWNT), which consists of nested single wall nanotubes, in a Russian doll fashion as shown in Fig. 1.3 (c). Again, the distance between walls of neighboring tubes is comparable to the interplanar distance of graphite. Carbon nanotubes also occur in more complex shapes such as junctions between nanotubes of two different chiralities (Fig. 1.3 (d)) and Y-junctions (Fig. 1.3 (e)) These carbon nanotube junctions are atomically precise in that each carbon atom preserves its sp^2 hybridization and thus makes bonds with its three nearest neighbors without introducing dangling bonds. The curvature needed to create these interesting shapes arises from pentagon–heptagon defects in the hexagonal network.

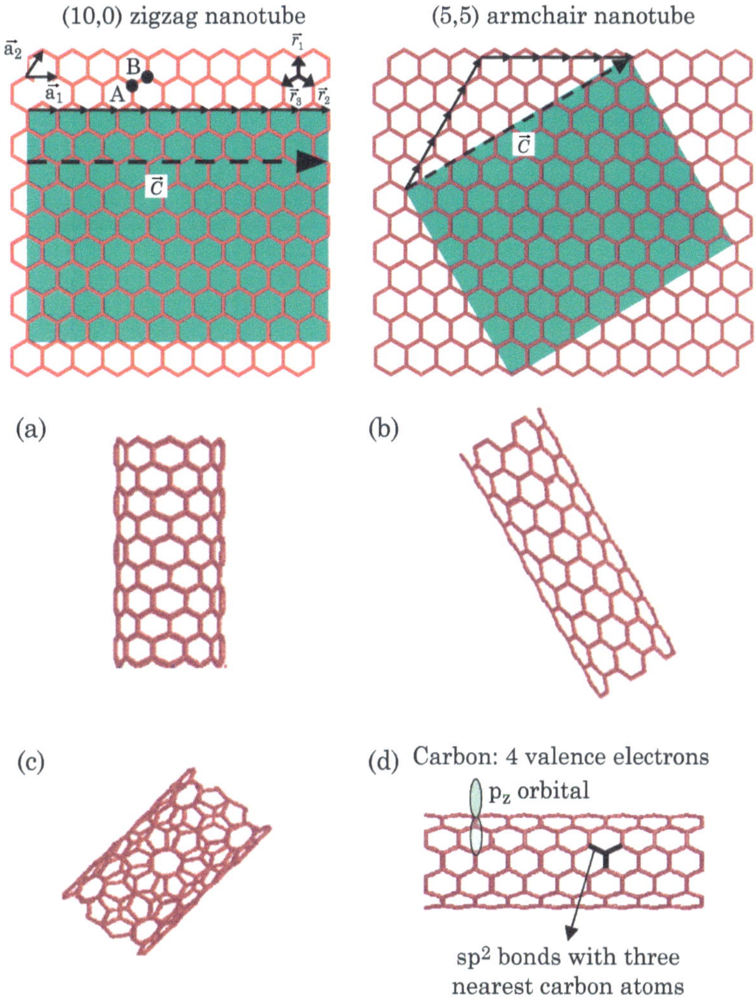

Figure 1.2 Sketch of a graphene sheet and procedure for generating single wall carbon nanotubes. $\vec{a}_1$ and $\vec{a}_2$ denote the lattice vectors of graphene, with $|a_1| = |a_2| = \sqrt{3}a$, where a is the carbon–carbon bond length. There are two atoms per unit cell marked as A and B. Single wall carbon nanotubes are obtained by cutting a strip in the graphene sheet and rolling it up such that each carbon atom is bonded to its three nearest neighbors. The creation of a $(n, 0)$ zigzag nanotube is shown in (a). (b) Creation of a (n, n) armchair nanotube. (c) Chiral nanotube. (d) The bonding structure of a nanotube. Carbon has four valence electrons. Three of these electrons are bonded to nearest neighbor carbon atoms by sp^2 bonding, in a manner similar to graphene. The fourth electron is a nonhybridized p_z orbital perpendicular to the cylindrical surface.

1.2 Electronic Properties of Carbon Nanotubes

Elemental carbon has six electrons with orbital occupancy $1s^2\ 2s^2\ 2p^2$, and thus has four valence electrons. The 2s and 2p orbitals can hybridize to form sp, sp^2 and sp^3 orbitals, and this leads to the different structures that carbon

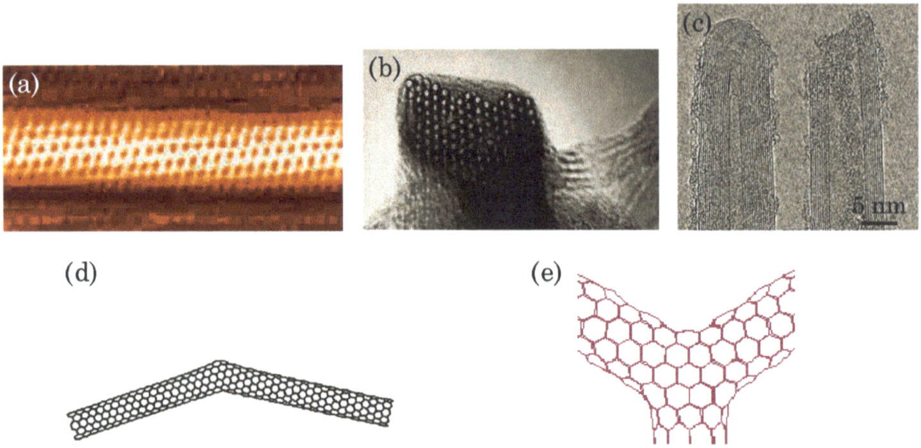

Figure 1.3 Forms of nanotubes: (a) scanning tunneling microscope image of a single wall carbon nanotube. (b) A bundle of single wall nanotubes. (c) Two multiwall nanotubes. (d) Junction between two single wall nanotubes of different chiralities. (e) Y-junction between single wall nanotubes. In (d) and (e), each carbon atom has only three nearest neighbors and there are no dangling bonds despite the presence of the junctions. Figure (a) from C. Dekker, (b) Ref. [3], (c) Ref. [4], (d) Ref. [5], and (e) from J. Han.

materials adopt. For example, the sp hybridization leads to linear carbon molecules, while the sp^3 hybridization gives the diamond structure. The sp^2 hybridization is responsible for the graphene and carbon nanotube structures. In graphene and nanotubes, each carbon atom has three $2sp^2$ electrons and one 2p electron. The three $2sp^2$ electrons form the three bonds in the plane of the graphene sheet (Fig. 1.4), leaving an unsaturated p_z orbital (Fig. 1.2 (d)). This p_z orbital, perpendicular to the graphene sheet and thus the nanotube surface, forms a delocalized π network across the nanotube, which is responsible for its electronic properties. These electronic properties can be well described starting from a tight-binding model for graphene, as we now discuss.

1.2.1 Graphene Electronic Structure

A carbon atom at position $\vec{r}_s$ has an unsaturated p_z orbital described by the wave function $\chi_{\vec{r}_s}(\vec{r})$. In the orthogonal tight-binding representation, the interaction between orbitals on different atoms vanishes unless the atoms are nearest neighbors. With H the Hamiltonian, this can be written as

$$\begin{aligned} \langle \chi_{\vec{r}_A} | H | \chi_{\vec{r}'_A} \rangle &= \langle \chi_{\vec{r}_B} | H | \chi_{\vec{r}'_B} \rangle = 0 \\ \langle \chi_{\vec{r}_A} | H | \chi_{\vec{r}_B} \rangle &= \langle \chi_{\vec{r}_B} | H | \chi_{\vec{r}_A} \rangle = \gamma \delta_{\vec{r}_A - \vec{r}_B = \vec{r}_i} \end{aligned} \tag{1.2}$$

where $\vec{r_i}$ is a vector connecting nearest-neighbors between the A and B sublattices (Fig. 1.2 (a)), and where we have set the on-site interaction energy

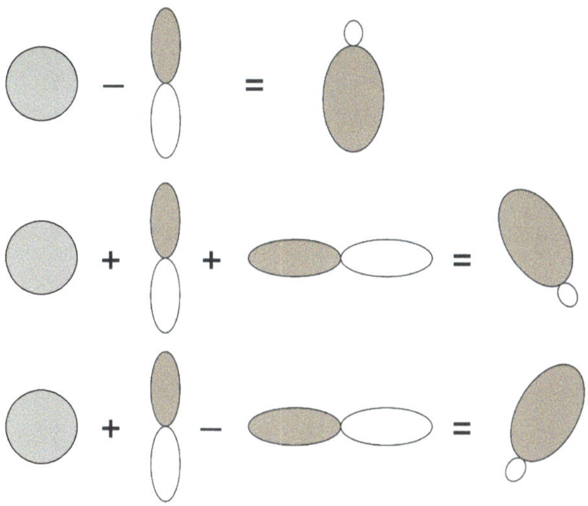

Figure 1.4 Simplified sketch of the combination of s and p orbitals that form the three sp^2 hybridized orbitals in the plane of the graphene sheet. The coefficients for each of the wavefunctions were omitted for simplicity and the final wavefunctions are sketched approximately.

to zero, without loss of generality. The tight-binding parameter γ represents the strength of the nearest-neighbor interactions. The A and B sublattices correspond to the set of all A and B atoms in Fig. 1.2 (a).

To calculate the electronic structure, we construct the Bloch wavefunction for each of the sublattices as

$$\phi_{s\vec{k}}(\vec{r}) = \sum_{\vec{r}_s} e^{i\vec{k}\cdot\vec{r}_s} \chi_{\vec{r}_s}(\vec{r}) \tag{1.3}$$

where $s = \mathrm{A}$ or B refers to each sublattice, and $\vec{r}_s$ refers to the set of points belonging to sublattice s. The total wavefunction is then a linear combination of these two functions,

$$\psi_{\vec{k}}(\vec{r}) = \frac{1}{\sqrt{2}} \left[\phi_{A\vec{k}}(\vec{r}) + \lambda_{\vec{k}} \phi_{B\vec{k}}(\vec{r})\right] \tag{1.4}$$

where $\lambda_{\vec{k}}$ is the mixing parameter to be determined below. The Hamiltonian matrix elements are calculated as follows:

$$\begin{aligned} \langle \phi_{s\vec{k}} | H | \phi_{s'\vec{k}} \rangle &= \int d^3r \sum_{\vec{r}_s} e^{-i\vec{k}\cdot\vec{r}_s} \chi^*_{\vec{r}_s}(\vec{r}) H \sum_{\vec{r}_{s'}} e^{i\vec{k}\cdot\vec{r}_{s'}} \chi_{\vec{r}_{s'}}(\vec{r}) \\ &= \sum_{\vec{r}_s} e^{-i\vec{k}\cdot\vec{r}_s} \sum_{\vec{r}_{s'}} e^{i\vec{k}\cdot\vec{r}_{s'}} \int d^3r \chi^*_{\vec{r}_s}(\vec{r}) H \chi_{\vec{r}_{s'}}(\vec{r}). \end{aligned} \tag{1.5}$$

From Eq. (1.2) we have

$$\int d^3r \chi^*_{\vec{r}_s}(\vec{r}) H \chi_{\vec{r}_{s'}}(\vec{r}) = \gamma \delta_{\vec{r}_s - \vec{r}_{s'} = \vec{r}_i} \tag{1.6}$$

and the Hamiltonian matrix elements are therefore

$$\begin{aligned} \langle \phi_{A\vec{k}} | H | \phi_{A\vec{k}} \rangle &= \langle \phi_{B\vec{k}} | H | \phi_{B\vec{k}} \rangle = 0 \\ \langle \phi_{A\vec{k}} | H | \phi_{B\vec{k}} \rangle &= \langle \phi_{B\vec{k}} | H | \phi_{A\vec{k}} \rangle^* = H_{AB} = \gamma \sum_{j=1,2,3} e^{i\vec{k}\cdot\vec{r}_j}, \end{aligned} \tag{1.7}$$

leading to the Schrödinger equation $H\psi = E\psi$ in matrix form

$$\begin{pmatrix} E & -H_{AB} \\ -H^*_{AB} & E \end{pmatrix} \begin{pmatrix} 1 \\ \lambda_{\vec{k}} \end{pmatrix} = 0. \tag{1.8}$$

Diagonalization of this matrix leads to the solution

$$\begin{aligned} E &= \pm |H_{AB}| \\ &= \pm\gamma \sqrt{1 + 4\cos\left(\frac{3}{2}k_y a\right)\cos\left(\frac{\sqrt{3}}{2}k_x a\right) + 4\cos^2\left(\frac{\sqrt{3}}{2}k_x a\right)}. \end{aligned} \tag{1.9}$$

This band structure for graphene is plotted in Fig. 1.5 as a function of k_x and k_y. The valence and conduction bands meet at six points $[(\pm 4\pi/3\sqrt{3}a, 0); (\pm 2\pi/3\sqrt{3}a, \pm 2\pi/3a)]$ at the corner of the first Brillouin zone. Graphene is thus a peculiar material: bands cross at the Fermi level, but the Fermi surface consists only of points in k-space, and the density of states at these so-called Fermi points vanishes. Graphene can be described as a gapless semiconductor, or a as a semi-metal with zero overlap. Because of the symmetry of the graphene lattice, the Brillouin zone has $2\pi/3$ rotational symmetry, and there are only two nonequivalent Fermi points.

The values of $\lambda_{\vec{k}}$ that correspond to the two branches in Eq. (1.9) are

$$\lambda^+_{\vec{k}} = \frac{H^*_{AB}}{|H_{AB}|} \quad \text{and} \quad \lambda^-_{\vec{k}} = -\frac{H^*_{AB}}{|H_{AB}|}. \tag{1.10}$$

The lower branch, corresponding to lower energies, has opposite sign of the wavefunction on the two atoms of the unit cell while the wavefunction for the upper branch has the same sign on the two atoms of the unit cell. Thus the low energy branch has bonding character while the higher energy branch has antibonding character.

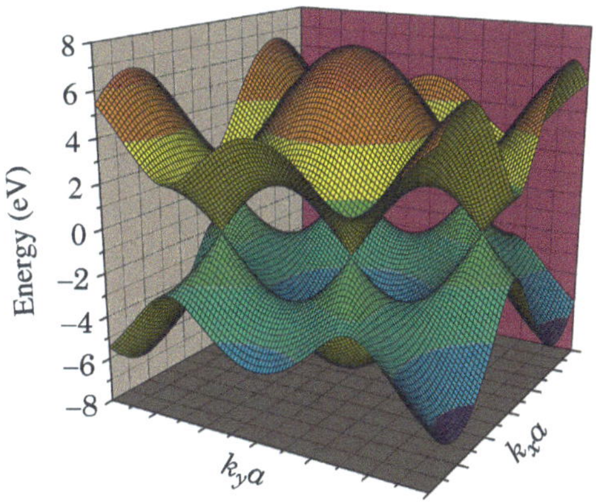

Figure 1.5 Electronic structure of graphene calculated within a tight-binding model with only π electrons. The conduction and valence bands meet at six points at the corner of the first Brillouin zone.

Near the Fermi points, the band structure can be approximated as [6]

$$E \approx \frac{3\gamma a}{2} \left| \vec{k} - \vec{k}_F \right| \tag{1.11}$$

and is thus isotropic and linear. This relation will be useful to obtain the bandgap of semiconducting nanotubes as described in the next section.

1.2.2 Carbon Nanotube Electronic Structure

To obtain the electronic structure of carbon nanotubes, we start from the band structure of graphene and quantize the wavevector in the circumferential direction:

$$\vec{k} \cdot \vec{C} = k_x C_x + k_y C_y = 2\pi p \tag{1.12}$$

where $\vec{C}$, the circumferential vector, is shown in Fig. 1.2 and p is an integer. Eq. (1.12) provides a relation between k_x and k_y defining lines in the (k_x, k_y) plane. Each line gives a one-dimensional energy band by slicing the two-dimensional band structure of graphene shown in Fig. 1.5. The particular values of C_x, C_y and p determine where the lines intersect the graphene band structure, and thus, each nanotube will have a different electronic structure. Perhaps the most important aspect of this construction is that nanotubes can be metallic or semiconducting, depending on whether or not the lines pass through the graphene Fermi points. This concept is illustrated in Fig. 1.6 where the first Brillouin zone of graphene is shown as a shaded hexagon with the Fermi points at the six corners. In the left panel, the lines of quantized circumferential

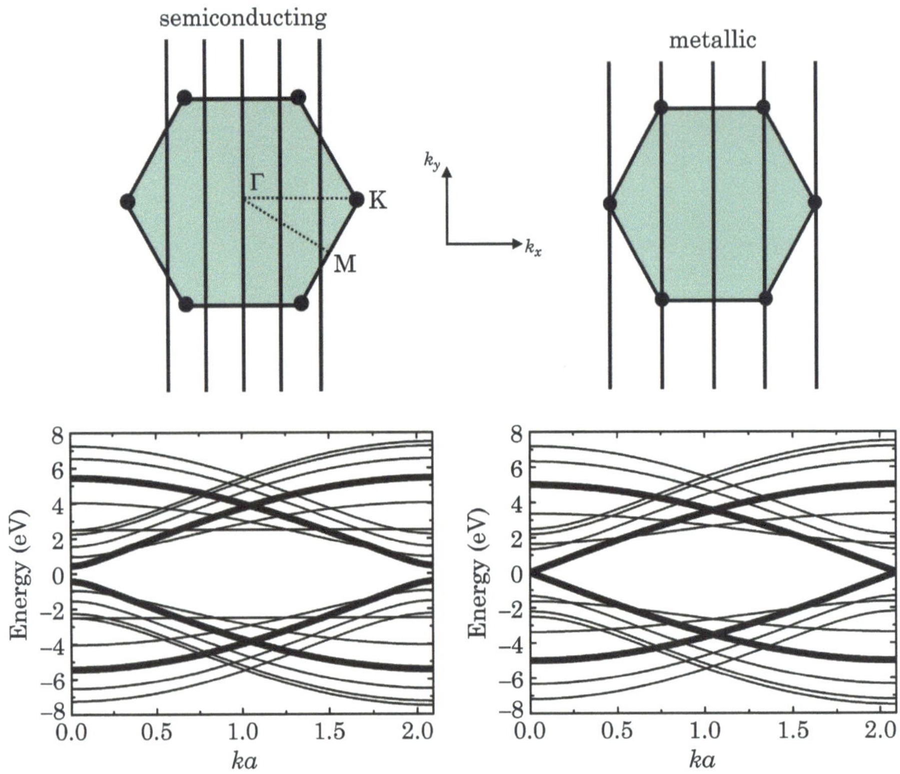

Figure 1.6 Illustration of the first Brillouin zone of graphene, and the allowed wavevector lines leading to semiconducting and metallic nanotubes. Examples of band structures for semiconducting and metallic zigzag nanotubes are displayed at the bottom of the figure. The thick lines indicate the bands that cross or come closest to the Fermi level, taken as the zero of energy in these figures.

wavevectors do not intersect the graphene Fermi points, and the nanotube is semiconducting, with a bandgap determined by the two lines that come closer to the Fermi points. The right panel illustrates a situation where the lines pass through the Fermi points, leading to crossing bands at the nanotube Fermi level, and thus metallic character.

We can express mathematically the electronic band structure of nanotubes by defining components of the wavevector perpendicular and parallel to the tube axis. By expressing k_x and k_y in terms of these components and substituting in Eq. (1.9), we obtain

$$\frac{E^2}{\gamma} = 1 + 4\cos\left(\frac{3C_x ka}{2C} - \frac{3\pi paC_y}{C^2}\right)\cos\left(\frac{\sqrt{3}C_y ka}{2C} + \frac{\sqrt{3}\pi paC_x}{C^2}\right) + 4\cos^2\left(\frac{\sqrt{3}C_y ka}{2C} + \frac{\sqrt{3}\pi paC_x}{C^2}\right) \quad (1.13)$$

where k is the wavevector in the axial direction, $C_x = a\sqrt{3}\,(n + m/2)$ and $C_y = 3am/2$. Band structures for semiconducting and metallic nanotubes computed from this expression are shown in Fig. 1.6. The index p in the above expression takes the values $p = 1, 2, \ldots, N/2$ where N is the number of carbon atoms in the nanotube unit cell. The value of N is given by [7]

$$N = \frac{4\left(m^2 + mn + n^2\right)}{d_R} \tag{1.14}$$

where $d_R = 3q$ if $n - m = 3qI$ and $d_R = q$ if $n - m \neq 3qI$ with I an integer. Here q is the greatest common divisor of n and m. Since each value of p gives two subbands (the positive and negative signs in Eq. (1.9)), the total number of bands is N. Because the nanotube band structure is symmetric about the Fermi level within the orthogonal tight-binding model considered here, half of the bands will be below the Fermi level and half of the bands will be above the Fermi level. And often, many of the bands will be degenerate, so the number of independent bands is less than N. Two special cases are worth mentioning. In the case of zigzag nanotubes$(n, 0)$ we have $q = n$, $d_R = n$ and $N = 4n$. In the case of armchair nanotubes(n, n) we have $q = n$, $d_R = 3n$ and $N = 4n$. The fact that the zigzag and armchair nanotubes have the same number of atoms per unit cell arises because the structure of both of these nanotubes consists of parallel rings of atoms—zigzag nanotubes have n atoms per ring with four rings per unit cell, while armchair nanotubes have $2n$ atoms per ring with two rings per unit cell.

As discussed above, the condition for nanotubes to be metallic is for some of the allowed lines $k_y = \frac{2\pi p}{C_y} - \frac{C_x}{C_y} k_x$ to cross one of the Fermi points of graphene. This leads to the general condition $|n - m| = 3I$ where I is an integer. We now derive this relation, considering each of the two inequivalent Fermi points in turn.

Case 1: $\vec{k}_F = (4\pi/3\sqrt{3}a, 0)$

One of the lines will cross this Fermi point when

$$k_y = \frac{2\pi p}{C_y} - \frac{C_x}{C_y} k_x = \frac{2\pi p}{C_y} - \frac{C_x}{C_y} \frac{4\pi}{3\sqrt{3}a} = 0 \tag{1.15}$$

and using the expressions $C_x = a\sqrt{3}\,(n + m/2)$ and $C_y = 3am/2$ we obtain

$$(2n + m) = 3p \quad \text{or} \quad -n + m = 3\,(p + n) \tag{1.16}$$

and since both p and n are integers, $p + n = I$ is also an integer, and the condition is

$$-n + m = 3I. \tag{1.17}$$

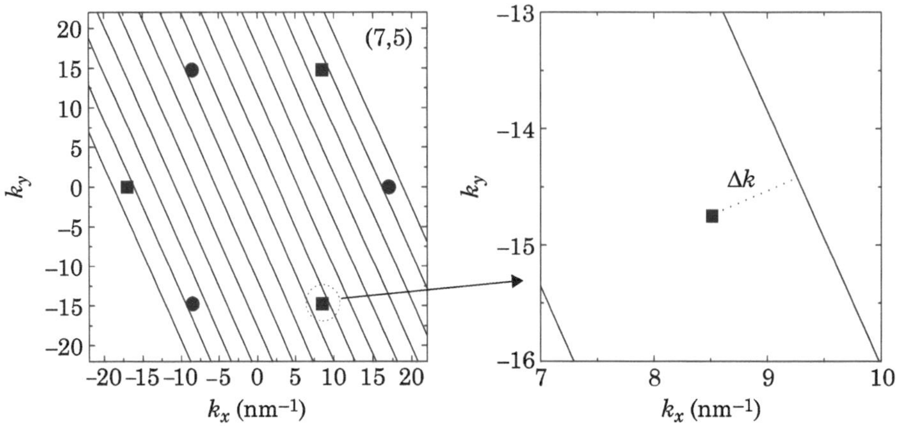

Figure 1.7 The left panel shows the first Brillouin zone of graphene with the two sets of inequivalent Fermi points indicated by circles and squares. The right panel shows the distance Δk between one of the allowed wavevector lines and a Fermi point.

Case 2: $\vec{k}_F = (-4\pi/3\sqrt{3}a, 0)$

We repeat the above analysis to obtain the condition

$$k_y = \frac{2\pi p}{C_y} - \frac{C_x}{C_y}k_x = \frac{2\pi p}{C_y} + \frac{C_x}{C_y}\frac{4\pi}{3\sqrt{3}a} = 0 \tag{1.18}$$

giving

$$-2n - m = 3p \quad \text{or} \quad n - m = 3I. \tag{1.19}$$

The two conditions for the two sets of nonequivalent Fermi points can be summarized as

$$|n - m| = 3I. \tag{1.20}$$

It is worth mentioning that because armchair nanotubes have $n = m$, they always satisfy the condition (1.20) and are therefore metallic.

Nanotubes for which this condition does not hold are semiconducting, and this represents 2/3 of all nanotubes that can be generated using the procedure of Fig. 1.2 (there is an infinite number of such nanotubes). The relationship between bandgap and diameter [8,9] can be obtained by finding the line that comes closest to a graphene Fermi point. The distance between the Fermi points and each line (Fig. 1.7) can be calculated from

$$\Delta k = \left| \left(\vec{k}_F - \vec{k}\right) \cdot \frac{\vec{C}}{\left|\vec{C}\right|} \right|. \tag{1.21}$$

Using the relations $\vec{k} \cdot \vec{C} = 2\pi p$ and $\left|\vec{C}\right| = 2\pi R$ we obtain

$$\Delta k = \frac{1}{R}\left|\frac{\vec{k}_F \cdot \vec{C}}{2\pi} - p\right|. \tag{1.22}$$

There are six cases to consider:

Case 1: $\vec{k}_F = (4\pi/3\sqrt{3}a, 0)$

$$\Delta k = \frac{1}{R}\left|\frac{\vec{k}_F \cdot \vec{C}}{2\pi} - p\right| = \frac{1}{R}\left|\frac{2C_x}{3\sqrt{3}a} - p\right| = \frac{1}{3R}\left|3p - (2n + m)\right|. \tag{1.23}$$

Case 2: $\vec{k}_F = (2\pi/3\sqrt{3}a, 2\pi/3a)$

$$\begin{aligned}\Delta k &= \frac{1}{R}\left|\frac{\vec{k}_F \cdot \vec{C}}{2\pi} - p\right| = \frac{1}{R}\left|\frac{C_x}{3\sqrt{3}a} + \frac{C_y}{3a} - p\right| \\ &= \frac{1}{3R}\left|3p - (n + 2m)\right|.\end{aligned} \tag{1.24}$$

Case 3: $\vec{k}_F = (-2\pi/3\sqrt{3}a, 2\pi/3a)$

$$\begin{aligned}\Delta k &= \frac{1}{R}\left|\frac{\vec{k}_F \cdot \vec{C}}{2\pi} - p\right| = \frac{1}{R}\left|-\frac{C_x}{3\sqrt{3}a} + \frac{C_y}{3a} - p\right| \\ &= \frac{1}{3R}\left|3p + (n - m)\right|.\end{aligned} \tag{1.25}$$

Case 4: $\vec{k}_F = (-4\pi/3\sqrt{3}a, 0)$

$$\Delta k = \frac{1}{R}\left|\frac{\vec{k}_F \cdot \vec{C}}{2\pi} - p\right| = \frac{1}{R}\left|-\frac{2C_x}{3\sqrt{3}a} - p\right| = \frac{1}{3R}\left|3p + (2n + m)\right|. \tag{1.26}$$

Case 5: $\vec{k}_F = (-2\pi/3\sqrt{3}a, -2\pi/3a)$

$$\begin{aligned}\Delta k &= \frac{1}{R}\left|\frac{\vec{k}_F \cdot \vec{C}}{2\pi} - p\right| = \frac{1}{R}\left|-\frac{C_x}{3\sqrt{3}a} - \frac{C_y}{3a} - p\right| \\ &= \frac{1}{3R}\left|3p + (n + 2m)\right|.\end{aligned} \tag{1.27}$$

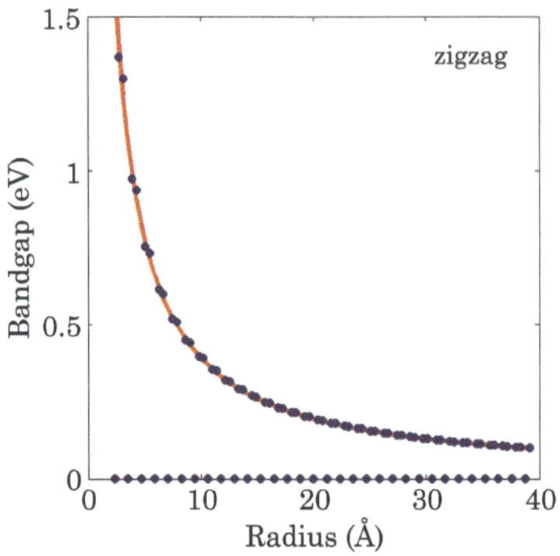

Figure 1.8 Bandgap versus radius for zigzag nanotubes according to Eq. (1.29). The bandgap decreases inversely with increase in diameter. The points with zero bandgap correspond to metallic nanotubes which satisfy $n = 3I$, where I is an integer. Note that when curvature effects are introduced all metallic nanotubes (except armchair) develop a small bandgap. Figure from Ref. [1].

Case 6: $\vec{k}_F = (2\pi/3\sqrt{3}a, -2\pi/3a)$

$$\Delta k = \frac{1}{R}\left|\frac{\vec{k}_F \cdot \vec{C}}{2\pi} - p\right| = \frac{1}{R}\left|\frac{C_x}{3\sqrt{3}a} - \frac{C_y}{3a} - p\right| = \frac{1}{3R}\left|3p - (n-m)\right|. \quad (1.28)$$

Because $|n - m| = 3I \pm 1$ for a semiconducting nanotube, one or more of the above relations will give a value of 1 for the term in absolute values, giving the smallest value for $\Delta k = 1/(3R)$. Using the approximate linear dependence of the graphene band structure near the Fermi points given in Eq. (1.11), we obtain the bandgap

$$E_g \approx 2 \times \frac{3a\gamma}{2}\left|\vec{k} - \vec{k}_F\right| = 3a\gamma\Delta k = \frac{a\gamma}{R} = \frac{2a\gamma}{d} \quad (1.29)$$

where d is the nanotube diameter. The bandgaps of zigzag nanotubes are shown in Fig. 1.8 as a function of their radius.

A particular example helps in understanding how the above relations are used in practice. We consider the (17,0) zigzag nanotube, which has a diameter of 1.32 nm, typical of many nanotubes used experimentally. This nanotube has $|n - m| = 17 \neq 3I$ and is therefore semiconducting. For values of $p = \pm 1, \ldots, \pm 17$ the six cases above lead to the bandgaps:

Cases 1 and 4: $E_g = \frac{a\gamma}{R}\left|3p \pm 34\right|$
The minimum bandgap is thus $E_g = a\gamma/R$ for $p = \pm 11$.

Cases 2, 3, 5 and 6: $E_g = \frac{a\gamma}{R}\left|3p \pm 17\right|$

The minimum bandgap is thus $E_g = a\gamma/R$ for $p = \pm 6$.

Therefore, the bands $p = \pm 6$ and $p = \pm 11$ are degenerate. Note that because each value of p gives one valence and one conduction band, the band degeneracy is one-half of the number of bands that give the same bandgap, i.e. in this case it is equal to 2. Such band degeneracies have important consequences for electronic transport, as will be discussed later.

Having expressions for the wavefunctions of electrons in the valence and conduction bands for carbon nanotubes is useful when calculating material properties. To obtain these wavefunctions, we first consider the matrix elements

$$H_{AB} = \gamma \sum_{j=1,2,3} e^{ik_{\parallel} r_{\parallel}^{j}} e^{ik_{\perp} r_{\perp}^{j}} \tag{1.30}$$

where the subscripts refer to vector components parallel and perpendicular to the nanotube axis. In this coordinate system the circumferential vector is $\vec{C} = (0, C)$ and it makes an angle χ with the reference x–y axis. We express the bond vectors in the new coordinate system by using a rotation of the axes:

$$\begin{aligned}
\vec{r}_1 &= a\,(0,1)\begin{pmatrix} \cos\chi & \sin\chi \\ -\sin\chi & \cos\chi \end{pmatrix} = a\,(-\sin\chi, \cos\chi) \\
\vec{r}_2 &= \frac{a}{2}\left(\sqrt{3}, -1\right)\begin{pmatrix} \cos\chi & \sin\chi \\ -\sin\chi & \cos\chi \end{pmatrix} \\
&= \frac{a}{2}\left(-\sqrt{3}\cos\chi - \sin\chi, -\sqrt{3}\sin\chi + \cos\chi\right) \\
\vec{r}_3 &= -\frac{a}{2}\left(\sqrt{3}, 1\right)\begin{pmatrix} \cos\chi & \sin\chi \\ -\sin\chi & \cos\chi \end{pmatrix} \\
&= -\frac{a}{2}\left(\sqrt{3}\cos\chi - \sin\chi, \sqrt{3}\sin\chi + \cos\chi\right).
\end{aligned} \tag{1.31}$$

Inserting these expressions into Eq. (1.30) and relabeling $k_{\parallel} = k$ we have

$$\begin{aligned}
H_{AB}(J,k) = \gamma e^{-ika\sin\chi} e^{i\frac{Ja}{R}\cos\chi} \Bigg[1 + 2e^{i\frac{ka}{2}\sin\chi} e^{i\frac{Ja}{2R}\cos\chi} \\
\times \cos\left(\frac{\sqrt{3}}{2} ka\cos\chi + \frac{\sqrt{3}}{2}\frac{Ja}{R}\sin\chi\right)\Bigg]
\end{aligned} \tag{1.32}$$

where we have used the quantization condition $k_{\perp} = 2\pi J/C = J/R$. The important point is that, in contrast to graphene, the matrix elements acquire a dependence on the band index J. The electronic structure is thus of the form

$$E_{Jk}^{h} = \pm\left|H_{AB}(J,k)\right| \tag{1.33}$$

where the label $h = c, v$ for the conduction (top sign) and valence bands (bottom sign). Similarly, we have

$$\lambda_{Jk}^{h} = \mp \frac{H_{AB}^{*}(J,k)}{|H_{AB}(J,k)|} \tag{1.34}$$

and the wavefunctions are

$$\psi_{Jk}^{h}(\vec{r}) = \frac{1}{\sqrt{2}} \left[\phi_{AJk}(\vec{r}) + \lambda_{Jk}^{h} \phi_{BJk}(\vec{r}) \right] \tag{1.35}$$

with

$$\phi_{sJk}(\vec{r}) = \sum_{\phi_s, z_s} e^{ikz_s} e^{iJ\phi_s} \chi_{\vec{r}_s}(\vec{r}) . \tag{1.36}$$

An effective mass model can be obtained for the nanotube electronic structure by considering an expansion of the energy dispersion near the minimum of the conduction band. (Because the nanotube electronic structure is symmetric around the Fermi level in the tight-binding approximation used in this book, the electron and hole effective masses are the same.) To obtain the effective mass representation, we expand the band structure near the Fermi points for small values of the axial wavevector k:

$$\begin{aligned} E_c &\approx \frac{3a\gamma}{2} \left| \vec{k} - \vec{k}_F \right| = \frac{3a\gamma}{2} \sqrt{\Delta k^2 + k^2} \\ &\approx \frac{E_g}{2} + \frac{9a\gamma R}{4} k^2 = \frac{E_g}{2} + \frac{\hbar^2}{2m^*} k^2 \end{aligned} \tag{1.37}$$

giving the effective mass

$$m^* = \frac{2\,\hbar^2}{9a\gamma R} . \tag{1.38}$$

Fig. 1.9 shows the effective mass as a function of nanotube radius, indicating that it is significantly less than the free electron mass. It should be noted that each band in the nanotube electronic structure will have its own effective mass; the derivation of the effective mass above is only valid for the bands giving the smallest bandgap since it relies on expansions near the Fermi level.

The model above is also useful to understand the density of states of nanotubes. The density of states can be expressed as

$$D(E) = \frac{\sqrt{3}a^2}{2\pi R} \sum_i \int dk\, \delta(k - k_i) \left| \frac{\partial \varepsilon}{\partial k} \right|^{-1} \tag{1.39}$$

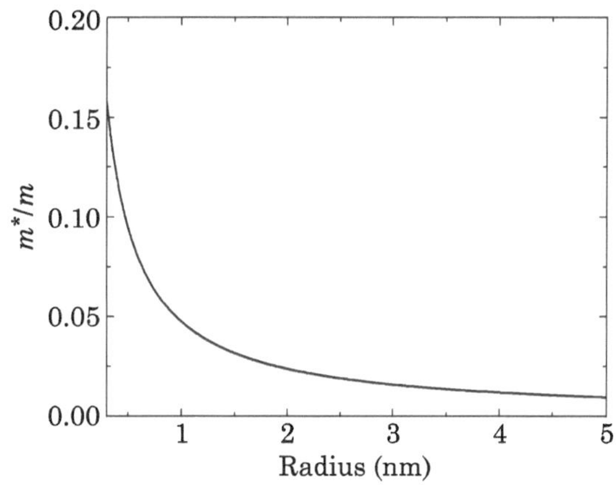

Figure 1.9 Ratio of nanotube effective mass to the bare electron mass for the first subband, as a function of nanotube radius.

where $\varepsilon(k_i) = E$. Fig. 1.10 shows the density of states calculated from Eqs. (1.39) and (1.13) for (11,0) and (12,0) nanotubes. The unique feature, as compared to bulk materials, is the presence of singularities at the band edges. To understand the basic shape of the density of states, we use the expansion of the dispersion relation around the Fermi point to get

$$\frac{\partial \varepsilon}{\partial k} = \frac{3a\gamma}{2}\frac{\partial}{\partial k}\sqrt{\Delta k^2 + k^2} = 3a\gamma\frac{k}{\sqrt{\Delta k^2 + k^2}} = 3a\gamma\frac{\sqrt{E^2 - \varepsilon_m^2}}{E} \tag{1.40}$$

where $\varepsilon_m = |3m + 1|\frac{a\gamma}{2R}$ for semiconducting tubes and $\varepsilon_m = |3m|\frac{a\gamma}{2R}$ for metallic tubes. Using this relation, we obtain the density of states per carbon atom [10]

$$D(E) = \frac{a\sqrt{3}}{\pi^2 R\gamma}\sum_{m=1}^{N}\frac{|E|}{\sqrt{E^2 - \varepsilon_m^2}}. \tag{1.41}$$

In the case of metallic tubes, the $m = 0$ band gives a nonzero density of states at the Fermi level, with $D(E_F) = \frac{a\sqrt{3}}{\pi^2 R\gamma}$. The expression for the density of states shows van Hove singularities when $E = \pm\varepsilon_m$, which is characteristic of quasi-one-dimensional materials (Fig. 1.10).

The presence of these singularities in the density of states, the dependence of the bandgap on diameter, and the presence of both metallic and semiconducting nanotubes has been verified by scanning tunneling microscopy of individual nanotubes [11], as Fig. 1.11 shows.

Finally, it is important to note that there are some deviations in the electronic properties of nanotubes from the simple π-orbital graphene picture described above, due to curvature. As a result of curvature (i) the hopping integrals

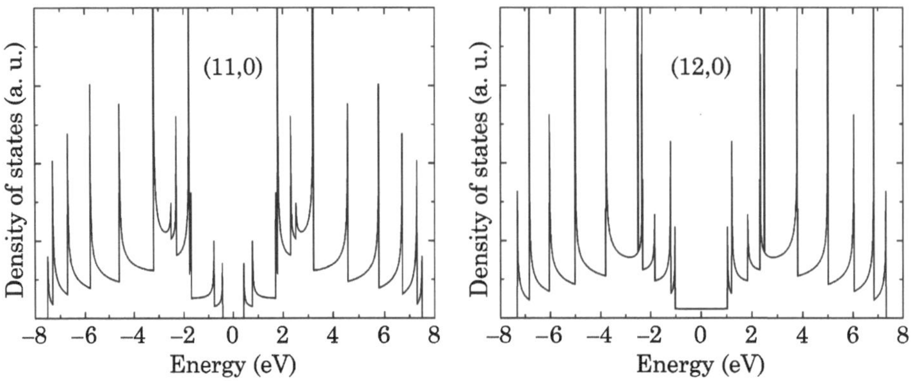

Figure 1.10 Density of states for (11,0) and (12,0) nanotubes computed from tight binding showing van Hove singularities typical of quasi-one-dimensional systems. Figure from Ref. [1].

describing the three bonds between nearest neighbors are not identical and (ii) σ–π hybridization and charge self-consistency become important. Since curvature becomes larger with decrease in nanotube diameter, deviations from the simple π-orbital graphene picture become more important in small diameter nanotubes. Nanotubes satisfying $|n-m| = 3I$ develop a small curvature-induced bandgap, and are hence quasi-metallic. Armchair nanotubes are an exception because of their special symmetry, and they remain metallic for all diameters. The bandgap of quasi-metallic nanotubes is small and varies inversely as the square of nanotube diameter. For example, while a semiconducting nanotube with a diameter of 10 Å has a bandgap of 0.7 eV, a quasi-metallic nanotube with a comparable diameter has a bandgap of only 40 meV. In graphene, hybridization between π and σ orbitals is absent. In contrast, the curvature of a nanotube can induce σ–π hybridization, which can have an important influence on the electronic properties of small diameter nanotubes. Reference [12] found that, while tight-binding calculations predict small diameter (4,0) and (5,0) zigzag nanotubes to be semiconducting with bandgaps exceeding 1 eV, ab initio calculations show that they are semi-metallic. Similarly, while tight-binding calculations predict that the (6,0) zigzag nanotube is semi-metallic with a bandgap of approximately 200 meV, ab initio calculations indicate that they are metallic [12,13]. Generally, small diameter nanotubes require a more refined treatment beyond the simple tight-binding graphene model.

1.2.3 Carrier Concentration in Intrinsic Carbon Nanotubes

The concentration of free carriers at thermal equilibrium for undoped carbon nanotubes can be calculated much the same way as for traditional bulk materials. For the tight-binding model where the nanotube valence and

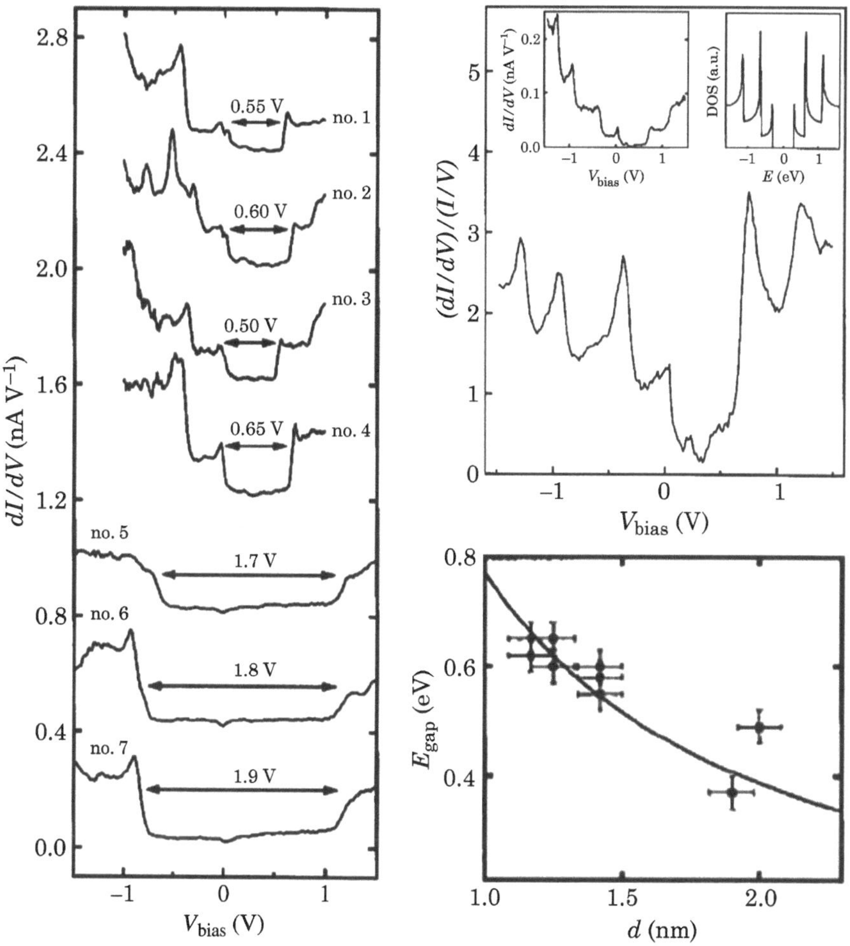

Figure 1.11 Electronic structure of carbon nanotubes measured with a scanning tunneling microscope (STM). The left panel shows the tunneling conductance between the STM tip and the gold surface on which the nanotubes sit. The upper set of curves (no. 1–4) corresponds to semiconducting nanotubes and the lower set of curves (no. 5–7) shows metallic nanotubes. The top right panel shows the normalized tunneling conductance (a measure of the density of states) with the peaks indicative of the van Hove singularities. The bottom right panel shows that the bandgap decreases with diameter, as expected from the tight-binding framework. Figure from Ref. [11].

conduction bands are symmetric, the Fermi level will be located at midgap, and the free carrier concentration in the conduction band due to the first subband can be expressed as

$$n = \int_{E_c}^{\infty} D(E)f(E)dE = \frac{a\sqrt{3}g}{\pi^2 R\gamma} \int_{E_c}^{\infty} \frac{E}{\sqrt{E^2 - E_c^2}} \frac{1}{1 + e^{E/k_B T}} dE \quad (1.42)$$

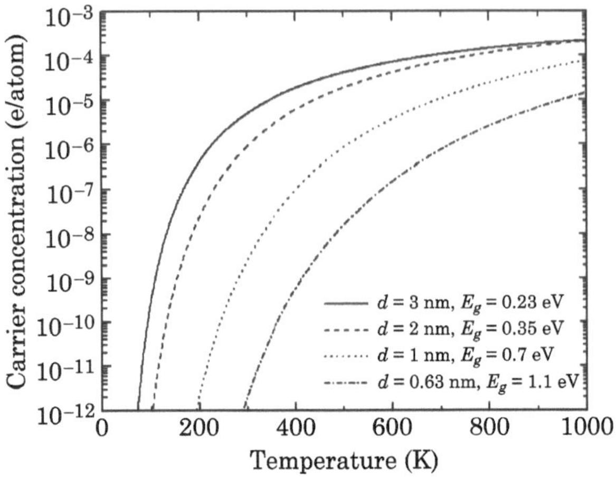

Figure 1.12 Intrinsic carrier concentration as a function of temperature for nanotubes of different radii, assuming a nondegenerate first subband.

where $f(E)$ is the Fermi distribution, T is the temperature, k_B is Boltzmann's constant, g is equal to the degeneracy of the first subband, and n is in units of electrons/atom. In writing this expression we set the Fermi level as the reference energy. Fig. 1.12 shows the calculated intrinsic carrier concentration as a function of temperature for nanotubes of different diameters. The carrier concentration is much larger than that of intrinsic silicon, which has carrier concentrations of about 10^{-14} electrons/atom at room temperature. A major part of this difference is due to the smaller bandgap of typical nanotubes. However, the figure also shows that for the same bandgap as that of silicon (1.1 eV), nanotubes have a much higher carrier concentration; the origin of this effect is the diverging density of states at the band edge (for a bulk material, the density of states goes to zero at the band edge). To further illustrate this behavior we consider the nondegenerate case where $k_B T \ll E_g$ to obtain

$$n_{NT} \approx \frac{9\sqrt{3}a^2 m^* \sqrt{E_g k_B T}}{4\pi^{3/2}\,\hbar^2} \exp\left(-\frac{E_g}{2k_B T}\right) \tag{1.43}$$

which is to be compared with the similar expression for bulk materials

$$n_{\text{bulk}} \approx \frac{2}{N_v}\left(\frac{m^* k_B T}{2\pi\,\hbar^2}\right)^{3/2} \exp\left(-\frac{E_g}{2k_B T}\right) \tag{1.44}$$

where N_v is the volume density of atoms. The ratio of these two equations for the same bandgap and effective mass is

$$\frac{n_{NT}}{n_{\text{bulk}}} = \frac{9\sqrt{6}}{4}\frac{a^2\hbar N_v\sqrt{E_g}}{\sqrt{m^* k_B T}} \tag{1.45}$$

which is about 100 at room temperature. Thus, the presence of the van Hove singularity at the band edge leads to a much larger density of intrinsic carriers.

1.2.4 Doped Carbon Nanotubes

The ability to dope semiconducting materials is important for the realization of electronic devices. In nanotubes, doping can be accomplished in different ways: substitution of B or N atoms in the lattice, insertion of atoms inside of the nanotubes, electrostatically, and charge transfer from adsorbates or substrates. These approaches and related devices will be discussed in Chapter 4. Here we are interested in establishing the basic equations that determine the position of the Fermi level in doped nanotubes. This can be accomplished with the help of Eq. (1.43) where we replace the intrinsic carrier concentration with the doping fraction f (electrons/atom)

$$f \approx \frac{9\sqrt{3}a^2 m^* \sqrt{E_g k_B T}}{4\pi^{3/2}\,\hbar^2} \exp\left(\frac{E_F - E_c}{k_B T}\right). \tag{1.46}$$

The position of the Fermi level with respect to the conduction band edge is thus

$$E_F - E_c = k_B T \ln\left(\frac{2\pi^{3/2}\gamma^{1/2}R^{3/2}f}{\sqrt{3}a^{3/2}g\sqrt{k_B T}}\right). \tag{1.47}$$

Fig. 1.13 shows the behavior of this function at room temperature for nanotubes of different sizes, assuming a band degeneracy of one (for a doubly degenerate first band, the Fermi level is lower by $k_B T \ln 2$). Of note is the doping at which the Fermi level reaches the band edge, with a value of about 10^{-3} electrons/atom for nanotubes with radii in the 0.5–1 nm range. In fact, this doping is given by

$$f^* = \frac{\sqrt{3}a^{3/2}g\sqrt{k_B T}}{2\pi^{3/2}\gamma^{1/2}R^{3/2}} = 5.2 \times 10^{-3} g \frac{\sqrt{k_B T}}{R^{3/2}} \tag{1.48}$$

where R is in nanometers and $k_B T$ is in eV. At room temperature a good rule of thumb is $f^* \approx 10^{-3}/R^{3/2}$.

1.2.5 Temperature Dependence of Bandgap

The derivation of the electronic band structure of carbon nanotubes in the previous sections assumed that it was independent of temperature. However,

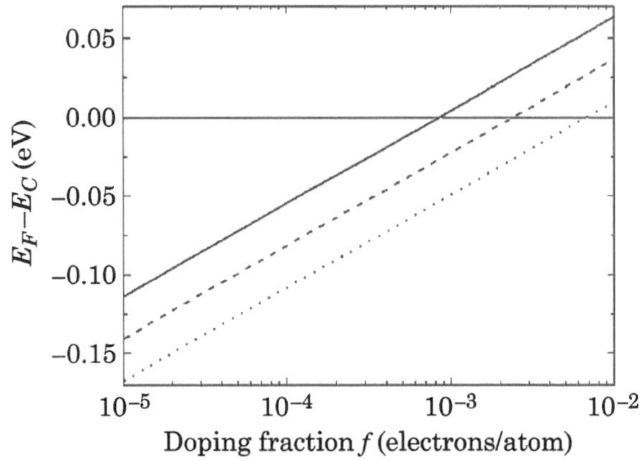

Figure 1.13 Position of Fermi level with respect to the conduction band edge as a function of the doping fraction for three nanotubes. The lines from top to bottom correspond to nanotubes of radius 1 nm, 0.5 nm and 0.25 nm.

in bulk semiconductor materials, it is well known that electron–phonon interactions and temperature-dependent structural parameters lead to a reduction of the bandgap with temperature. In fact, for silicon, experimental and theoretical work has shown that the bandgap follows the temperature dependence

$$\Delta E_g^{\mathrm{Si}}(T) = -4.73 \times 10^{-4} \frac{T^2}{T + 636 \text{ K}}. \tag{1.49}$$

Thus, for silicon, the bandgap monotonically decreases with temperature, and at room temperature is reduced by about 45 meV. This reduction can be significant when calculating, for example, carrier concentrations. It is therefore important to ascertain the role of temperature in renormalizing the bandgap in carbon nanotubes. For nanotubes, photoluminescence [14] and Raman spectroscopy [15] indicate in general that the bandgap is not strongly temperature dependent. Fig. 1.14 shows the photoluminescence spectrum from a single carbon nanotube as a function of temperature. The blueshift is clearly visible in the data, and at room temperature the bandgap is reduced by less than 8 meV from the small temperature limit. For single isolated nanotubes, molecular dynamics calculations have shown that the bandgap is essentially unchanged with temperature [16,17]. Thus expansion of the nanotubes and the ensuing reduction of the bandgap are not expected to play a significant role. It is expected that the temperature dependence originates mostly from the electron–phonon interaction. Tight-binding calculations incorporating electron–phonon interactions indicate that the variation of the bandgap with temperature is fairly small, on the order of 10 meV at room temperature

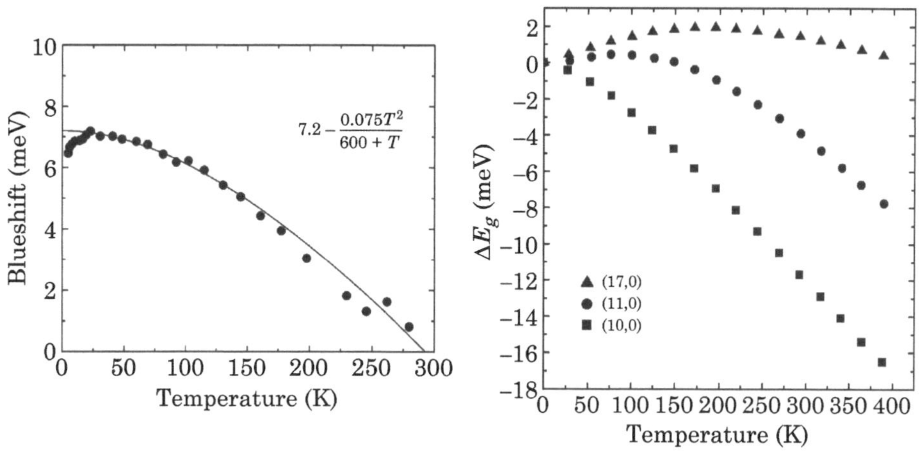

Figure 1.14 Left: experimentally measured photoluminescence shift of the band edge peak of a carbon nanotube with temperature. Figure after Ref. [14]. Right: calculated temperature dependence of the bandgap of carbon nanotubes. Figure after Ref. [18].

(Fig. 1.14), in good agreement with experiment. Thus corrections to carrier concentrations, position of Fermi level, etc. are expected to be relatively small compared to bulk semiconductors, and can usually be ignored.

The temperature dependence of the nanotube bandgap can be approximated by an equation of the form [18]

$$\Delta E_g(T) = \frac{\alpha_1 \Theta_1}{e^{\Theta_1/T} - 1} + \frac{\alpha_2 \Theta_2}{e^{\Theta_2/T} - 1} \tag{1.50}$$

where α_1, α_2, Θ_1, Θ_2 are constants representing two effective phonons that govern the temperature dependence. The presence of two modes is necessary to reproduce the nonmonotonic temperature dependence calculated in some nanotubes, with the first mode describing the low temperature behavior and the second mode the high temperature properties. The behavior of the bandgap with temperature can be well reproduced with this simplified model, provided that the constants in Eq. (1.50) be chirality and diameter dependent [18].

1.3 Phonon Spectra

Knowledge of the phonon spectrum is important to understand many aspects of materials such as electron–phonon interactions and thermal transport. In this section, we present the phonon spectra of carbon nanotubes, based on zone-folding of the graphene phonon spectrum, much like was done for the electronic structure. The goal of this section is to illustrate the basic phonon features of carbon nanotubes that will be useful in later chapters and sketch the procedure for obtaining them. A more thorough and detailed discussion of phonons in carbon nanotubes can be found in reference [7].

The simplest model describing phonons assumes that the atoms are connected by springs, with different spring constants depending on which two atoms are considered. Thus in the harmonic approximation, the change in energy due to atomic displacements $\vec{u}_i$ around the equilibrium positions (where the subscript i represents atom i in the lattice) is

$$E = \frac{1}{2} \sum_{i,j,\alpha,\beta} u_i^\alpha K_{ij}^{\alpha\beta} u_j^\beta \tag{1.51}$$

where the Greek indices indicate the three directions x, y, z. The dynamical equation for the displacements of atoms of mass M is

$$M \frac{\partial^2 u_i^\alpha}{\partial t^2} = -\frac{\partial E}{\partial u_i^\alpha} = -\sum_{j,\beta} K_{ij}^{\alpha\beta} u_j^\alpha. \tag{1.52}$$

The phonon modes correspond to simultaneous displacements of all of the atoms with the same frequency ω, and this can be expressed in Fourier space as

$$\hat{u}_i^\alpha \left(\vec{k}, t\right) = \frac{1}{\sqrt{N_v}} \sum_{\vec{r}_i} u_i^\alpha e^{i\left(\vec{k}\cdot\vec{r}_i - \omega t\right)} \tag{1.53}$$

where N_v is the number of atoms in the volume under consideration and $\vec{k}$ is a wave vector in the first Brillouin zone. The dynamical equation for the Fourier components is

$$\left(M\omega^2 - \sum_{j,\beta} K_{ij}^{\alpha\beta} \right) \hat{u}_i^\alpha + \sum_{j,\beta} K_{ij}^{\alpha\beta} e^{i\vec{k}\cdot\Delta\vec{R}_{ij}} \hat{u}_j^\beta = 0 \tag{1.54}$$

where $\Delta\vec{R}_{ij} = \vec{r}_i - \vec{r}_j$. This equation can be written in terms of a so-called dynamical matrix $\overleftrightarrow{D}$:

$$\overleftrightarrow{D}\, \overrightharpoon{u}_{\vec{k}} = 0 \tag{1.55}$$

where the vector $\vec{u}_{\vec{k}}$ is a column vector with components ${}^t(\hat{u}_1^x, \hat{u}_1^y, \hat{u}_1^z, \hat{u}_2^x, \hat{u}_2^y, \hat{u}_2^z, \ldots \hat{u}_N^x, \hat{u}_N^y, \hat{u}_N^z)$. The dynamical matrix provides $3N$ equations for the N atoms in the first Brillouin zone, and the matrix D is thus of dimension $3N \times 3N$. This matrix has blocks of size 3×3 representing the possible combination

Table 1.1 Coupling Parameters for Graphene up to the fourth neighbor, in units of eV/Å². Adapted from Ref. [7]

First neighbor	Second neighbor	Third neighbor	Fourth neighbor
$\phi_R^1 = 22.81$	$\phi_R^2 = 5.50$	$\phi_R^3 = 1.88$	$\phi_R^4 = -1.20$
$\phi_{IPT}^1 = 15.31$	$\phi_{IPT}^2 = -2.02$	$\phi_{IPT}^3 = -3.28$	$\phi_{IPT}^4 = 1.43$
$\phi_{OPT}^1 = 6.14$	$\phi_{OPT}^2 = -0.25$	$\phi_{OPT}^3 = 0.094$	$\phi_{OPT}^4 = -0.36$

of cartesian coordinates. Labeling the submatrices with indices $i, j = 1, \ldots N$ gives the expression

$$D_{ij}^{\alpha\beta} = \left(M\omega^2 - \sum_l K_{il}^{\alpha\beta} \right) \delta_{ij}\delta_{\alpha\beta} + \sum_l K_{il}^{\alpha\beta} e^{i\vec{k}\cdot\Delta\vec{R}_{il}}. \tag{1.56}$$

The procedure for calculating the phonon modes is to solve Eq. (1.56) for the eigenvectors and the dispersion relation $\omega(\vec{k})$; nontrivial solutions are obtained when the determinant of $\overleftrightarrow{D}$ is equal to zero. To proceed further, the coupling constants K need to be specified. While in principle all atoms in the unit cell are coupled, the calculation of the phonon modes is simplified by truncating the interaction to a few neighbor atoms. For graphene, it has been shown that interactions up to fourth neighbor atoms need to be included to reproduce the experimental results [7]. In general, for two atoms that lie on the x-axis, the coupling constants are taken to be of the form

$$K^n = \begin{pmatrix} \phi_R^n & 0 & 0 \\ 0 & \phi_{IPT}^n & 0 \\ 0 & 0 & \phi_{OPT}^n \end{pmatrix} \tag{1.57}$$

where the subscript n represents the nth neighbor and the constants ϕ_R^n, ϕ_{IPT}^n and ϕ_{OPT}^n represent the coupling parameters in the radial, in-plane tangential, and out-of-plane tangential directions, respectively. Coupling matrices K for the other atoms can be obtained by rotating the matrix (1.57), while values for the twelve coupling parameters are obtained by fitting to experimentally measured dispersion relations, with values given in Table 1.1.

The calculated phonon dispersion for graphene using these parameters is shown in Fig. 1.15, showing the good agreement with experimental data. Since in graphene there are two atoms per unit cell each with three degrees of freedom, there are six phonon bands in total. Three of these branches are acoustic branches corresponding to out-of-plane, in-plane tangential, and in-plane radial modes, as indicated in the figure.

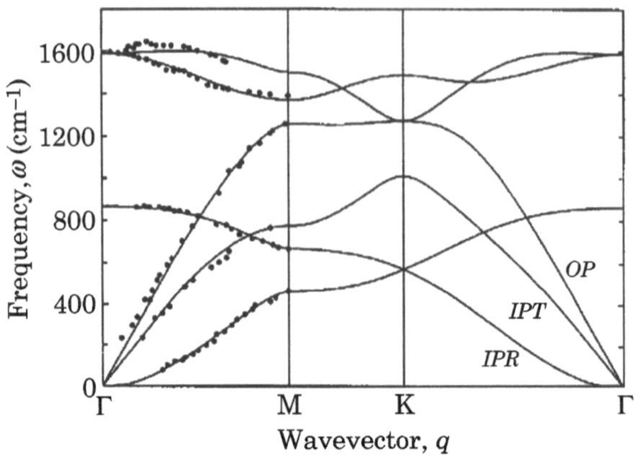

Figure 1.15 Phonon dispersion of graphene along high-symmetry directions, calculated using fourth neighbor interactions. The circles indicate experimental data points. Figure from Ref. [19].

The phonon dispersion of carbon nanotubes can be obtained from the graphene phonon dispersion by quantization of the phonon wavevector around the nanotube circumference, just like the electron momentum is quantized to obtain the electronic structure. The quantization condition is

$$\vec{k} \cdot \vec{C} = k_x C_x + k_y C_y = 2\pi p \tag{1.58}$$

where $C_x = a\sqrt{3}\,(n + m/2)$ and $C_y = 3am/2$. Each of the $N/2$ values of p will intersect the six bands of graphene, giving a total of $3N$ phonon bands, consistent with the 3 degrees of freedom for each of the N atoms of the unit cell.

The calculated phonon dispersion relation for a (10,10) nanotube is shown in Fig. 1.16. Because there are 40 atoms per unit cell in a (10,10) nanotube, there is a total of 120 phonon bands (54 are doubly degenerate and 12 are nondegenerate). In this nanotube, there are four acoustic modes: two degenerate transverse modes (T), a twist mode (TW) and one longitudinal mode (L), as indicated in the figure. It is worth noting that there is an optical radial breathing mode (RBM) at 20 meV. The inset in Fig. 1.16 shows the low-energy phonon density of states of the nanotube (solid line) and that of graphite (dashed line) and graphene (dot–dashed line). The characteristics of the nanotube phonon density of states are its constant value at low energies and the presence of singularities at the band edges. Examples of phonon modes in the (10,10) nanotube are illustrated in Fig. 1.17. Finally, it should be noted that the zone-folding scheme to obtain the nanotube phonon dispersion from that of graphene misses some phonon modes particular to carbon nanotubes. A detailed discussion of this point can be found in [20].

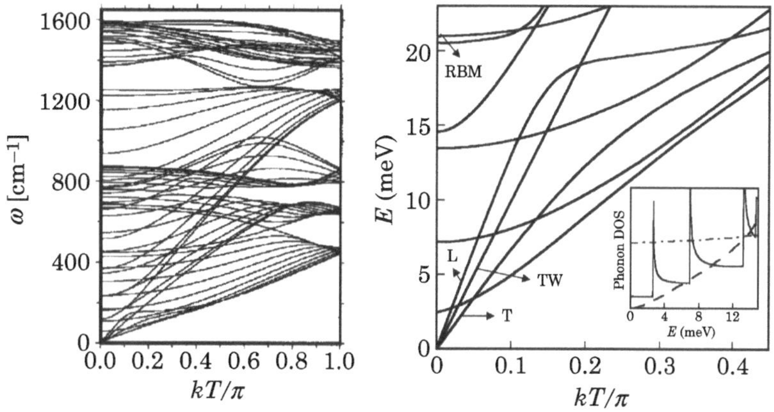

Figure 1.16 Calculated phonon dispersion for a (10,10) carbon nanotube. Left: full spectrum. Right: low energy spectrum. The acoustic modes are denoted with L for longitudinal, TW for twist and T for transverse. The radial breathing mode (optical) is denoted with RBM. The inset in the right panel shows the density of phonon states. Figures adapted from Ref. [20].

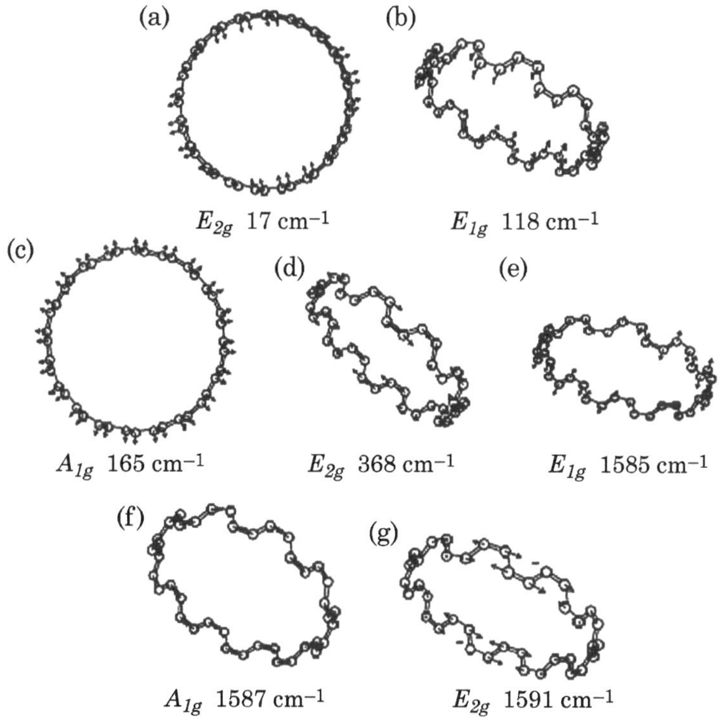

Figure 1.17 Examples of phonon modes in a (10,10) carbon nanotube. Mode (c) is the radial breathing mode. Figure from Ref. [20].

References

1. M.P. Anantram and F. Léonard, "Physics of carbon nanotube devices", *Rep. Prog. Phys.*, Vol. 69, p. 507, 2006.
2. J. Tersoff and R.S. Ruoff, "Structural properties of a carbon-nanotube crystal", *Phys. Rev. Lett.*, Vol. 73, p. 676, 1994.
3. A. Thess, R. Lee, P. Nikolaev, H. Dai, P. Petit, J. Robert, C. Xu, Y.H. Lee, S.G. Kim, A.G. Rinzler, D.T. Colbert, G.E. Scuseria, D. Tománek, J.E. Fischer and R.E. Smalley, "Crystalline ropes of carbon nanotubes", *Science*, Vol. 273, p. 483, 1996.
4. J.-M. Bonard, J.-P. Salvetat, T. Stöckli, L. Forró and A. Châtelain, "Field emission from carbon nanotubes: perspectives for applications and clues to the emission mechanism", *Applied Physics* A : *Materials Science & Processing*, Vol. 69, p. 245, 1999.
5. A.A. Farijian, H. Mizuseki and Y. Kawazoe, "Electronic transport properties of a metal–semiconductor carbon nanotube heterojunction", *Physica* E, Vol. 22, p. 675, 2004.
6. C.T. White and T.N. Todorov, "Carbon nanotubes as long ballistic conductors", *Nature*, Vol. 393, p. 240, 1998.
7. R. Saito, G. Dresselhaus and M.S. Dresselhaus, *Physical properties of carbon nanotubes*, Imperial College Press, London, UK, 2005.
8. C.T. White, D.H. Robertson and J.W. Mintmire, "Helical and rotational symmetries of nanoscale graphitic tubules", *Phys. Rev.* B, Vol. 47, p. 5485, 1993.
9. C.T. White and J.W. Mintmire, "Fundamental properties of single-wall carbon nanotubes", *J. Phys. Chem.* B, Vol. 109, p. 52, 2005.
10. J.W. Mintmire and C.T. White, "Universal density of states for carbon nanotubes", *Phys. Rev. Lett.*, Vol. 81, p. 2506, 1998.
11. J.W.G. Wildöer, L.C. Venema, A.G. Rinzer, R.E. Smalley and C. Dekker, "Atomic structure of atomically resolved carbon nanotubes", *Nature*, Vol. 391, p. 59, 1998.
12. I. Cabria, J.W. Mintmire and C.T. White, "Metallic and semiconducting narrow carbon nanotubes", *Phys. Rev.* B, Vol. 67, p. 121406, 2003.
13. X. Blase, L.X. Benedict, E.L. Shirley and S.G. Louie, "Hybridization effects and metallicity in small radius carbon nanotubes", *Phys. Rev. Lett.*, Vol. 72, p. 1878, 1994.
14. J. Lefebvre, P. Finnie and Y. Homma, "Temperature-dependent photoluminescence from single-walled carbon nanotubes", *Phys. Rev.* B, Vol. 70, p. 045419, 2004.
15. S.B. Cronin, Y. Yin, A. Walsh, R.B. Capaz, A. Stolyarov, P. Tangney, M.L. Cohen, S.G. Louie, A.K. Swan, M.S. Ünlü, B.B. Goldberg and M. Tinkham, "Temperature dependence of the optical transition energies of carbon nanotubes: the role of electron–phonon coupling and thermal expansion", *Phys. Rev. Lett.*, Vol. 96, p. 127403, 2006.
16. N.R. Raravikar, P. Keblinski, A.M. Rao, M.S. Dresselhaus, L.S. Schadler and P.M. Ajayan, "Temperature dependence of radial breathing mode Raman frequency of single-walled carbon nanotubes", *Phys. Rev.* B, Vol. 66, p. 235424, 2002.
17. Y.-K. Kwon, S. Berber and D. Tomanek, "Thermal contraction of carbon fullerenes and nanotubes", *Phys. Rev. Lett.*, Vol. 92, p. 015901, 2004.
18. R.B. Capaz, C.D. Spataru, P. Tangney, M.L. Cohen and S.G. Louie, "Temperature dependence of the band gap of semiconducting carbon nanotubes", *Phys. Rev. Lett.*, Vol. 94, p. 036801, 2005.
19. R.A. Ishi, L. Venkataraman, M.S. Dresselhaus and G. Dresselhaus, "Phonon modes in carbon nanotubules", *Chem Phys. Lett.*, Vol. 209, p. 77, 1993.
20. M.S. Dresselhaus and P.C. Eklund, "Phonons in carbon nanotubes", *Adv. in Phys.*, Vol. 49, p. 705, 2000.

2 Metallic Carbon Nanotubes for Current Transport

2.1 Introduction

Metallic carbon nanotubes have attracted significant attention because their ability to carry high current densities is unparalleled in the family of emerging nanowires. Ballistic transport has been observed, and values for the conductance that approach the theoretical limit have been measured at small biases. They hold promise as interconnects in both nanoelectronics and molecular electronics because of their low resistance and outstanding mechanical properties (Fig. 2.1). A problem with interconnects in the semiconductor industry is breakdown of copper wires due to electromigration when current densities exceed $10^6\,A/cm^2$. Preliminary work has shown that arrays of carbon nanotubes can be integrated with silicon technology and hold promise as vertical vias to carry more than an order of magnitude larger current densities than conventional ones [1,2], and reference [3] demonstrated that multiwall nanotubes carry current densities approaching $10^9\,A/cm^2$. Single wall metallic nanotubes are also important from the perspective of molecular electronics, where they can be used either as interconnects or nanoscale contacts.

The unique band structure of metallic carbon nanotubes, which is partly responsible for their superb current carrying capacity, differs from the electronic band structure of conventional metals. We will now discuss essential aspects of the electronic band structure of metallic carbon nanotubes that are necessary to understand their current–voltage characteristics. A more in-depth discussion is provided in the following sections. The band structure and transmission versus energy of a single wall armchair metallic nanotube is shown in Fig. 2.2. The band structure shows various subbands arising from quantization of the wave vector around the circumference of the nanotube. The total transmission at a given electron energy is equal to the electron transmission probability times the number of channels, and for perfect transmission without scattering (i.e. transmission probability equal to 1), is simply equal to the number of channels. Thus, the total transmission shows steps when a subband opens or closes. The magnitude of the change in transmission at these steps corresponds to the subband degeneracy. For example, the subbands shown by the dashed lines are not degenerate, and each contributes one channel for transmission. Hence the total transmission at the Fermi energy ($E = 0$) is equal to two. At about 0.85 eV, the transmission jumps from two to six units because the subbands shown in solid grey in Fig. 2.2 have a degeneracy of two.

The total number of subbands increases with increasing nanotube diameter because the number of quantum numbers arising from quantization of the electron wavefunction around the nanotube circumference becomes larger.

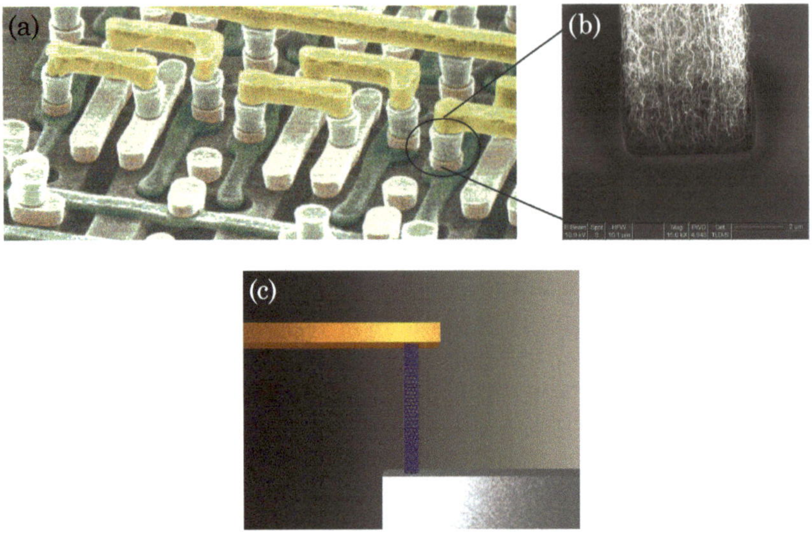

Figure 2.1 (a) Wiring in an integrated circuit. The horizontal interconnects are shown in yellow and the vertical interconnects are shown in gray. Figure from IBM Journal of Research and Development. (b) An example of a multiwall carbon nanotube array grown in a silicon oxide via, which can potentially be used as a vertical interconnect. Figure from Ref. [1]. (c) A pictorial representation of a single wall nanotube used as a vertical interconnect.

Because the conductance and valence bands have mirror symmetry, the Fermi energy is at the energy where the subbands denoted by the dashed lines cross (which occurs at the zero of energy in Fig. 2.2), independent of nanotube diameter. These subbands are called crossing subbands. However, the location of the wave vector k where the crossing occurs changes with nanotube chirality. There are only two subbands per spin at the Fermi energy, independent of nanotube diameter and chirality. The energy of the first semiconducting subbands (solid grey lines of Fig. 2.2) decreases inversely with increase in nanotube diameter. For example, a metallic nanotube with a diameter of 1.5 nm has its first semiconducting subband 0.85 eV away from the Fermi energy, while for a nanotube of 18 nm diameter that subband is only 0.0625 eV from the Fermi level. Note that the semiconducting subbands are also referred to as noncrossing subbands.

From a basic physics perspective, metallic nanotubes have been of immense interest to researchers studying electron–electron interactions in condensed matter systems because they exhibit the physics of Luttinger liquid behavior. We will however restrict ourselves to properties of nanotubes that emerge from their single particle physics in this book and refer the interested reader to references [4–6] to learn about their Luttinger liquid properties. In the remainder of this section, we will discuss the physics dictating the current carrying properties of metallic nanotubes in low and high bias regimes in

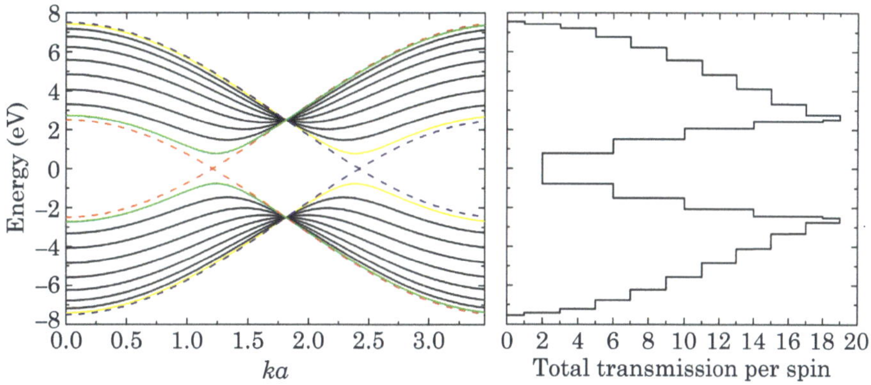

Figure 2.2 Left: energy versus wave vector ka for a (10,10) armchair metallic nanotube. Right: energy versus total transmission per spin degree of freedom for the (10,10) nanotube.

Sections 2.2 and 2.3 respectively, and discuss the quantum capacitance, kinetic inductance and speed of signal propagation in Section 2.4.

2.2 Low Bias Transport

2.2.1 Electronic Transport in Ballistic Conductors

In systems with a low density of defects such as carbon nanotubes, the seminal work of Landauer [7] established that the conductance can be related to the electron transmission probability across the conductor. As illustrated in Fig. 2.3, this can be understood by considering the flow of electrons between two leads separated by the conductor. In the case where the electrons have perfect transmission between the leads, the current flowing to the right is

$$I_R = \int_{-\infty}^{\infty} eD(E)v(E)f\left(E - E_F^L\right)dE \tag{2.1}$$

and the current flowing to the left is

$$I_L = -\int_{-\infty}^{\infty} eD(E)v(E)f\left(E - E_F^R\right)dE \tag{2.2}$$

where $D(E)$ is the density of states in units of (states/eV/nm), $v(E)$ is the electron velocity and $f(E)$ is the Fermi function with Fermi levels E_F^R in the right lead and E_F^L in the left lead. These equations are simply expressing the fact that the current at energy E is the product of the number of charges $eD(E)f(E - E_F)$ and their velocity $v(E)$. The total current is the sum of the

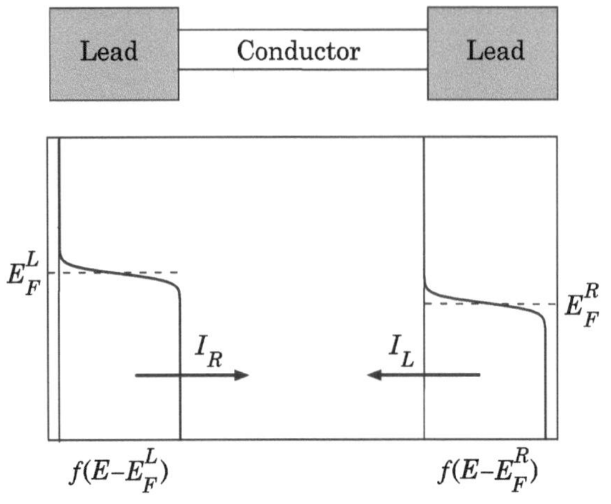

Figure 2.3 Sketch of a ballistic conductor between two leads as well as the finite-temperature Fermi distributions in each lead. The current through the conductor can be considered as the sum of right and left currents.

right and left currents

$$I = I_R + I_L = \int_{-\infty}^{\infty} eD(E)v(E)\left[f\left(E - E_F^L\right) - f\left(E - E_F^R\right)\right] dE. \tag{2.3}$$

The difference of the Fermi functions implies that most of the current will flow between the two Fermi levels (this is exactly true at zero temperature where the Fermi distributions are step functions).

Further insight into Eq. (2.3) can be obtained by considering simple expressions for the density of states and the electron velocity. For one-dimensional materials, the density of states is given by

$$D(E) = \frac{1}{\pi}\left|\frac{\partial \varepsilon}{\partial k}\right|^{-1} \tag{2.4}$$

and the velocity is

$$v\,(E) = \frac{1}{\hbar}\frac{\partial \varepsilon}{\partial k}. \tag{2.5}$$

Substituting these expression in Eq. (2.3) gives the current

$$I = \frac{2e}{h}\int_{-\infty}^{\infty} \left[f\left(E - E_F^L\right) - f\left(E - E_F^R\right)\right] dE. \tag{2.6}$$

For small biases $eV = E_F^L - E_F^R$ this expression reduces to

$$I = \left[\frac{2e^2}{h}\int_{-\infty}^{\infty}\left(-\frac{\partial f}{\partial E}\right)dE\right]V. \tag{2.7}$$

The small bias conductance$G = I/V$ is thus given by

$$G = \frac{2e^2}{h}\int_{-\infty}^{\infty}\left(-\frac{\partial f}{\partial E}\right)dE \tag{2.8}$$

and at zero temperature we obtain the quantum of conductance

$$G_0 = \frac{2e^2}{h}. \tag{2.9}$$

There are two generalizations of the above derivation that need to be considered when describing transport through real systems. The first is that in general there may be several modes that contribute to the current, and each mode will contribute one quantum of conductance. The second point is that, because of scattering processes in the conductor, the electron transmission probability may be less than unity. Putting this together gives the final expressions for the current

$$I = \frac{2e}{h}\sum_m \int_{-\infty}^{\infty} T_m(E)\left[f\left(E - E_F^L\right) - f\left(E - E_F^R\right)\right]dE, \tag{2.10}$$

and the total conductance (including spin) is given as

$$G = \frac{2e^2}{h}\sum_m \int_{-\infty}^{\infty} T_m(E)\left(-\frac{\partial f}{\partial E}\right)dE \tag{2.11}$$

where $T_m(E)$ is the transmission probability at energy E for channel m.

Fig. 2.2 shows the total transmission probability for a pristine (10,10) nanotube as a function of energy. Because the nanotube is free of defects, the transmission probability is equal to 1 for each channel, i.e. the total transmission consists of integer values representing the number of subbands at that energy. Since the factor $\partial f/\partial E$ is significant only at energies near the Fermi energy, and noting that for metallic carbon nanotubes the number of subbands at the Fermi energy is two, the linear response conductance is then given by

$$G = \frac{2e^2}{h} \times 2 = \frac{4e^2}{h} = \frac{1}{6.5\ \mathrm{k}\Omega}. \tag{2.12}$$

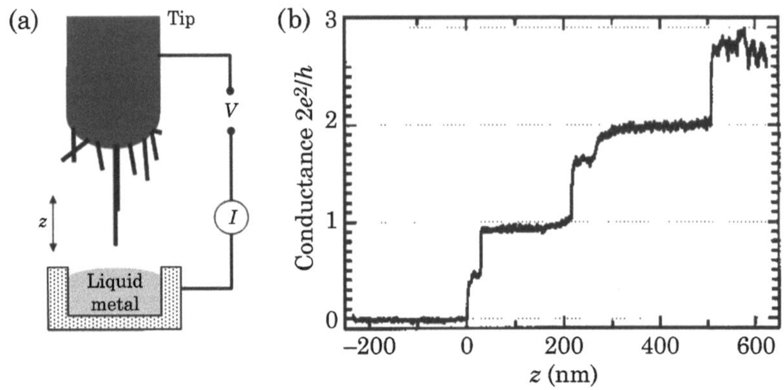

Figure 2.4 (a) Schematic diagram of the experimental setup to measure the conductance of carbon nanotubes in contact with liquid metal. A long nanotube bundle protrudes from the tip of a scanning probe microscope. The nanotube is lowered to a liquid metal surface. After contact is established, the current is measured as the tip is moved into the liquid metal, so that the conductance can be determined as a function of the position of the nanotube contact. (b) Conductance as a function of dipping length into liquid mercury contact. The quantized conductance of $2e^2/h$ corresponds to a single nanotube making contact to mercury. The other quantized conductance values correspond to two and three nanotubes making contact to the metal. Figures from, and after Ref. [8].

We have explicitly assumed that the nanotube diameter is small enough so that $\partial f/\partial E$ is negligible at energies corresponding to the semiconducting subbands. Experiments have measured small bias conductances between $2e^2/h$ and $4e^2/h$. A classic experiment that measured the small bias conductance of multiwall nanotubes involved two unconventional contacts, a scanning probe microscope tip and a liquid metal [8,9]. The experimental setup is shown in Fig. 2.4, with the liquid metal being mercury, cerrolow or gallium. In all cases, the small bias conductance is $2e^2/h$, which is half of the maximum possible value (Eq. (2.12)) for a single nanotube. The conductance increases in steps of $2e^2/h$ with increase in the number of nanotubes making contact to the liquid metal. Modeling of this experimental effect has been challenging and a clear explanation is pending. Reference [10] found that inter-wall interactions modify the density of states near the Fermi level leading to a conductance of $2e^2/h$. On the other hand, modeling of metal-armchair nanotube contacts using a jellium model shows that only one of the modes at the Fermi energy couples to the metal; the other mode does not couple well to the metal due to wave vector mismatch [11,12] or reflection at the nanotube-metal interface [13]. For metallic zigzag nanotubes, both subbands couple well to the metal and hence the nanotube is capable of carrying twice as much current, close to the theoretical maximum of $4e^2/h$[12].

In contrast to the multiwall nanotube experiment just discussed, values of conductance closer to the theoretical maximum of $4e^2/h$ have been measured in single wall nanotubes as shown in Fig. 2.5 [14]. For these experiments, a

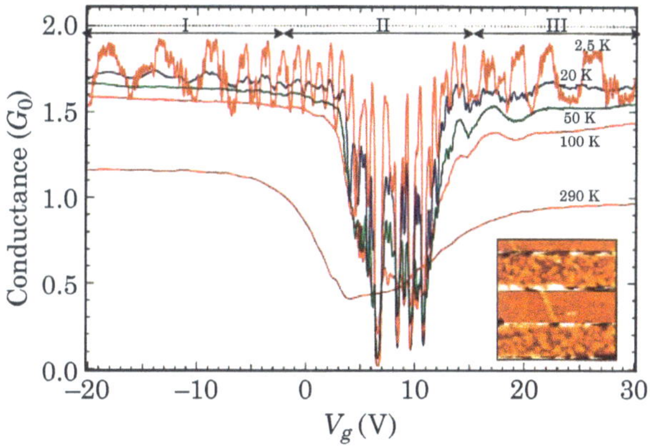

Figure 2.5 Conductance versus gate voltage for a metallic single wall nanotube. The inset shows the nanotube lying between four metal contacts. As the temperature is lowered, the conductance approaches $2G_0$, where $G_0 = 1/(12.9\ \text{k}\Omega)$. The nanotube diameter is about 1.5 nm, and the distance between the contacts is 800 nm. Figure from Ref. [14].

metallic carbon nanotube is positioned between titanium contacts, and the source–drain conductance is measured as a function of the applied gate voltage. The room temperature curve shows that there is little modulation of the nanotube conductance with the gate voltage, indicating the metallic character of the nanotube. As the temperature is lowered, the maximum conductance approaches $4e^2/h$. Motivated by this experimental work, ab initio methods have been used to study the properties of contacts to metallic nanotubes [15–18]. In agreement with experiments, the theoretical studies show that titanium forms the best contact when compared to other metals such as Au, Al, Ni, Fe and Co. However, experiment and theory [19] both indicate that Pd is the best material for making contacts to nanotubes. The topic of electrical contacts will be discussed at length in the next chapter.

The results discussed above show that metallic nanotubes are excellent conducting wires, which can attain near perfect experimentally measured conductance. At first sight this is surprising because surfaces, disorder, defects, and phonons are all sources of scattering that usually lead to a decrease in conductance in conventional materials. In nanotubes these processes are ineffective, and the reasons for this are:

1. the mean free paths for scattering with acoustic phonons are longer than a micron [20]. This occurs because the Fermi velocity is large, and the electron–phonon interaction is relatively weak.

2. the nanotube has a crystalline surface without disordered boundaries. In a silicon field-effect transistor, there is significant scattering of electrons due to the disordered nature of the Si/SiO_2 interface.

3. the crossing points of metallic nanotubes correspond to the reciprocal lattice wave vector $|K| = 4\pi/3|a_1|$ of the underlying graphene sheet, where $|a_1|$ is the length of the real space lattice vector of graphene (Fig. 1.2 (a)). The

two different crossing subbands of metallic nanotubes (Fig. 2.2) correspond to these K points in the graphene sheet. As a result, any potential that is long-ranged compared to a_1 will not effectively couple the two crossing subbands because of the lack of wave vector components in reciprocal space [21].

4. metallic nanotubes are robust to weak disorder. The localization lengths due to weak disorder are much longer than those in other thin conducting wires [22,23].

We now discuss in more detail points (1) and (4).

Electron–Phonon Scattering: To understand the long mean free path at low bias due to electron–phonon scattering (point (1) above) requires a calculation of the scattering rate. This can be accomplished by using a second quantization approach and Fermi's golden rule [20]. The Hamiltonian describing electron–phonon interactions is written as

$$H^{\alpha}_{\mathrm{e-ph}} = \sum_{\vec{k},\vec{q}} U^{\alpha}_{\vec{k},\vec{q}} c^{\dagger}_{\vec{k}+\vec{q}} c_{\vec{k}} \left(b^{\alpha}_{\vec{q}} + b^{\alpha\dagger}_{-\vec{q}} \right) \tag{2.13}$$

where $c^{\dagger}_{\vec{k}}$ and $c_{\vec{k}}$ are electron creation and annihilation operators, and $b^{\alpha\dagger}_{\vec{q}}$ and $b^{\alpha}_{\vec{q}}$ are creation and annihilation operators for phonons in phonon band α. At low bias, electrons do not have enough energy to interact with optical phonons, and we thus consider scattering with acoustic phonons only. The coupling $U^{\alpha}_{\vec{k},\vec{q}}$ for acoustic phonons is given by

$$U^{\alpha}_{\vec{k},\vec{q}} = \sqrt{\frac{\hbar}{2L\rho\Omega^{\alpha}_{\vec{q}}}} U_{\alpha} q \tag{2.14}$$

where ρ is the linear mass density, L is the length of the nanotube, and U_{α} is the deformation potential (more about this below). Energy and momentum conservation require that an electron with energy $E_{\vec{k}}$ and momentum $\vec{k}$ interacting with a phonon of frequency $\Omega^{\alpha}_{\vec{q}}$ and momentum $\vec{q}$ scatter to a state

$$E_{\vec{k}+\vec{q}} = E_{\vec{k}} \pm \hbar\Omega^{\alpha}_{\vec{q}} \tag{2.15}$$

where $E_{\vec{k}+\vec{q}}$ is the energy of the electron in the final state with momentum $\vec{k}+\vec{q}$, and the positive (negative) sign corresponds to phonon absorption (emission). According to Fermi's golden Rule, the scattering rate is given by

$$\frac{1}{\tau^{\alpha}_{ac}} = \frac{2\pi}{\hbar} \sum_{\vec{q}} |\langle f| H_{\mathrm{e-ph}} |i\rangle|^2 \times \left[\delta\left(E_{\vec{k}+\vec{q}} - E_{\vec{q}} - \hbar\Omega^{\alpha}_{\vec{q}} \right) + \delta\left(E_{\vec{k}+\vec{q}} - E_{\vec{q}} + \hbar\Omega^{\alpha}_{\vec{q}} \right) \right] \tag{2.16}$$

which includes contributions from phonon absorption and emission. Using a notation where $|E_{\vec{k}}, N_{\vec{q}}^{\alpha}\rangle$ represents an electron at energy E with momentum $\vec{k}$ in a phonon bath with $N_{\vec{q}}^{\alpha}$ phonons of wavevector $\vec{q}$ in band α, the matrix elements for phonon absorption are evaluated as follows:

$$\begin{aligned}\langle f|\, H_{\mathrm{el-ph}}^{\alpha}\,|i\rangle &= \langle E_{\vec{k}+\vec{q}}, N_{\vec{q}}^{\alpha}-1|U_{\vec{k},\vec{q}}^{\alpha} c_{\vec{k}+\vec{q}}^{\dagger} c_{\vec{k}} \left(b_{\vec{q}}^{\alpha}+b_{-\vec{q}}^{\alpha\dagger}\right)|E_{\vec{k}}, N_{\vec{q}}^{\alpha}\rangle \\ &= \langle E_{\vec{k}+\vec{q}}, N_{\vec{q}}^{\alpha}-1|U_{\vec{k},\vec{q}}^{\alpha} c_{\vec{k}+\vec{q}}^{\dagger} c_{\vec{k}} \left(\sqrt{n_{\vec{q}}^{\alpha}}\,|E_{\vec{k}}, N_{\vec{q}}^{\alpha}-1\rangle\right. \\ &\left.+\sqrt{n_{\vec{q}}^{\alpha}+1}|E_{\vec{k}}, N_{\vec{q}}^{\alpha}+1\rangle\right) \\ &= \langle E_{\vec{k}+\vec{q}}, N_{\vec{q}}^{\alpha}-1|U_{\vec{k},\vec{q}}^{\alpha}\left(\sqrt{n_{\vec{q}}^{\alpha}}|E_{\vec{k}+\vec{q}}, N_{\vec{q}}^{\alpha}-1\rangle\right. \\ &\left.+\sqrt{n_{\vec{q}}^{\alpha}+1}|E_{\vec{k}}, N_{\vec{q}}^{\alpha}+1\rangle\right) \\ &= \sqrt{n_{\vec{q}}^{\alpha}}U_{\vec{k},\vec{q}}^{\alpha} \end{aligned} \tag{2.17}$$

and a calculation for phonon emission gives the same result. The scattering time becomes

$$\begin{aligned}\frac{1}{\tau_{ac}^{\alpha}} &= \frac{2\pi}{\hbar}\sum_{q} n_{\vec{q}}^{\alpha}\left(U_{\vec{k},\vec{q}}^{\alpha}\right)^{2} \\ &\times\left[\delta\left(E_{\vec{k}+\vec{q}}-E_{\vec{q}}-\hbar\Omega_{\vec{q}}^{\alpha}\right)+\delta\left(E_{\vec{k}+\vec{q}}-E_{\vec{q}}+\hbar\Omega_{\vec{q}}^{\alpha}\right)\right].\end{aligned} \tag{2.18}$$

Using Eq. (2.14), we obtain the expression

$$\begin{aligned}\frac{1}{\tau_{ac}^{\alpha}} &= 2\frac{2\pi}{\hbar}\sum_{q}\frac{\hbar U_{\alpha}^{2} n_{\vec{q}}^{\alpha}}{2L\rho\Omega_{\vec{q}}^{\alpha}} q^{2}\delta\left(E_{\vec{k}+\vec{q}}-E_{\vec{q}}-\hbar\Omega_{\vec{q}}^{\alpha}\right) \\ &= 2\frac{2\pi}{\hbar}\frac{\hbar U_{\alpha}^{2}}{2L\rho}\int\frac{L}{2\pi\Omega_{\vec{q}}^{\alpha}} D\left(E_{\vec{k}+\vec{q}}\right) \\ &\times n_{\vec{q}}^{\alpha} q^{2}\delta\left(E_{\vec{k}+\vec{q}}-E_{\vec{q}}-\hbar\Omega_{\vec{q}}^{\alpha}\right) dq\end{aligned} \tag{2.19}$$

where $D(E)$ is the density of electronic states and n is the phonon occupation number. For acoustic phonon temperatures much less than room temperature we have

$$n_{\vec{q}}^{\alpha} = \frac{1}{e^{\hbar\Omega_{\vec{q}}^{\alpha}/k_{B}T}-1} \approx \frac{k_{B}T}{\hbar\Omega_{\vec{q}}^{\alpha}} \tag{2.20}$$

and thus

$$\frac{1}{\tau_{ac}^{\alpha}} = \frac{k_B T U_{\alpha}^2}{\pi \hbar^2 \rho v_{ac}^{\alpha 2} v_F} \tag{2.21}$$

where $v_{ac}^{\alpha} = \Omega_q^{\alpha}/q$ is the velocity of acoustic phonons. The mean free path for electrons traveling at the Fermi velocity($v_F = 3a\gamma/2\hbar$) is thus

$$l_{ac}^{\alpha} = v_F \tau_{ac}^{\alpha} = \frac{\pi \hbar^2 \rho v_{ac}^{\alpha 2} v_F^2}{k_B T U_{\alpha}^2}. \tag{2.22}$$

To estimate the mean free path we need to obtain a value for the deformation potential. To do this, we consider the perturbations in the energy eigenvalues for the nanotube electronic structure under atomic displacements $\vec{R}_{\alpha}$ due to phonon α. For a wavefunction ψ_n with energy E_n, the Hellman–Feynman theorem yields

$$\frac{\partial E_n}{\partial R_{\alpha i}} = \langle \psi_n | \frac{\partial H}{\partial R_{\alpha i}} | \psi_n \rangle \tag{2.23}$$

where $R_{\alpha i}$ is the component of the vector $\vec{R}_{\alpha}$ in the direction of the phonon polarization. For small displacements $\vec{u}_{\alpha}$ around the equilibrium atomic positions we have

$$\delta E_n = \langle \psi_n | \frac{\partial H}{\partial R_{\alpha i}} | \psi_n \rangle \, u_{\alpha i}. \tag{2.24}$$

The time-averaged deviation in the energy eigenvalues is

$$\langle \delta E_n^2 \rangle = \left(\langle \psi_n | \frac{\partial H}{\partial R_{\alpha i}} | \psi_n \rangle \right)^2 \langle u_{\alpha i}^2 \rangle \tag{2.25}$$

and using the expression for the vibrational modes

$$\vec{u}_{\alpha}(t) = \frac{1}{\sqrt{N_v}} \sum_{\vec{k}} \hat{u}^{\alpha} \left(\vec{k} \right) e^{i(\vec{k}\cdot\vec{r} - \omega t)} \tag{2.26}$$

we obtain

$$\langle \delta E_n^2 \rangle = \left(\langle \psi_n | \frac{\partial H}{\partial R_{\alpha i}} | \psi_n \rangle \right)^2 \frac{1}{N_v} \left(\hat{u}_i^{\alpha} \right)^2. \tag{2.27}$$

At thermal equilibrium, the amplitude of the vibrational modes is given by

$$(\hat{u}_i^\alpha)^2 = \frac{k_B T}{2L\rho\Omega_\alpha^2}, \tag{2.28}$$

therefore leading to the expression

$$\left\langle \delta E_n^2 \right\rangle^{1/2} = \sqrt{\frac{k_B T}{2L\rho N_v \Omega_\alpha^2}} \left\langle \psi_n \right| \frac{\partial H}{\partial R_{\alpha i}} \left| \psi_n \right\rangle. \tag{2.29}$$

The deformation potential is related to the matrix element through the definition

$$U_\alpha q = \left\langle \psi_n \right| \frac{\partial H}{\partial R_{\alpha i}} \left| \psi_n \right\rangle. \tag{2.30}$$

Therefore, Eqs. (2.29) and (2.30) allow the calculation of the deformation potential when the shift of the energy levels due to phonons is provided. We obtain an approximate value for the deformation potential for carbon nanotubes by considering the change in bandgap associated with strain (the impact of strain on carbon nanotube electronic structure will be discussed in more detail in the chapter on electromechanical devices). Theoretical work [24] has indicated that the bandgap change in armchair carbon nanotubes due to uniaxial strain is given by

$$\Delta E_g = \frac{3\gamma}{1+\nu}\sigma \tag{2.31}$$

where σ is the strain and $\nu = 0.2$ is the nanotube Poisson ratio (we consider armchair nanotubes because they are metallic). If we equate the strain with the dilatation of the volume due to the phonon atomic displacements

$$\sigma \sim \left\langle (\nabla \cdot \vec{u})^2 \right\rangle^{1/2} = \sqrt{\frac{k_B T}{2L\rho N_v} \frac{q}{\Omega_\alpha}} \tag{2.32}$$

and equate the change in bandgap in Eq. (2.31) with the average shift of the electronic levels obtained in Eq. (2.29), we obtain

$$U_\alpha = \frac{3\gamma}{1+\nu}. \tag{2.33}$$

Thus for a typical value of the overlap integral $\gamma = 2.5$ eV the maximum deformation potential is about 6 eV.

Returning to the mean free path, we consider a (10,10) zigzag nanotube which has 40 atoms per unit cell, with the unit cell 0.246 nm long. This

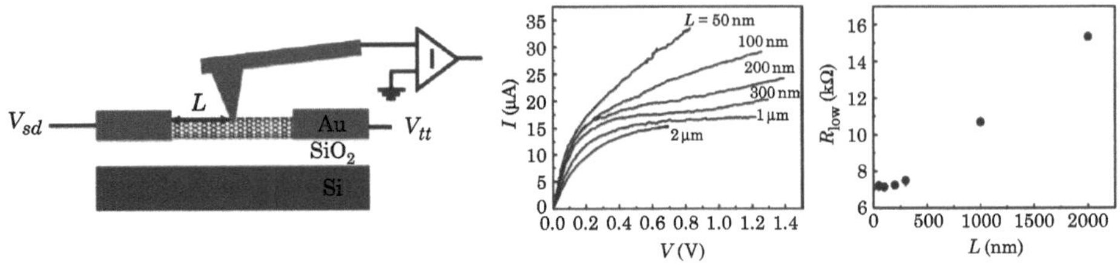

Figure 2.6 Sketch of experimental setup and results of studies of electron–phonon scattering in metallic carbon nanotubes. Current–voltage curves (middle panel) are measured between one of the contacts and the tip of an atomic force microscope (AFM); changing the position of the AFM tip along the tube gives the resistance as a function of the nanotube length (right panel). Figure from Ref. [20].

gives a mass density of 3.24×10^{-15} kg/m, and from Fig. 1.16 the acoustic phonon velocity is about 10^4 m/s. With the estimated value for the deformation potential of 6 eV we obtain

$$l_{ac} \approx 1.8\ \mu\text{m} \tag{2.34}$$

indicating that the mean free path is very long and that scattering with acoustic phonons is relatively inefficient.

This prediction for the long mean free path at small bias has been verified by measuring the current flowing through a metallic nanotube between a contact electrode and the tip of an atomic force microscope (Fig. 2.6). By using the tip of the atomic force microscope as a movable electrode, the resistance as a function of length is obtained, which can be related to the mean free path through the relation

$$R = \frac{h}{4e^2}\frac{L}{l} \tag{2.35}$$

where L is the length of the nanotube between the contact and the tip and l is the mean free path. Fitting this expression to the linear region of the data in the right panel of Fig. 2.6 gives a mean free path of 1.6 μm, in reasonable agreement with the theoretical estimate (differences may be due to the presence of other scattering mechanisms that reduce the mean free path, or to values of the parameters used to estimate the mean free path from the theoretical expression).

Robustness to Weak Disorder: For a thin conducting wire, weak disorder gives rise to weak localization, and a conductance that depends exponentially on the wire length $G \sim e^{-L/L_0}$ where L_0 is the localization length. The same feature can be seen in metallic carbon nanotubes, except that the decay length L_0 is found to be extremely long. This is illustrated by considering random disorder in on-site potential at every atomic site and calculating the conductance using a quantum transport approach [22]. Fig. 2.7 shows the calculated conductance as a function of nanotube length for three values of the strength of the random

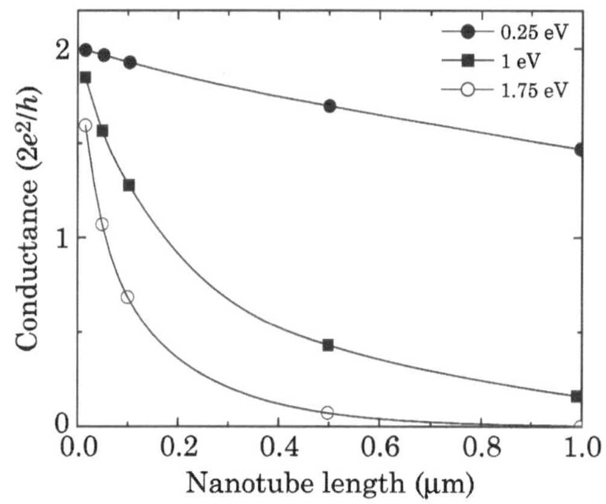

Figure 2.7 Effect of disorder on the conductance of a (10,10) metallic nanotube, for three values of the strength of the disorder. The key feature here is that the decay of the conductance occurs over a long length scale. Figure after Ref. [22].

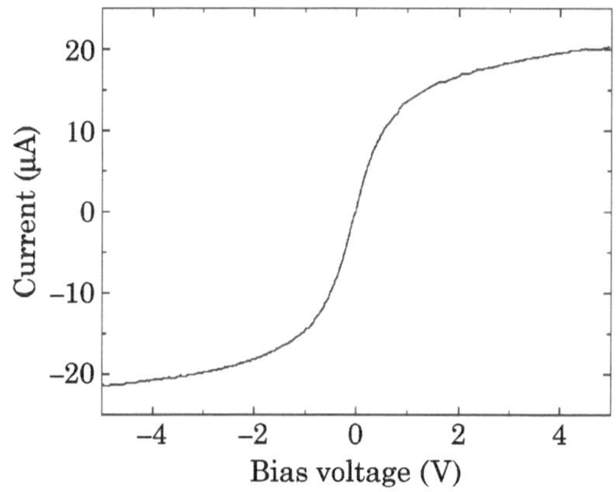

Figure 2.8 Current versus applied bias of a metallic nanotube at different temperatures. The differential conductance is largest at zero bias and saturates to much lower values at high biases. Figure after Ref. [25].

disorder. It is clear that there is an exponential-like drop of the conductance as the length of the nanotube increases. However, even for the cases of disorder strengths of 1 eV and 1.75 eV (comparable to the overlap integral $\gamma = 2.5$ eV), the decay length is found to be 3.4 μm and 1.4 μm, respectively; and even larger values are obtained if one considers weaker disorder or greater separation between the disordered sites, as for example would occur in the case of impurities. Thus, metallic carbon nanotubes are less sensitive to weak localization than conventional metallic wires. The origin of this behavior lies in the fact that the wavefunctions in carbon nanotubes span the nanotube circumference, so the strength of the scattering potential is "averaged" over the circumference, reducing the impact on the conductance. A consequence of this effect is that the localization length decreases as the diameter of the nanotube decreases [22,23].

2.3 High Bias Transport

At high biases, the current carrying capacity of carbon nanotubes is significantly affected by electron–phonon scattering. Fig. 2.8 shows the experimentally measured current–voltage characteristics of a small diameter nanotube. The conductance is largest at zero bias and decreases with increase in bias, a signature of increased electron–phonon scattering. To treat the regime of high-bias transport in metallic nanotubes, an approach based on the Boltzmann equation has been proposed [25]. The Boltzmann equation approach describes the time evolution and spatial dependence of the electron distribution functions $f_L(E,x)$ and $f_R(E,x)$, which represent left- and right-moving electrons. In the presence of scattering processes and a uniform electric field these equations are

$$\begin{aligned} \frac{\partial f_L}{\partial t} + v_F \frac{\partial f_L}{\partial x} + \frac{1}{\hbar}\frac{eV}{L}\frac{\partial f_L}{\partial k} &= \left[\frac{\partial f_L}{\partial t}\right]_{\text{scattering}} \\ \frac{\partial f_R}{\partial t} - v_F \frac{\partial f_R}{\partial x} - \frac{1}{\hbar}\frac{eV}{L}\frac{\partial f_R}{\partial k} &= \left[\frac{\partial f_R}{\partial t}\right]_{\text{scattering}} . \end{aligned} \tag{2.36}$$

Three sources of scattering are included to describe high bias transport in metallic carbon nanotubes: elastic scattering by defects, backscattering by phonons, and forward scattering by phonons. The elastic scattering is given by the expression

$$\left[\frac{\partial f_L}{\partial t}\right]_{\text{elastic}} = \frac{v_F}{l_e}\left(f_L - f_R\right) \tag{2.37}$$

where l_e is the elastic mean free path. The backscattering collisions with phonons lead to a rate of change of the occupation function

$$\begin{aligned} \left[\frac{\partial f_L(E)}{\partial t}\right]_{bp} = \frac{v_F}{l_{bp}} &\{[1 - f_L(E)]\, f_R(E + \hbar\Omega) \\ &- [1 - f_R(E - \hbar\Omega)]\, f_L(E)\} \end{aligned} \tag{2.38}$$

while the forward scattering with phonons is

$$\begin{aligned} \left[\frac{\partial f_L(E)}{\partial t}\right]_{fp} = \frac{v_F}{l_{fp}} &\{[1 - f_L(E)]\, f_L(E + \hbar\Omega) \\ &- [1 - f_L(E - \hbar\Omega)]\, f_L(E)\} . \end{aligned} \tag{2.39}$$

These equations are supplemented by boundary conditions at the contacts

$$\begin{aligned} f_R(E)|_{x=0} &= t_L^2 f_0(E - \mu_L) + \left(1 - t_L^2\right) f_L(E)|_{x=0} \\ f_L(E)|_{x=L} &= t_R^2 f_0(E - \mu_R) + \left(1 - t_R^2\right) f_R(E)|_{x=L} \end{aligned} \tag{2.40}$$

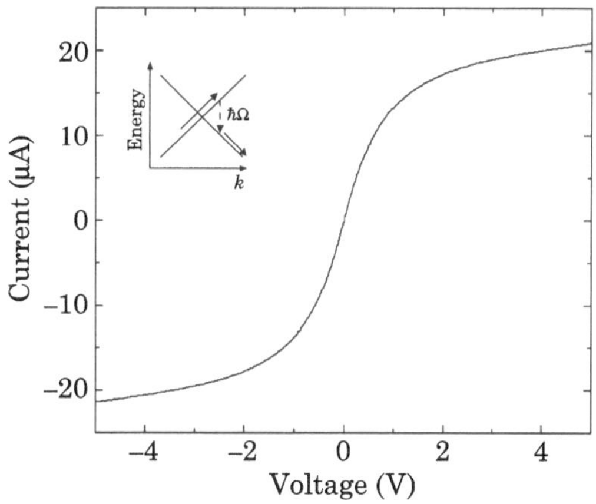

Figure 2.9 Calculated current versus voltage for a metallic carbon nanotube using the Boltzmann transport equation and electron–phonon scattering. The inset shows the electron–phonon scattering process whereby electrons with energy larger than the phonon energy emit a phonon and backscatter. Figure after Ref. [25].

where f_0 is the equilibrium Fermi distribution and $t_{L,R}$ are the transmission coefficients at the contacts. Once the distribution functions are determined by solving the Boltzmann equations with the boundary conditions, the current is calculated from

$$I = \frac{4e^2}{h} \int (f_L - f_R)\, dE \tag{2.41}$$

where the distribution functions can be evaluated at any (but the same) point x in the steady-state. Fig. 2.9 shows the numerically calculated [25] current versus voltage for a one micron long metallic carbon nanotube including electron–phonon scattering with 150 meV phonons, and with parameters $t_{L,R}^2 = 0.5$, $l_e = 300$ nm, $l_{pb} = 10$ nm, and $l_{pf} = \infty$. The excellent agreement with experiment indicates that the mean free path for optical phonon scattering is around 10 nm, and is dominated by scattering with phonons in the 150 meV range.

Because the mean free path for scattering with optical phonons is small, the conductance at high bias decreases appreciably in nanotubes that are much longer than this mean free path. If one assumes that all electrons incident from the left contact with energy 160 meV larger than the drain side Fermi energy are reflected by phonon emission, the maximum current that flows in a long nanotube (many mean free paths) at large biases is then approximately

$$I = \frac{4e^2}{h} 160 \text{ mV} = 25\ \mu\text{A}. \tag{2.42}$$

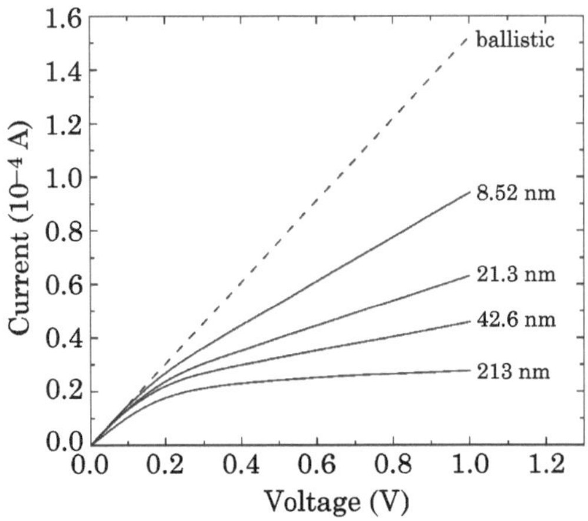

Figure 2.10 Computed current–voltage characteristics in the ballistic limit (dashed line) and with electron–phonon scattering for various lengths. For the longest nanotube considered (213 nm), the current is close to 25 μA, as suggested by Eq. (2.42). The current approaches the ballistic limit as the nanotube length decreases. Figure after Ref. [27].

A number of experiments have reported currents comparable to 25 μA in long nanotubes [20,25,26]. Recent simulations of the current–voltage characteristics in the ballistic limit and with electron–phonon interactions have also indicated that the scattering with optical phonons occurs over a length scale of a few tens of nanometers, as shown in Fig. 2.10. At small biases, the conductancedI/dV is nearly $4e^2/h$, independent of the nanotube length, indicating ballistic charge transport in the crossing subbands. As the bias increases, the current-carrying capacity and differential conductance are length dependent. The longest nanotube considered (length of 213 nm), is considerably longer than the mean free path of about 10 nm. The computed current for this nanotube is about 25 μA at a bias of 1 V, in agreement with Eq. (2.42). As the length of the nanotube decreases, the current carrying capacity increases and approaches the ballistic limit (dashed line) in Fig. 2.8.

It is worth mentioning that the experimentally measured mean free paths for optical phonon scattering are nearly five times smaller than the theoretical predictions. Reference [20] theoretically estimated the mean free path due to optical and zone boundary scattering to be about 50 nm but found that the experimental data could be explained only if a net mean free path of 10 nm was assumed. The reason for this disparity is unclear. One possibility is that the emitted phonons cannot easily dissipate to the environment, resulting in an excess of hot phonons, and the smaller experimentally observed mean free path.

In contrast to small diameter nanotubes, large diameter multiwall nanotubes show an increase in differential conductance with applied bias [8,28,29]. Fig. 2.11 shows the experimentally measured current and conductance versus bias for a nanotube with diameter of 15.6 nm [28]. The low bias conductance is

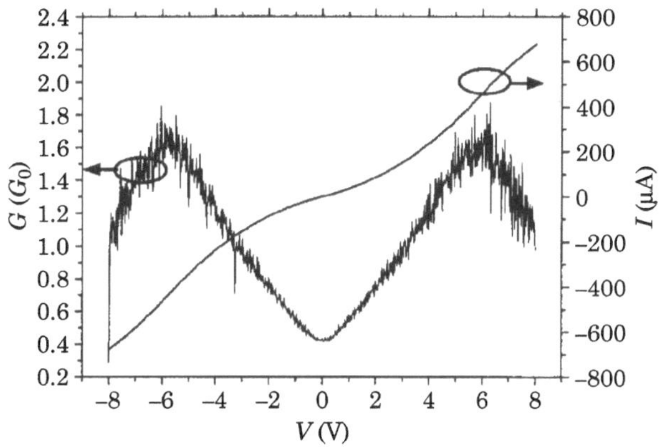

Figure 2.11 Observed I–V curve of a single multiwall carbon nanotube in the bias range from −8 to 8 V (right axis). The conductance around zero bias is $0.4G_0$, and increases linearly until an applied bias of 5.8 V where it decreases. The multiwall nanotube has more than 15 shells, and diameters and lengths of approximately 15.6 and 500 nm respectively. Figure from Ref. [28].

$0.4G_0$ instead of the maximum of $2G_0$. More importantly, the conductance increases with applied bias, a feature also seen in Ref. [8]. This is qualitatively different from the case of small diameter nanotubes discussed above where the conductance decreases with increase in bias (Fig. 2.8). There are many potential reasons for the increase in conductance with bias seen in these large diameter multiwall nanotubes. One possibility is that the inner walls of the multiwall nanotube start to carry current as the bias increases. However, recent theoretical work found that this mechanism is unlikely [30]. The most likely explanation for the increase in conductance with applied bias is Zener tunneling between noncrossing valence and conduction bands [31]. This process is illustrated in Fig. 2.12. Consider an electron incident into a noncrossing valence subband of the nanotube from the left contact. This electron can either tunnel to the noncrossing conductance subband with the same symmetry (dashed arrow) or be Bragg reflected back into the left contact (dotted arrow). The barrier for Zener tunneling in the noncrossing subband is ΔE_{NC}, and the width of the tunneling barrier depends on the potential profile in the nanotube. Since the barrier height ΔE_{NC} increases with decrease in nanotube diameter, it turns out that the noncrossing subbands of small diameter metallic nanotubes do not carry significant current [27,31]. On the other hand, for large diameter nanotubes, the barrier for tunneling ΔE_{NC} is much smaller, and as a result, the probability for tunneling increases with increase in nanotube diameter. Self-consistent calculations of the current–voltage characteristics of short nanotubes indeed show a significant diameter dependence of the conductance arising from tunneling into noncrossing/semiconducting subbands [27,31].

Finally, we discuss the electrostatic potential drop in carbon nanotubes at low and high biases. We will limit the discussion here to perfect coupling between

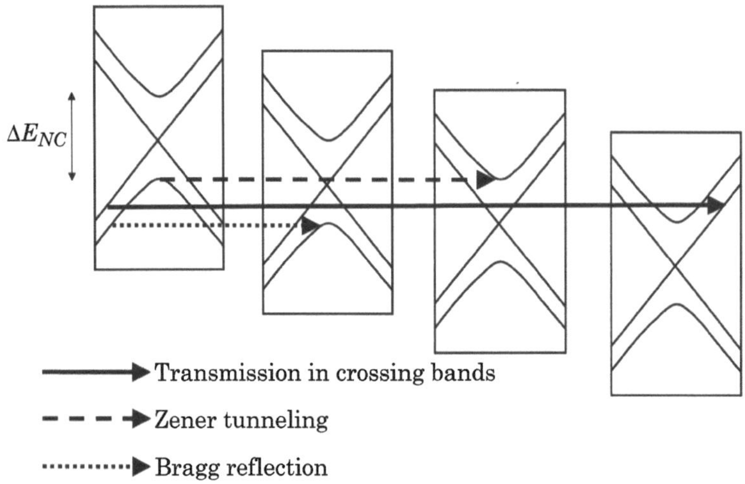

Figure 2.12 Each rectangular box is a plot of energy versus wave vector, with the subband bottom equal to the electrostatic potential. Only a few subbands are shown for clarity. The three processes shown are direct transmission (solid line), Bragg reflection (dotted line), and intersubband tunneling (dashed line). Figure after Ref. [31].

the nanotube and contacts. The nanotube conductance is then determined by the number of subbands carrying current and scattering due to electron–phonon interaction inside the nanotube. Note that an additional resistance at the nanotube-contact interface will cause the applied bias to drop across this resistance, in addition to the drop across the nanotube.

At low bias, smaller than the energy of optical and zone boundary phonons (160 meV), electron–phonon scattering is suppressed, and hence defect-free nanotubes are essentially ballistic. In this low bias limit, the applied bias primarily drops across the two ends of the nanotube as shown in Fig. 2.13 (a). Interestingly, even though the nanotube is ballistic, the electric field near the contact depends on the tube diameter. The electric field at the center of the nanotube increases with increase in diameter because the density of states per atom decreases with increase in diameter, as shown for example in Eq. (1.41). This makes screening in the larger diameter nanotubes less effective. When the applied bias increases, allowing the emission of optical and zone boundary phonons, the electrostatic potential drops uniformly over the length of the nanotube provided that the nanotube length is many times the mean free path. The potential drop in Fig. 2.13 (b) corresponds to this case.

2.4 Capacitance and Inductance

In this section, we discuss the concepts of capacitance and inductance of a carbon nanotube, and then describe how they impact the velocity of

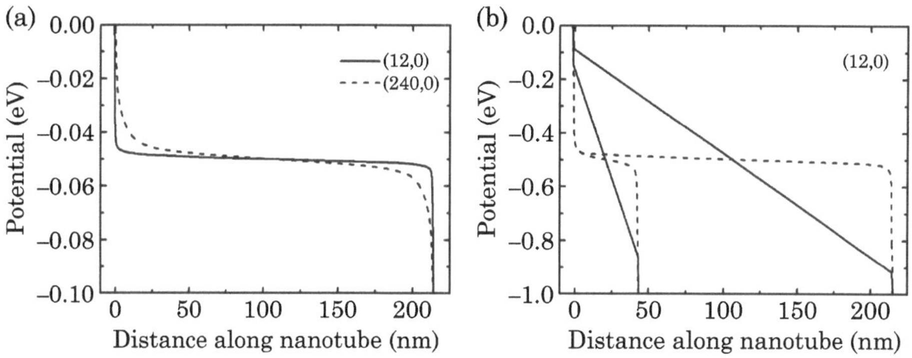

Figure 2.13 Calculated electrostatic potential along the nanotube axis. (a) Low bias potential for (12,0) and (240,0) nanotubes, which have diameters of 0.94 and 18.8 nm, respectively. The applied bias is 100 mV. The screening for the large-diameter nanotube is significantly poorer. The nanotube length is 213 nm. (b) The potential as a function of position is shown for (12,0) nanotubes of lengths 42.6 and 213 nm in the presence of scattering (solid line), with the potential profile in the ballistic limit (dashed line) shown for comparison. Figure after Ref. [27].

electromagnetic signal propagation in nanotubes. In conductors, there are two sources of energy associated with the addition of charge (capacitance) and current flow (inductance): classical and material-specific. The former follows from classical electrodynamics and leads to the common electrostatic capacitance C_c and magnetic inductance L_m discussed in basic physics textbooks. These classical values are obtained by assuming that the conductor is a perfect metal with an infinite density of states near the Fermi level. The second source is a contribution that arises from the fact that for real conductors the density of states is not infinite at the Fermi level. This leads to corrections to the classical values of C_c and L_m, which are often referred to as the "quantum capacitance" and "quantum inductance". However, since these corrections are not generally quantized or do not generally arise from quantization conditions, the terms "intrinsic capacitance" and "intrinsic inductance" will be used here.

2.4.1 Classical Capacitance

The classical capacitance between two conductors is a geometry-specific quantity. The geometry we consider is a metallic nanotube of diameter d at a distance h above a grounded conducting plate (Fig. 2.14). To obtain the capacitance, we assume that there is a fixed, uniformly distributed charge σ per unit surface area of the nanotube, and use an image charge construction to obtain the potential

$$V = -\frac{\sigma d}{2\varepsilon} \ln \frac{4h}{d} \tag{2.43}$$

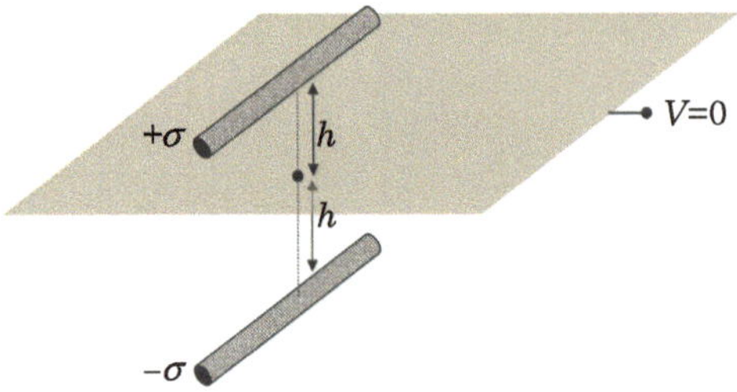

Figure 2.14 Positively charged metallic nanotube over a ground plane, and its image charge.

where ε is the dielectric constant of the medium surrounding the nanotube. Using the definition of the capacitance $Q = CV$, the classical capacitance per unit nanotube length is

$$C_c = \frac{2\pi\varepsilon}{\ln(4h/d)}. \tag{2.44}$$

For a nanotube with $d = 1.5$ nm and $h = 100$ nm, the classical capacitance in vacuum is

$$C_c = 0.01 \text{ aF/nm}. \tag{2.45}$$

Note that the derivation of the classical capacitance assumed a *fixed* charge on the nanotube, which did not depend on the potential V. This assumption is equivalent to having a fixed Fermi level position, which can only occur when the density of states at the Fermi level is infinite. However, real metals have a finite density of states at the Fermi level, and a shift of the Fermi level gives rise to induced charge and an additional intrinsic capacitance, as we now discuss.

2.4.2 Intrinsic Capacitance

We present two derivations of the intrinsic capacitance. For the first derivation, we consider a nanotube of length L with discrete energy levels separated in momentum space by $\delta k = 2\pi/L$. The energy needed to add an electron to this one-dimensional conductor at the Fermi energy is

$$\delta E = \delta k \left.\frac{\partial E}{\partial k}\right|_{E=E_F} = \frac{1}{2\pi D(E_F)}\delta k \tag{2.46}$$

where $\frac{\partial E}{\partial k} = \frac{1}{2\pi D(E_F)}$ is evaluated at the Fermi energy E_F and $D(E_F)$ is the density of states per unit length at energy E. Eq. (2.46) can be written as

$$\delta E = \frac{1}{D(E_F)L} \tag{2.47}$$

and equating δE to the expression for capacitive energy e^2/LC, the intrinsic capacitance C_i per unit length is

$$C_i = e^2 D(E_F). \tag{2.48}$$

A second method of obtaining the expression for the intrinsic capacitance is to compute the change in charge density due to a change in potential. Consider a nanotube in equilibrium at an electrostatic potential V_0. If the electrostatic potential changes to $V_0 + \delta V$, then the change in charge per unit nanotube length is given by

$$\delta Q = e \int dED(E)[f(E - E_F) - f(E - e\delta V - E_F)]. \tag{2.49}$$

Assuming that the density of states does not change appreciably around the Fermi energy. We obtain

$$\delta Q = e^2 D(E_F)\delta V, \tag{2.50}$$

and the intrinsic capacitance is then given by

$$C_i = \frac{\delta Q}{\delta V} = e^2 D(E_F) \tag{2.51}$$

which is the same as Eq. (2.48). For a metallic carbon nanotube, the density of states per unit length at the Fermi energy is $D(E_F) = \frac{a\sqrt{3}}{\pi^2 R\gamma} = \frac{4}{\pi\hbar v_f}$. Using this, we find that the intrinsic capacitance per unit length is [32,33]

$$C_i = \frac{4e^2}{\pi\hbar v_F} = 0.4 \text{ aF/nm}. \tag{2.52}$$

This value for the capacitance is a factor of 10 larger than the classical capacitance obtained from Eq. (2.45) if we consider a nanotube in vacuum 100 nm above a ground plane. Since the classical and gate capacitance add like capacitors in series, the total capacitance will be dominated by the lower of the two values. In most cases, the classical capacitance dominates, but there are situations where this is not necessarily the case. For example, we will see in Chapter 8 that liquid-gating is one situation where the intrinsic capacitance dominates. (As an additional note, the true classical limit where only the classical capacitance is used emerges naturally from the above treatment. Indeed, the intrinsic capacitance is proportional to the density of states at the Fermi level as given by Eq. (2.51). For an infinite density of states (the classical limit), the series capacitance is entirely determined by the classical capacitance.)

2.4.3 Classical Inductance

The magnetic inductance of a nanotube carrying a steady current I over a plane (Fig. 2.14) is calculated using the method of images and Ampère's law, giving the magnetic field

$$B\hat{\phi} = \left(\frac{\mu I}{2\pi\,|z-h|} - \frac{\mu I}{2\pi\,|z+h|}\right)\hat{\phi} \tag{2.53}$$

where $\hat{\phi}$ is the circumferential unit vector, z is the coordinate perpendicular to the ground plane and μ is the permeability of the material in which the nanotube is embedded. The magnetic flux per unit length through a loop made of the two wires connected at positive and negative infinity is

$$\Phi = \int \vec{B}\cdot d\vec{a} = \frac{\mu I}{\pi}\ln\frac{4h}{d} \tag{2.54}$$

and using the relation $\Phi = L_m I$ we obtain the magnetic inductance per unit length

$$L_m = \frac{\mu}{2\pi}\ln\frac{4h}{d}. \tag{2.55}$$

For $d = 1.5$ nm and $h = 100$ nm, the magnetic inductance is $L_m \sim 2.2 \times 10^{-3}$ pH/nm.

2.4.4 Intrinsic Inductance

The physical origin of the intrinsic inductance L_K is excess kinetic energy associated with current flow [6,32,33]. This is illustrated in Fig. 2.15, which shows ballistic electrons flowing in the shaded band between the source Fermi level and the drain Fermi level. Here $E = E_F^R$ is the equilibrium Fermi energy. The average energy of excess electrons in the bias window is

$$E = \tfrac{1}{2}\left(E_F^L - E_F^R\right) \tag{2.56}$$

while the number of electrons in the bias window is

$$N = \frac{D(E_F)}{2}\left(\frac{E_F^L - E_F^R}{2}\right) \tag{2.57}$$

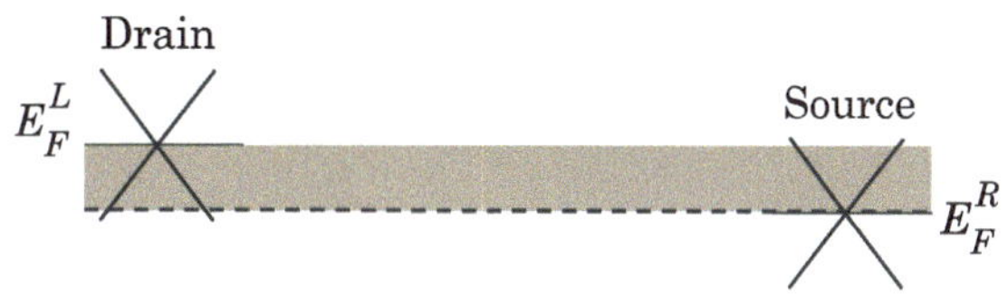

Figure 2.15 Transport of electrons in the shaded band between the drain and source Fermi energies leads to an excess kinetic energy, which gives rise to kinetic inductance in addition to magnetic inductance.

which includes factors of 1/2 because only right moving carriers contribute to the current. From the above two expressions, we obtain the excess kinetic energy

$$E_K = \frac{1}{8} D(E_F) \left(E_F^R - E_F^L\right)^2 . \tag{2.58}$$

For ballistic transport in the nanotube the current is given by

$$I = \frac{4e}{h} \left(E_F^L - E_F^R\right) \tag{2.59}$$

and the excess kinetic energy is of the form

$$E_K = D(E_F) \frac{\pi^2 \hbar^2 I^2}{32e^2} . \tag{2.60}$$

Equating (2.60) to an inductive energy $L_K I^2/2$, the kinetic inductance is

$$L_K = \frac{\pi^2 \hbar^2 D(E_F)}{16e^2} . \tag{2.61}$$

Using the expression for the density of states at the Fermi energy of metallic carbon nanotubes, we obtain the result

$$L_K = \frac{\pi \hbar}{4e^2 v_F} . \tag{2.62}$$

Substituting the Fermi velocity$v_F = 8 \times 10^5$ m/s in the above expression, the kinetic inductance of a single metallic carbon nanotube is

$$L_K = 4 \text{ pH/nm}. \tag{2.63}$$

The kinetic inductance of a single nanotube is about a thousand times larger than the classical inductance and cannot be neglected in modeling of carbon nanotube interconnects. However, at biases larger than 160 mV, scattering due to optical phonons decreases the conductance and leads to relaxation of the incident carriers, causing a decrease of the kinetic inductance for nanotubes much longer than the mean free path.

2.4.5 Electromagnetic Wave Propagation

The large value of kinetic inductance has an important effect on the speed at which signals are propagated in a transmission line consisting of a single nanotube. In general the velocity of electromagnetic wave propagation is given by

$$v = \sqrt{\frac{1}{LC}} \tag{2.64}$$

where L is the total inductance and C is the total capacitance. The wave velocity for signal transmission in the nanotube transmission line is therefore

$$v = \sqrt{\frac{1}{L_K + L_m}\left(\frac{1}{C_c} + \frac{1}{C_i}\right)}. \tag{2.65}$$

For a single nanotube the kinetic inductance dominates over the magnetic inductance; thus in a situation where the intrinsic capacitance is much larger than the classical capacitance the wave velocity for propagation of an electromagnetic signal is (using Eqs. (2.52) and (2.62))

$$v \approx \sqrt{\frac{1}{L_K C_i}} = v_F \tag{2.66}$$

i.e. the Fermi velocity of electrons. A different situation arises in the case of a large number of nanotubes in parallel where the classical inductance dominates. The expression for the velocity of signal propagation in that case reduces to

$$v \approx \sqrt{\frac{1}{L_m C_c}}. \tag{2.67}$$

The last expression is the velocity of propagation of a system where the intrinsic capacitance and inductance are neglected.

References

1. F. Kreupl, A.P. Graham, G.S. Duesberg, W. Steinhogl, M. Liebau, E. Unger and W. Honlein, "Carbon nanotubes in interconnect applications", *Microelectron. Eng.*, Vol. 64, p. 399, 2002.

2. Q. Ngo, D. Petranovic, S. Krishnan, A.M. Cassell, Q. Ye, J. Li, M. Meyyappan and C.Y. Yang, "Electron transport through metal-multiwall carbon nanotube interfaces", *IEEE Trans. Nanotechnol.*, Vol. 3, p. 311, 2004.
3. B.Q. Wei, R. Vajtai and P.M. Ajayan, "Reliability and current carrying capacity of carbon nanotubes", *Appl. Phys. Lett.*, Vol. 79, p. 1172, 2001.
4. M. Bockrath, D.H. Cobden, J. Lu, A.G. Rinzler, R.E. Smalley, L. Balents and P.L. McEuen, "Luttinger-liquid behaviour in carbon nanotubes", *Nature*, Vol. 397, p. 598, 1999.
5. E. Graugnard, P.J. dePablo, B. Walsh, A.W. Ghosh, S. Datta and R. Reifenberger, "Temperature dependence of the conductance of multiwalled carbon nanotubes", *Phys. Rev.* B, Vol. 64, p. 125407, 2001.
6. R. Tarkiainen, M. Ahlskog, J. Penttilä, L. Roschier, P. Hakonen, M. Paalanen and E. Sonin, "Multiwalled carbon nanotube: Luttinger versus Fermi liquid", *Phys. Rev.* B, Vol. 64, p. 195412, 2001.
7. R. Landauer, "Spatial variation of currents and fields due to localized scatterers in metallic conduction", *IBM J. Res. Dev.*, Vol. 1, p. 223, 1957.
8. S. Frank, P. Poncharal, Z.L. Wang and W.A. de Heer, "Carbon nanotube quantum resistors", *Science*, Vol. 280, p. 1744, 1998.
9. P. Poncharal, C. Berger, Y. Yi, Z.L. Wang and W.A. de Heer, "Room temperature ballistic conduction in carbon nanotubes", *J. Phys. Chem.* B, Vol. 106, p. 12104, 2002.
10. S. Sanvito, Y.K. Kwon, D. Tomanek and C.J. Lambert, "Fractional quantum conductance in carbon nanotubes", *Phys. Rev. Lett.*, Vol. 84, p. 1974, 2000.
11. M.P. Anantram, S. Datta and Y. Xue, "Coupling of carbon nanotubes to metallic contacts", *Phys. Rev.* B, Vol. 61, p. 14219, 2000.
12. M.P. Anantram, "Which nanowire couples better electrically to a metal contact: Armchair or zigzag nanotube?", *Appl. Phys. Lett.*, Vol. 78, p. 2055, 2001.
13. H.J. Choi, J. Ihm, Y-G Yoon and S.G. Louie, "Possible explanation for the conductance of a single quantum unit in metallic carbon nanotubes", *Phys. Rev.* B, Vol. 60, p. R14009, 1999.
14. J. Kong, E. Yenilmez, T.W. Tombler, W. Kim, H. Dai, R.B. Laughlin, L. Liu, C.S. Jayanthi and S.Y. Wu, "Quantum interference and ballistic transmission in nanotube electron waveguides", *Phys. Rev. Lett.*, Vol. 87, p. 106801, 2001.
15. A.N. Andriotis, M. Menon and G.E. Froudakis, "Various bonding configurations of transition-metal atoms on carbon nanotubes: Their effect on contact resistance", *Appl. Phys. Lett.*, Vol. 76, p. 3890, 2000.
16. C.K. Yang, J. Zhao and J.P. Lu, "Binding energies and electronic structures of adsorbed titanium chains on carbon nanotubes", *Phys. Rev.* B, Vol. 66, p. 041403(R), 2002.
17. Y. Liu, "Ab initio study of Ti-contacted single-walled carbon nanotube", *Phys. Rev.* B, Vol. 68, p. 193409, 2003.
18. J.J. Palacios, A.J. Pérez-Jiménez, E. Louis, E. SanFabián and J.A. Vergés, "First-principles phase-coherent transport in metallic nanotubes with realistic contacts", *Phys. Rev. Lett.*, Vol. 90, p. 106801, 2003.
19. N. Nemec, D. Tománek and G. Cuniberti, "Contact dependence of carrier injection in carbon nanotubes: An ab initio study", *Phys. Rev. Lett.*, Vol. 96, p. 076802, 2006.
20. J. Park, S. Rosenblatt, Y. Yaish, V. Sazonova, H. Ustunel, S. Braig, T.A. Arias, P.W. Brouwer and P.L. McEuen, "Electron–phonon scattering in metallic single-walled carbon nanotubes", *Nano Lett.*, Vol. 4, p. 517, 2004.
21. T. Ando and T. Nakanishi, "Impurity scattering in carbon nanotubes – absence of back scattering", *J. Phys. Soc. Jpn.*, Vol. 67, p. 1704, 1998.
22. M.P. Anantram and T.R. Govindan, "Conductance of carbon nanotubes with disorder: a numerical study", *Phys. Rev.* B, Vol. 58, p. 4882, 1998.
23. C.T. White and T.N. Todorov, "Carbon nanotubes as long ballistic conductors", *Nature*, Vol. 393, p. 240, 1998.

24. L. Yang and J. Han, "Electronic structure of deformed carbon nanotubes", *Phys. Rev. Lett.*, Vol. 85, p. 154, 2000.
25. Z. Yao, C.L. Kane and C. Dekker, "High-field electrical transport in single-wall carbon nanotubes", *Phys. Rev. Lett.*, Vol. 84, p. 2941, 2000.
26. P.G. Collins, M. Hersam, M. Arnold, R. Martel and Ph. Avouris, "Current saturation and electrical breakdown in multiwalled carbon nanotubes", *Phys. Rev. Lett.*, Vol. 86, p. 3128, 2001.
27. A. Svizhenko and M.P. Anantram, "Effect of scattering and contacts on current and electrostatics in carbon nanotubes", *Phys. Rev.* B, Vol. 72, p. 085430, 2005.
28. Y.X. Liang, Q.H. Li and T.H. Wang, "Current saturation in multiwalled carbon nanotubes by large bias", *Appl. Phys. Lett.*, Vol. 84, p. 3379, 2004.
29. B. Bourlon, D.C. Glattli, B. Plaçais, J.M. Berroir, C. Miko, L. Forró and A. Bachtold, "Geometrical dependence of high-bias current in multiwalled carbon nanotubes", *Phys. Rev. Lett.*, Vol. 92, p. 026804, 2004.
30. Y.-G. Yoon, P. Delaney and S.G. Louie, "Quantum conductance of multiwall carbon nanotubes", *Phys. Rev.* B, Vol. 66, p. 073407, 2002.
31. M.P. Anantram, "Current-carrying capacity of carbon nanotubes", *Phys. Rev.* B, Vol. 62, p. 4837, 2000.
32. M.W. Bockrath, Carbon nanotubes: Electrons in one dimension, Ph. D thesis, University of California, Berkeley, 1999.
33. P.J. Burke, "Luttinger liquid theory as a model of the GHz electrical properties of carbon nanotubes", *IEEE Trans. Nanotechnol.*, Vol. 1, p. 129, 2002.

3 Physics of Nanotube/Metal Contacts

3.1 Introduction

Electrical contacts play a crucial role in electronic devices, and much work has been devoted to understanding and controlling the properties of contacts to traditional bulk semiconductors. The early work of Schottky, Mott and Bardeen established the fundamental principles that govern the properties of contacts to bulk semiconductors, and these concepts have been critical in the interpretation of experimental measurements, and in developing further refined models. In addition, many of the engineering solutions to improve contact quality are rooted in this basic understanding of contact properties. Because carbon nanotubes have fundamental properties that are much different from bulk semiconductor materials, much of the concepts developed for traditional contacts need to be re-visited, and entirely new ways to think about electrical contacts in the context of carbon nanotubes need to be considered. This "discovery" phase is similar to the one that occurred for bulk semiconductors that led to much progress in microelectronics. The need for new approaches becomes immediately clear if one considers the geometry of the contacts between metals and carbon nanotubes: while traditional contacts are essentially planar, nanotube contacts can show various structures. Fig. 3.1 shows a contact between a nanotube and SiC [1]. There, carbon nanotubes are deposited on Si and heating leads to the formation of a SiC film through the reaction $C + Si \rightarrow SiC$; the unreacted portion of the carbon nanotube protrudes from the film. Thus, in this situation, the carbon nanotube terminates at the SiC

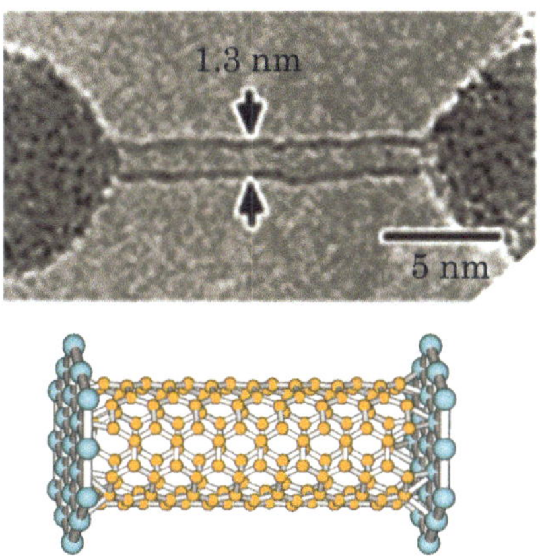

Figure 3.1 Top: transmission electron micrograph of a SiC/nanotube contact. Figure from Ref. [1]. Bottom: rendition of the contact showing that the nanotube does not penetrate into the metal, but rather is "end-bonded" to the metal surface. Figure from Ref. [2].

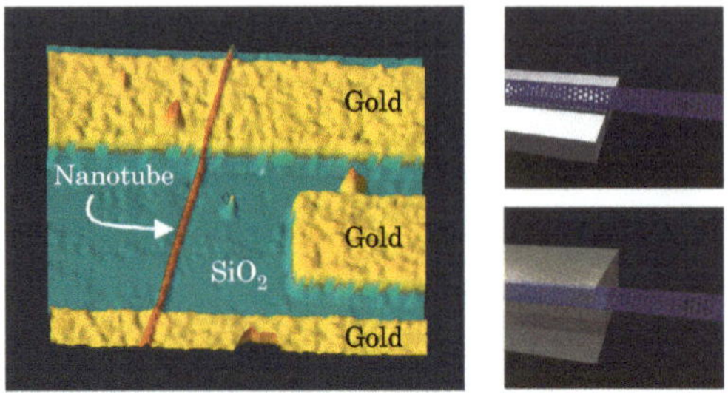

Figure 3.2 Left: atomic force microscope image of a carbon nanotube forming "side contacts" to gold electrodes. Figure from Ref. [3]. Right: sketch of two possible realizations of side contacts: the top figure shows a nanotube laying on a metal contact and the bottom figure shows a nanotube embedded in the metal.

surface, and the nanotube is said to be "end-bonded" to the metal. Even though this situation may seem at first very similar to a traditional contact where the semiconductor terminates and bonds with the metal, the small area of the contact and the quasi-one-dimensional nature of the carbon nanotube render the physics of this contact much different from that of traditional planar contacts. Fig. 3.2 shows a different situation where the nanotube is "side-contacted" by the metal. In this case, the interaction of the nanotube and the metal surface is predominantly of van der Waals character. This type of contact has two configurations: in the first case, the nanotube simply lays on the metal, while in the second case the nanotube is completely embedded in the metal. In turns out that these side-contacted contacts also have properties that are significantly different from those of bulk contacts. Thus, the different contact geometries and the reduced dimensionality of nanotubes can have a strong effect on the contact behavior, as we will discuss in this chapter.

3.2 End-Bonded Contacts

In traditional contacts between metals and semiconductors, a body of experimental and theoretical work has shown that Fermi level pinning usually dominates the contact behavior, leading to a Schottky barrier at the contact. Fig. 3.3 shows the measured [4] Schottky barrier height for contacts between Si and various metals. The measured values of the Schottky barriers vary by only 0.3 eV for different metals, despite the fact that the metal work functions change by almost 1 eV.

The simplest model for the barrier height consists of the bare alignment (i.e. without charge transfer) of the metal Fermi level and the semiconductor bandgap, as illustrated with the schematic in Fig. 3.3. In this model, the barrier height for electrons ϕ_{b0} is given by the difference between the the metal work

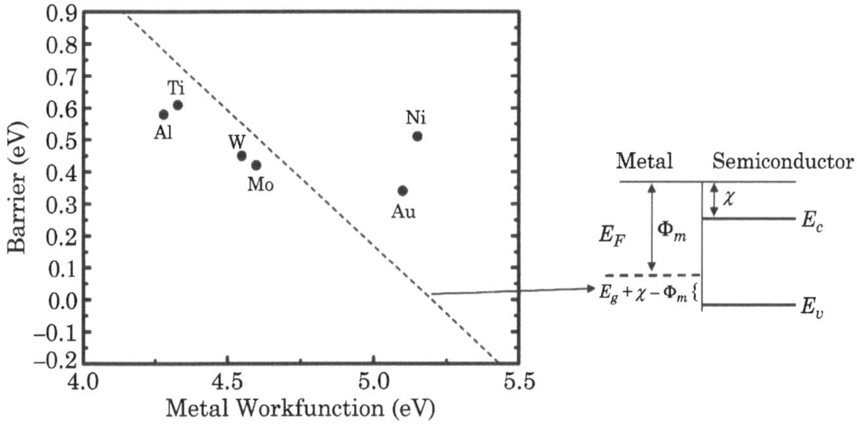

Figure 3.3 Measured Schottky barrier heights for traditional contacts between silicon and various metals. The dashed line represents the behavior predicted from Eq. (3.1), which corresponds to the direct band alignment between the metal Fermi level and the semiconductor valence and conduction bands, as shown in the schematic on the right. This simple model fails to explain the experimental data. See text for details.

function Φ_m and the semiconductor electron affinity χ,

$$\begin{aligned} \phi_{b0} &= \Phi_m - \chi \quad \text{for electrons} \\ \Delta_0 &= E_g + \chi - \Phi_m \quad \text{for holes} \end{aligned} \tag{3.1}$$

where the expression for the hole barrier height Δ_0 is also given, and can be obtained from the fact that $E_g = \phi_{b0} + \Delta_0$. As Fig. 3.3 indicates, this simple model fails to describe the experimentally measured barrier heights in planar bulk contacts since it would predict a strong linear dependence on the metal work function instead of the nearly constant behavior observed experimentally. The nearly constant barrier heights can be explained using the concept of Fermi level pinning. While an infinitely large semiconductor has a true electronic bandgap, when a surface is introduced at the metal/semiconductor interface, boundary conditions in the solutions of Schrödinger's equation must be taken into account, and lead to the appearance of electronic states in the semiconductor bandgap. These so-called metal-induced gap states (MIGS) decay exponentially away from the interface, and locally change the "neutrality" level in the semiconductor, i.e. the position of the Fermi level where the charge near the semiconductor surface vanishes. This situation is illustrated in Fig. 3.4. In general, the Fermi level of the isolated metal and the charge neutrality of the isolated semiconductor will not be aligned, and upon creation of an interface between the two materials, this leads to a charge on the semiconductor near the interface (the charge is localized near the interface because the MIGS decay rapidly away from the interface). The charge on the semiconductor is balanced by an image charge in the metal, of equal magnitude but opposite charge.

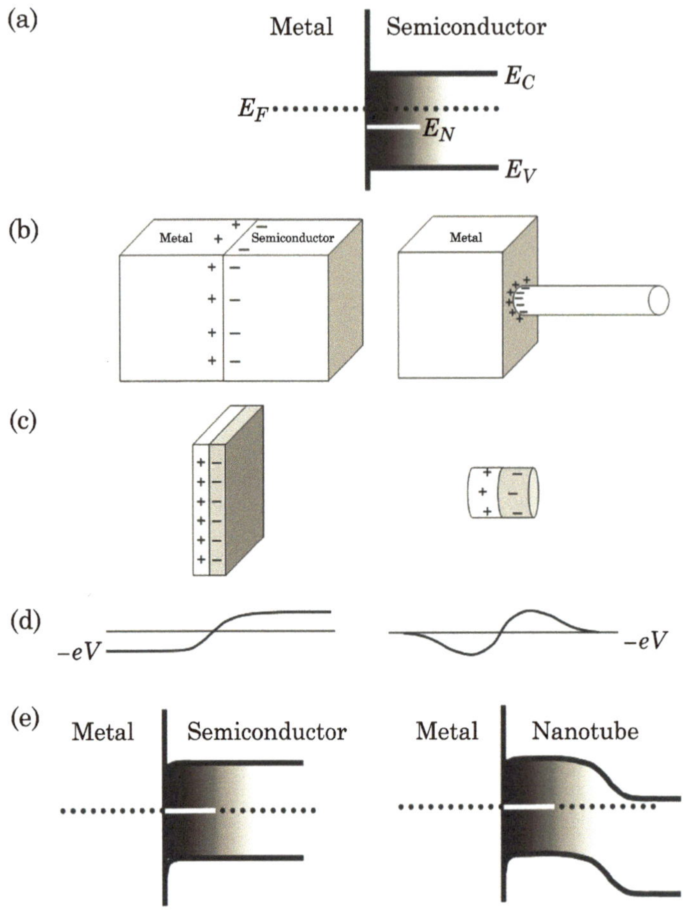

Figure 3.4 Schematics depicting Fermi level pinning in metal/semiconductor interfaces for the cases of bulk semiconductors and nanotubes end-bonded to metals. (a) shows the bare band alignment at the metal/semiconductor interface. The shaded area represents the metal-induced gap states (MIGS) that decay exponentially from the interface. E_N is the charge neutrality level for the semiconductor electronic structure including the MIGS. (b) and (c) show that the charge distribution at the interface consists of a dipole sheet for the bulk contact and a dipole ring for contacts to nanotubes. These charge distributions lead to the electrostatic potentials depicted in (d). As shown in (e), this electrostatic potential causes a bending of the bands. In the case of the bulk contact there is a constant potential shift far away from the interface; for the nanotube however, the band-bending decays away from the interface.

Thus, for a bulk metal/semiconductor interface, a dipole sheet is created at the interface; the electrostatic potential of an infinite dipole sheet is sketched in Fig. 3.4 (d), showing that it leads to a constant shift of the potential far away from the interface, despite the fact that the charge is localized in a thin region around the interface. As a consequence of this electrostatic potential the valence and conduction bands bend in the semiconductor until the metal Fermi level is aligned with the charge neutrality level due to MIGS. As will be discussed further below, additional band-bending due to doping of the semiconductor occurs on a much larger length scale (see Fig. 3.6).

The critical difference between conventional contacts and end-bonded nanotube contacts is due to the different geometry of the charge distribution generated at the interface. Instead of a dipole sheet, for a nanotube the charge distribution is that of a dipole ring, Fig. 3.4 (c). The key is that the electrostatic potential of a dipole ring falls off as the inverse distance squared (Fig. 3.4 (d)), and thus any potential shift near the interface quickly vanishes. We now discuss these effects in more mathematical detail.

Because the MIGS decay exponentially away from the interface, the charge due to MIGS on the semiconductor can be modeled as [5]

$$\sigma(z) = D_0(E_N - E_F)e^{-qz} \tag{3.2}$$

where D_0 is the density of MIGS, E_N is the position of the neutrality level at the interface, E_F is the metal Fermi level and q determines the distance over which the MIGS decay away from the interface. This model makes the explicit assumption that the MIGS are evenly distributed in energy. The factor $(E_N - E_F)$ shows that when the metal Fermi level is above (below) the semiconductor charge neutrality level, negative (positive) charge is generated because additional electronic states are filled (empty). Because the MIGS decay over a distance q^{-1}, this charge will be localized to a region of width q^{-1} near the interface.

The presence of this charge near the interface will change the electrostatic potential according to

$$V(z) = \int_{-\infty}^{\infty} K(z - z')\sigma(z')dz' \tag{3.3}$$

where $K(z)$ is the electrostatic kernel. For the charge dipole sheet with infinite cross-sectional area, the electrostatic kernel is

$$K_{\text{bulk}}(z - z') = 4\pi\varepsilon^{-1}\left|z - z'\right| \tag{3.4}$$

and the electrostatic potential in the semiconductor ($z > 0$) is calculated as

$$\begin{aligned} 4\pi\varepsilon V(z) = &- \int_{-\infty}^{0} (z - z')\,\sigma(z')\,dz' + \int_{0}^{z} (z - z')\,\sigma(z')\,dz' \\ &- \int_{z}^{\infty} (z - z')\,\sigma(z')\,dz'. \end{aligned} \tag{3.5}$$

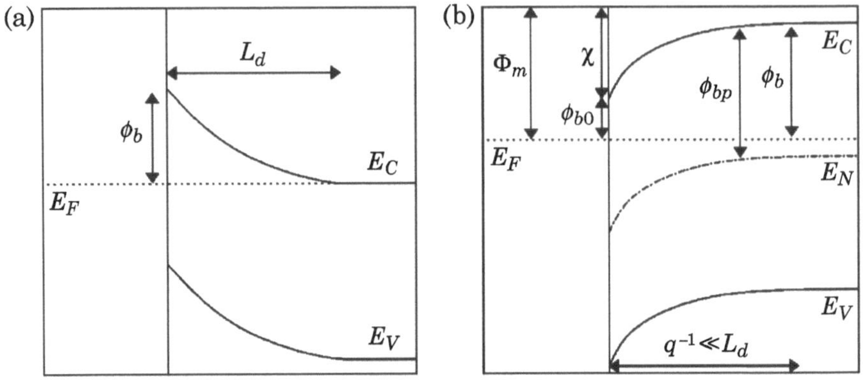

Figure 3.5 (a) The band-bending for a traditional semiconductor/metal junction. The band-bending occurs over a length scale L_d, which is the Debye length. (b) The near interface region, i.e. distances much smaller than L_d, is shown. Figure from Ref. [5].

The first term in this equation is the potential due to the image charge in the metal while the second and third terms arise from the charge in the semiconductor. Using the charge distribution from Eq. (3.2) we get

$$V_{\text{bulk}}(z) = (2\pi\varepsilon)^{-1} D_0 (E_N - E_F) q^{-2} \left(1 - e^{-qz}\right). \tag{3.6}$$

At distances $z \gg q^{-1}$ the potential attains the constant value $(2\pi\varepsilon)^{-1} D_0 (E_N - E_F) q^{-2}$ which brings the metal Fermi level closer to the charge neutrality level. In fact, the new Schottky barrier for electrons ϕ_b is

$$\phi_b = \phi_{b0} - e(2\pi\varepsilon)^{-1} q^{-2} D_0 (E_N - E_F) = \phi_{b0} + \alpha (\phi_{bp} - \phi_b) \tag{3.7}$$

where we defined $\alpha = e(2\pi\varepsilon)^{-1} q^{-2} D_0$, and where $\phi_{bp} = E_c - E_N$ is the Schottky barrier in the limit of a large density of MIGS, i.e. when the metal Fermi level is completely "pinned" at the charge neutrality level (see Fig. 3.5 for a schematic of the band-bending and a graphical definition of the various barrier heights). Solving the above equation for ϕ_b we obtain

$$\phi_b = \frac{1}{1+\alpha}\phi_{b0} + \frac{\alpha}{1+\alpha}\phi_{bp}. \tag{3.8}$$

For a typical semiconductor such as silicon, a conservative estimate of D_0 is $D_0 = 1$ state/(eV − atom) $= 8 \times 10^9$ C/(eV − m^3)while first principles calculations have obtained $q^{-1} = 0.3$ nm; from these values we have $\alpha \approx 12$. Thus, in general, the barrier height is well approximated by $\phi_b \approx \phi_{bp}$. If E_N lies in the middle of the semiconductor bandgap, the metal Fermi level will be located at midgap, thus giving rise to a barrier equal to half of the

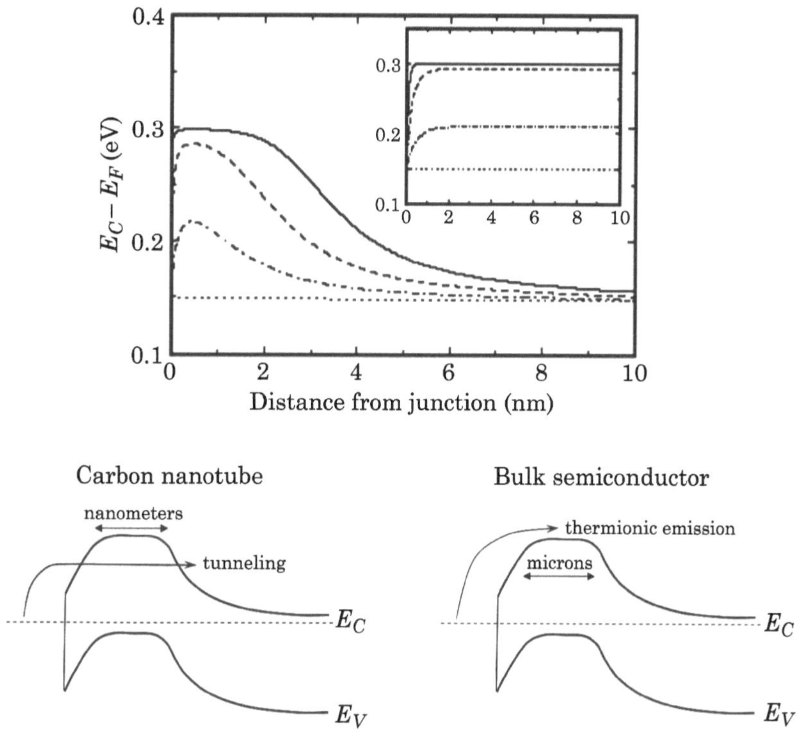

Figure 3.6 Top: calculated band-bending due to metal-induced gap states (MIGS) at an end-bonded nanotube/metal contact. The curves from bottom to top correspond to an increasing density of MIGS. The inset shows the band-bending with the same densities of MIGS for a planar contact. The bottom panels contrast the band-bending and the charge transport mechanisms for the carbon nanotube and the bulk semiconductor. Top figure from Ref. [5].

semiconductor bandgap, as shown in Fig. 3.5. This idealized limit agrees relatively well with the data presented in Fig. 3.3.

The presence of the Schottky barrier implies that electronic transport across the contact is dominated by thermionic emission over the Schottky barrier, significantly increasing the contact resistance. The key questions are: how is the physics modified for nanotube/metal contacts? What are the implications for electron transport? To address these questions, theoretical work [5] has considered the role of Fermi level pinning at end-bonded nanotube/metal contacts in more detail.

The key difference between end-bonded nanotube/metal contacts and planar contacts is that, while the model for the induced charge due to MIGS still applies (Eq. (3.2)) for the nanotube/metal contact, the MIGS charge takes the form of a dipole ring rather than a dipole sheet, as shown in Fig. 3.4. (Another way of saying this is that the electrostatic kernel in Eq. (3.3) is different for carbon nanotubes, as shown below.) This has a critical effect on the electrostatic potential and the resulting band-bending. While the electrostatic potential is a constant far from a dipole sheet, it decays as the second power of distance far from a dipole ring. Thus, in the nanotube contact, any potential shift near the interface will decay rapidly. Self-consistent calculations of Eqs. (3.2) and (3.3) with the electrostatic kernel

$$K_{NT}\left(z-z'\right)=\frac{R}{4\pi\varepsilon_0}\int_0^{2\pi}\frac{d\theta}{\sqrt{\left(z-z'\right)^2+2R^2\left(1-\cos\theta\right)}} \tag{3.9}$$

show that the potential shift essentially disappears within a few nanometers of the interface as shown in Fig. 3.6 [5]. Because the barrier is only nanometers wide, electrons can efficiently tunnel through the extra barrier due to Fermi level pinning, and it appears as if that extra barrier were inexistent. In contrast, for a traditional semiconductor/metal contact, the extra barrier due to Fermi pinning persists over distances of microns, until the band-bending due to doping in the semiconductor causes the bands to bend to achieve charge neutrality far from the interface (this length scale is determined by the Debye length). Thus, compared to planar contacts, Fermi level pinning is expected to play a minor role in end-bonded nanotube/metal contacts. An important consequence of this is that the type of metal used to contact the nanotube has a strong influence on the properties of the contacts.

Experimentally, formation of end-bonded nanotube/metal interfaces has only been observed in a few systems, with Fig. 3.1 showing one example. In another approach, carbon nanotubes were deposited on titanium contacts and annealed to high temperatures in order to form titanium carbide [6]. X-ray diffraction techniques were used to identify the formation of TiC, and transmission electron microscopy verified the abruptness of the TiC/nanotube junction. The experimental evidence points to the presence of Schottky barriers at these contacts (in general, contacts are intimately related to the behavior of nanotube transistors, as will be discussed in Chapter 4), and based on the theory discussed above, it is expected that these Schottky barriers arise because of the direct band alignment between the TiC Fermi level and the nanotube bandgap. Perhaps the simplest way to illustrate the presence of a Schottky barrier at these contacts is to measure the temperature dependence of the current, because in the presence of a Schottky barrier, the current is thermally activated above the barrier, and thus increases with increasing temperature. Fig. 3.7 shows the current–voltage characteristics of carbon nanotube transistors with TiC contacts. While such transfer characteristics will be discussed at length in Chapter 4, the point to note is that the current is found to increase with an increase in temperature, as the inset in the figure shows.

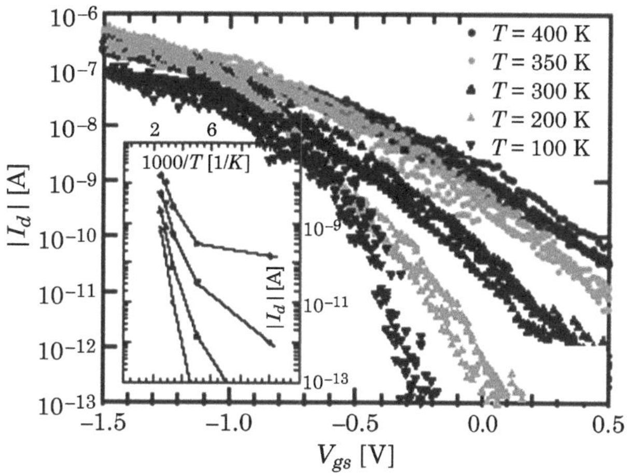

Figure 3.7 Main figure: current versus gate voltage in a carbon nanotube transistor with Ti contacts. Inset: measured current across a nanotube/Ti contact as a function of inverse temperature, indicating thermionic-like behavior over a Schottky barrier. Figure from Ref. [7].

3.3 Side Contacts

The most often utilized type of contact is the side-contact geometry shown in Fig. 3.2, as it can be obtained by depositing carbon nanotubes on pre-patterned metal electrodes, or depositing nanotubes on a substrate and depositing the metal on top of the nanotubes. In addition, both of these approaches can be combined to completely embed the nanotube in the metal. According to the discussion in Section 3.2, the barrier at nanotube/metal contacts (albeit for end-bonded contacts) should be well described by the difference between the metal work function and the nanotube electron affinity $\phi_{b0} = \Phi_m - \chi$. For a typical nanotube with bandgap of 0.6 eV, and for the nanotube midgap 4.5 eV below the vacuum level [8,9], metal work functions larger than 4.8 eV (or less than 4.2 eV) would thus lead to a negative Schottky barrier, i.e. the metal contacts the nanotube in the valence or conduction band, giving an ohmic contact. Thus, one may expect that Au (5.5 eV) and Pd (5.1 eV) would give ohmic contacts. Fig. 3.8 shows the measured conductance for a nanotube transistor with Pd contacts [10]. As seen in the figure, the device conductance is close to the maximum conductance of $4e^2/h$ thus indicating that little or no barrier exists at the contact. This can be further confirmed by studying the temperature dependence of the conductance. As the right panel of Fig. 3.8 shows, the conductance increases with a reduction in the temperature. This temperature dependence is opposite to that for Schottky barrier contacts as Fig. 3.7 illustrated.

Experiments using an atomic force microscope tip as an electrical scanning probe along the nanotube [11] have also observed ohmic behavior in Cr/Au

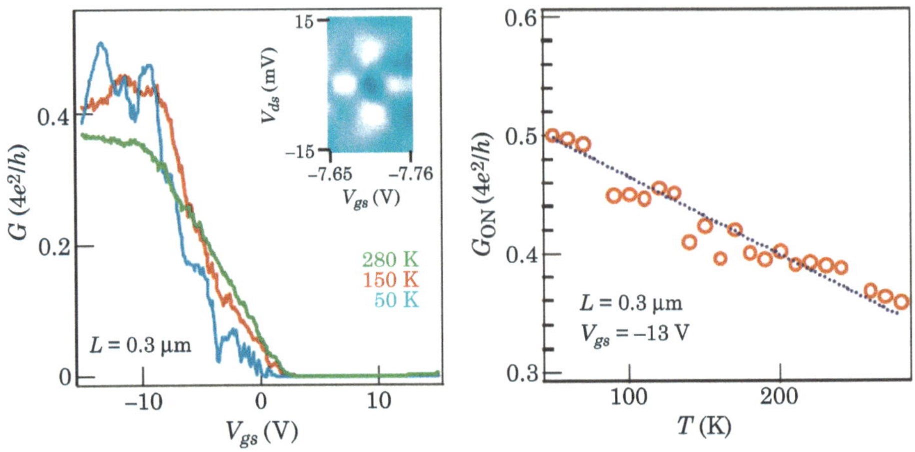

Figure 3.8 Left: measured conductance as a function of gate voltage in a nanotube transistor. The largest conductance measured is near the maximum possible value. The right panel shows the conductance as a function of temperature, with a behavior opposite to that in the Schottky barrier device of Fig. 3.7. Figure from Ref. [10].

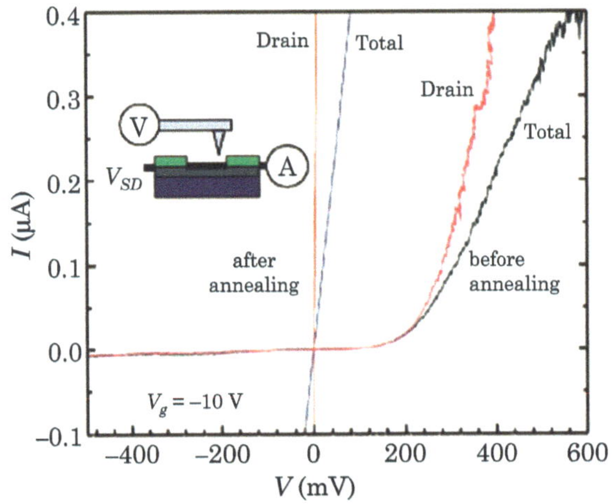

Figure 3.9 Current–voltage characteristics of a semiconducting nanotube contacting two metal electrodes, probed with the tip of an atomic force microscope. Figure from Ref. [11].

contacts. There, a thin Cr layer is first deposited as an adhesion layer, with Au deposited on top of the Cr. Fig. 3.9 shows the current between the tip and the contact electrode before and after annealing the contact. Clearly, the behavior is rectifying before annealing, indicating the presence of a Schottky barrier. After annealing, the current–voltage curve is linear showing that the contact is now ohmic. This result was also confirmed by cooling the device, which showed an increase in conductance. It is believed that upon annealing, Au diffuses through the thin Cr layer, changing the properties of the contact.

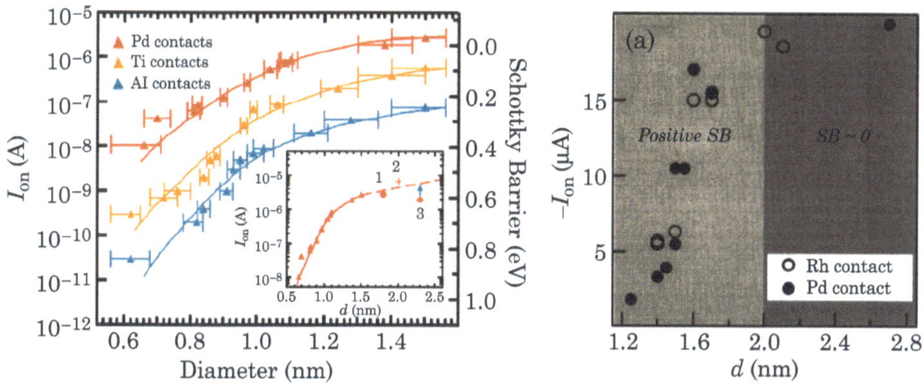

Figure 3.10 Experimental measurement of ON state current of nanotube transistors as a function of nanotube diameter. Figures from Refs [12] and [13].

Detailed measurements on a large number of devices have since indicated that the diameter of the carbon nanotube and the type of metal used for the contact play a critical role in determining the height of the Schottky barrier in side-contacted nanotubes [12–14]. Fig. 3.10 shows the measured current through metal/nanotube/metal devices with carbon nanotubes of different diameters. A clear increase of the current is observed as the nanotube diameter is increased, regardless of the type of metal used. In addition, it is found that Pd and Rh give the highest currents, followed by Ti and Al. Both of these results are in stark contrast to traditional metal/semiconductor contacts, where the diameter of the semiconductor does not come into play, and where the type of metal does not determine the Schottky barrier because of Fermi level pinning. In the previous section, we discussed why Fermi level pinning is expected to be less important in end-bonded contacts. However, the types of contacts used to obtain the experimental results in Fig. 3.10 are qualitatively different. To understand the properties of such contacts and explain the behavior observed in Fig. 3.10, we now discuss a theory of side contacts to carbon nanotubes.

We begin by describing the contact geometry considered here. Fig. 3.11 shows a sketch of a cross section of the contact consisting of a nanotube embedded in a metal. The metal forms a cylindrical cavity of radius $R + s$ where R is the nanotube radius and $s = 0.3$ nm is the distance between the nanotube and the metal (this value for s is typical of van der Waals interactions between nanotubes and surfaces).

In the simplest picture, the difference between the metal Fermi level E_F and the semiconductor valence band edge E_v (the barrier for holes) is simply given by (Fig. 3.11 (d))

$$\Delta_0 = E_g + \chi - \Phi_m \tag{3.10}$$

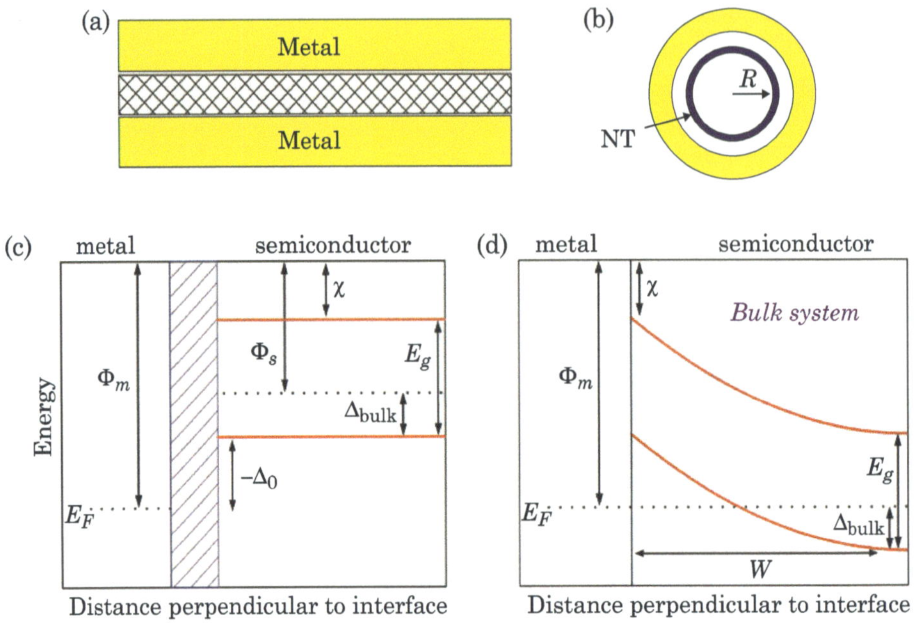

Figure 3.11 (a) shows a cross section of the contact along the length of the nanostructure. (b) shows a radial cross section. The separation between the nanotube and the metal surface is s (not shown in figure). (c) shows the band alignment before charge transfer. In a bulk contact,(d), band-bending over a distance W leads to a Schottky barrier Δ_{bulk}. Figure from Ref. [15].

where Φ_m is the metal work function, χ is the semiconductor electron affinity and E_g is the semiconductor bandgap. A positive value for Δ_0 indicates a Schottky barrier, while a negative value indicates an ohmic contact. Because the bandgap decreases with increasing diameter for nanotubes, the value of Δ_0 depends on the nanotube diameter. The behavior of Eq. (3.10) for undoped nanotubes is shown in Fig. 3.12 for a value of $\Phi_m - \Phi_{\text{nanotube}} = 0.4$ eV (typical of Pd), using the relation $E_g = 2a\gamma/d$ between bandgap and nanotube diameter d, and the parameters $a = 0.142$ nm for the C–C bond length, $\gamma = 2.5$ eV for the tight-binding overlap integral, and Φ_{NT} the nanotube work function assumed to be at midgap for an undoped nanotube. One problem with this picture (besides the fact that the physics is incomplete, as will be discussed below) is that Eq. (3.10) predicts large and negative values for Δ_0, signaling strong ohmic contacts. However, it is clear that such strong ohmic contacts are not observed experimentally.

In general, charge transfer between the metal and semiconductor leads to band re-alignment. At a bulk semiconductor junction (Fig. 3.11 (e)) this charge transfer leads to the Schottky barrier

$$\Delta_{\text{bulk}} = E_g + \chi - \Phi_s \tag{3.11}$$

where Φ_s is the semiconductor work function. This relationship arises because, in the bulk system, a band-bending of length W perpendicular to the metal–semiconductor interface is created until the band lineup in (3.11) is obtained. But for a side-contacted quasi-one-dimensional structure such as a nanotube or a nanowire, the semiconductor is only a few nanometers thick in the direction perpendicular to the metal–semiconductor interface; thus only a region of the order of the nanostructure cross-section can be depleted, giving partial band re-alignment. The value of Δ will then be somewhere between Δ_0 and Δ_{bulk} (for an undoped nanotube, $\Delta_{\text{bulk}} = E_g/2$, which would always give relatively high Schottky barriers).

Nanotubes are an extreme example of this situation, since the possible "depletion width" is the size of the nanotube wall; the charge transfer and image charge in the metal create two nested hollow cylinders with opposite charge, and an associated electrostatic potential. The charge and potential must of course be self-consistent. This behavior can be captured using analytical models for the charge and potential. The charge per unit area on the nanotube can be expressed as

$$\sigma = eN \int D_{NT}(E + eV_{NT})\, f(E - E_F)\, dE \tag{3.12}$$

where $D_{NT}(E)$ is the nanotube density of states [16] shifted by the self-consistent electrostatic potential on the nanotube eV_{NT}, $f(E - E_F)$ is the Fermi function, and $N = 4/\left(3\sqrt{3}a^2\right)$ is the atomic areal density. We assume a uniform distribution of the charge on the nanotube surface, and room temperature.

For the geometry of Fig. 3.11, solution of Poisson's equation gives the potential on the nanotube as

$$eV_{NT} = -\sigma\frac{eR}{\varepsilon_0}\ln\frac{R+s}{R} = -\frac{e^2}{C}\sigma \tag{3.13}$$

where ε_0 is the permittivity of free space and C is the capacitance per unit area between the metal and the nanotube. Eqs. (3.12) and (3.13) can be solved self-consistently for a given nanotube. In this model the electrostatic potential induced on the nanotube modifies the barrier to $\Delta = \Delta_0 - eV_{NT}$. Fig. 3.12 shows results of such calculations for parameters typical of Pd. Clearly, the behavior is different from the simple expressions in Eqs. (3.10) and (3.11). Indeed, the bulk limit $\Delta_{\text{bulk}} = E_g/2$ gives very large barriers, much too large to even appear on the scale of Fig. 3.12. These results suggest that there is a transition between Schottky and ohmic behavior at a nanotube diameter around 1.4 nm, in agreement with the experimental data for Pd contacts of Fig. 3.10.

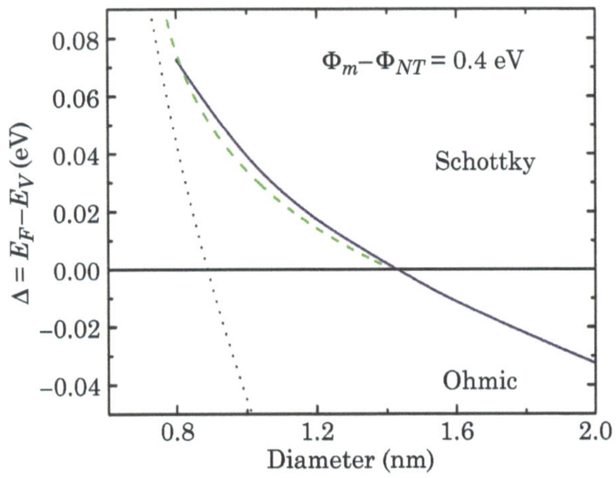

Figure 3.12 Schottky barrier Δ at nanotube–metal contacts for parameters typical of Pd. The dotted line is from Eq. (3.10), solid line is from Eqs. (3.12) and (3.13) solved numerically, and the dashed line is Eq. (3.20). Figure from Ref. [15].

To proceed further we focus on the small and positive Δ regime, and use an approximation of the integral in Eq. (3.12):

$$\begin{aligned}\sigma &= eND_0 \int_{E_c - eV_{NT}}^{\infty} \frac{E + eV_{NT}}{\sqrt{(E + eV_{NT})^2 - E_c^2}} \frac{1}{1 + \exp\left(\frac{E - E_F}{kT}\right)} dE \\ &\approx eND_0 \int_{E_c - eV_{NT}}^{\infty} \frac{E + eV_{NT}}{\sqrt{(E + eV_{NT})^2 - E_c^2}} \frac{1}{2} \exp\left(-\beta \frac{E - E_F}{kT}\right) dE\end{aligned} \tag{3.14}$$

where we replaced the Fermi distribution with an exponential function with a decay constant β. By fitting the Fermi distribution with this exponential form, we find that $\beta \approx 0.7$. Integration by parts gives

$$\sigma \approx \frac{eND_0}{2} \int_{E_c - eV_{NT}}^{\infty} \sqrt{(E + eV_{NT})^2 - E_c^2} \exp\left(-\beta \frac{E - E_F}{kT}\right) dE, \tag{3.15}$$

while the substitution of variables $x = E + eV_{NT} - E_c$ yields

$$\begin{aligned}\sigma &\approx \frac{eND_0}{2} \exp\left(-\frac{\beta E_c}{kT}\right) \int_0^{\infty} \sqrt{(x + E_g/2)^2 - (E_g/2)^2} \exp\left(-\frac{\beta x}{kT}\right) dE \\ &= \frac{eND_0}{2} \exp\left(-\frac{\beta E_c}{kT}\right) \int_0^{\infty} x^{1/2} \sqrt{x + E_g} \exp\left(-\frac{\beta x}{kT}\right) dE.\end{aligned} \tag{3.16}$$

The last integral can be approximated using the method of asymptotic expansion, in particular Watson's lemma, to give the result

$$\sigma \approx \frac{eNa\sqrt{3}}{2\sqrt{2\beta}\pi^{3/2}R\gamma}\sqrt{\frac{E_g kT}{2}}e^{-\beta\Delta/kT} \tag{3.17}$$

where we used the relation $\Delta = E_c - E_F$. This expression for the charge can be combined with Eq. (3.13) for the electrostatic potential to yield an equation for the Schottky barrier Δ:

$$-\alpha\sqrt{kTE_c}\exp\left(-\frac{\beta\Delta}{kT}\right) = \Delta_0 - \Delta \tag{3.18}$$

with $\alpha = \left(e^2Na\sqrt{3}\right) / \left(2\sqrt{2\beta}\pi^{3/2}R\gamma C\right)$.

This equation has solution

$$\Delta = \frac{kT}{\beta}W\left[\alpha\beta\sqrt{\frac{E_c}{kT}}\exp\left(-\beta\frac{\Delta_0}{kT}\right)\right] + \Delta_0 \tag{3.19}$$

where $W(x)$ is the Lambert W function. For large values of the argument this function has the expansion $W(x) \sim \ln x - \ln\ln x$, and this can be used to obtain the expression for the Schottky barrier

$$\Delta \approx \frac{kT}{\beta}\ln\left(\frac{\alpha\sqrt{\frac{E_g}{2kT}}}{\ln\alpha\sqrt{\frac{E_g}{2kT}} - \frac{\Delta_0}{kT}}\right). \tag{3.20}$$

The behavior of this function is plotted in Fig. 3.12 (dashed line), showing good agreement with the full calculation. The logarithmic dependence implies relatively slowly varying Δ, at least compared with Eq. (3.10). The nanotube diameter delimiting Schottky from ohmic behavior [15] is calculated by setting the argument of the logarithm equal to 1, giving

$$d \approx d_0\left(1 + \alpha\sqrt{\frac{kT}{\Phi_m - \Phi_{NT}}}\right), \tag{3.21}$$

where d_0 is the crossover diameter that would be obtained from Eq. (3.10). Thus the crossover diameter is increased from its bare value by $\delta d = \alpha\left[kT/\left(\Phi_m - \Phi_{NT}\right)\right]^{1/2} d_0$. Making ohmic contact to a wide range of nanotube diameters requires a small δd; this can be accomplished at low temperature, with a large metal work function, or with a large capacitance (giving a small α). Embedded contacts thus provide an advantage over planar contacts because of their larger capacitance.

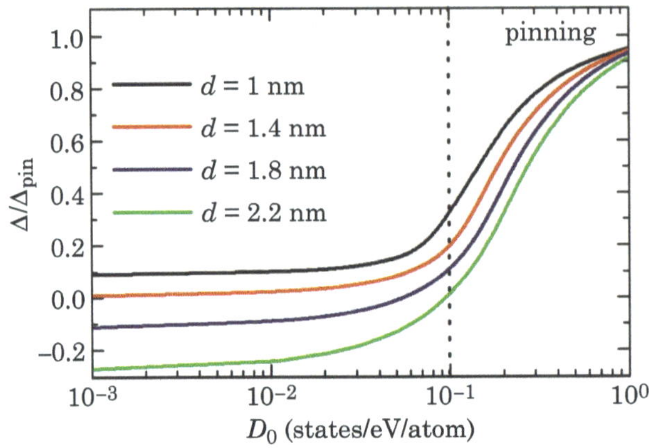

Figure 3.13 Schottky barrier as a function of the density of metal-induced gap states for several nanotubes side-contacted by a metal.

As we discussed in a previous section in this chapter, in a bulk metal–semiconductor contact, metal-induced gap states (MIGS) lead to Fermi level pinning, and modification of the Schottky barrier height to Δ_{pin}. We showed how Fermi level pinning is expected to play a minor role in end-bonded nanotube/metal contacts. One question is: how will Fermi level pinning affect side contacts to carbon nanotubes? To model the impact of MIGS in side contacts to nanotubes, we consider a pinning charge on the nanotube wall

$$\sigma_{\text{pin}} = D_0 N \left(E_F - E_N\right) \tag{3.22}$$

where the neutrality level E_N is assumed to be at midgap for simplicity. The absence of the exponentially decaying spatial dependence is due to the fact that MIGS decay exponentially in a direction perpendicular to the metal/semiconductor interface, and for a side-contacted nanotube this is the radial direction. Thus, pinning charge is only generated on the nanotube wall, and the decay of the MIGS wavefunctions across the small metal–nanotube gap is captured in the constant D_0. To calculate the impact of the MIGS, this pinning charge is added to Eq. (3.12) and the self-consistent calculations are repeated.

Fig. 3.13 shows the Schottky barrier calculated for several nanotubes as a function of the density of gap states ($\Delta_{\text{pin}} = E_g/2$ in these calculations). Clearly, there is a rapid onset of pinning at $D_0 \sim 0.1$ states/(atom eV); this value of D_0 is rather large considering the van der Waals bonding of nanotubes to surfaces, and that atomistic calculations [17,18] have obtained much smaller values. Thus, as in end-bonded contacts, it is expected that Fermi level pinning will play a minor role in side-contacts to nanotubes.

The reduced impact of Fermi level pinning on Schottky barrier heights can be seen in the experimental data presented in Fig. 3.10, where the current

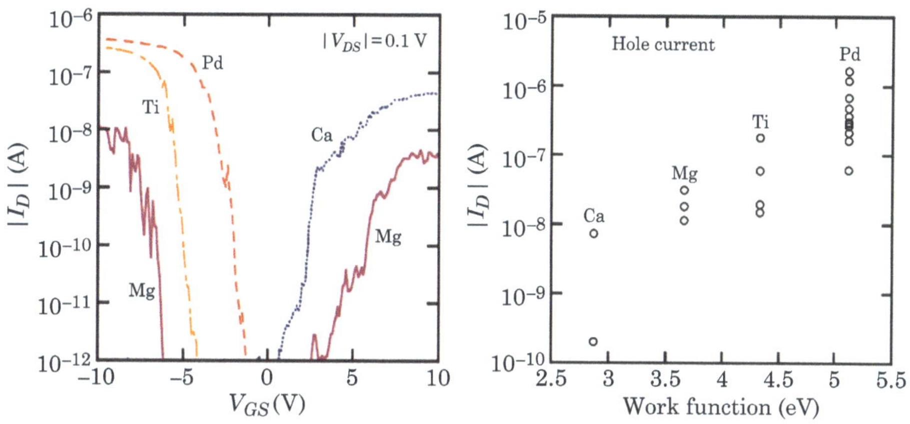

Figure 3.14 Left: current–voltage characteristics of carbon nanotube field-effect transistors for different contact metals. Right: hole current for several devices as a function of the work function of the contact metal. Figures from Ref. [14].

depends on the type of metal used. In addition to the results presented in that figure, the role of the metal work function has been probed over a large range of metal work functions [14] by fabricating devices with contacts made of Ca (2.9 eV work function), Ti (4.4 eV), Mg (3.6 eV) and Pd (5.1 eV). Fig. 3.14 presents experimental measurements of the transfer characteristics of carbon nanotube field-effect transistors with contacts made of these different metals. As can be seen in the figure, the hole current (negative gate voltages) is found to increase as the metal work function increases; the converse is true for the electron current. This behavior clearly points to the ability to tailor the type of contact by using different metals.

3.4 Contacts to Metallic Carbon Nanotubes

Given that the dependence of the current on diameter in semiconducting carbon nanotubes can be well-described by a metal-dependent Schottky barrier concept, it would seem that electrical contacts to metallic nanotubes should be of high quality because of the absence of a Schottky barrier. While such high quality contacts have been reported, it turns out that the quality of the contact in side-contacted metallic nanotubes also depends on the diameter of the metallic carbon nanotubes. This behavior is illustrated in Fig. 3.15, where the conductance of metal/(metallic nanotube)/metal devices is plotted as a function of the nanotube diameter. Much like is seen for the case of semiconducting nanotubes, the conductance is found to decrease as the nanotube diameter is decreased. Based on these measurements, it is proposed that diameter-dependent tunneling barriers are present between the metal and the nanotubes. Evidence for the presence of these barriers [13] comes from low-

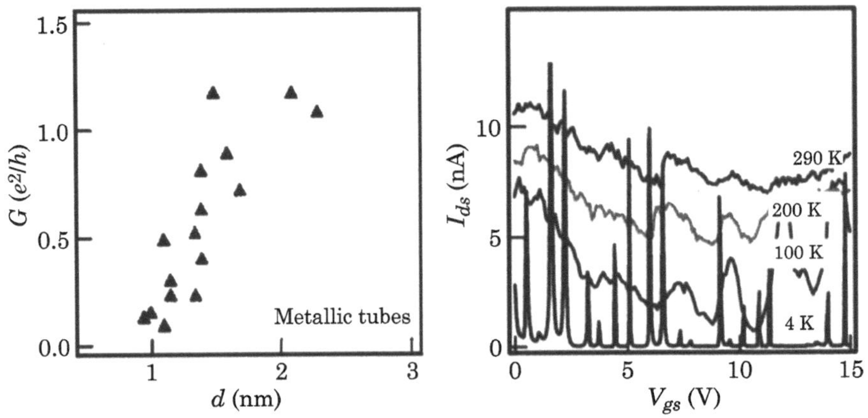

Figure 3.15 Measured conductance of metallic carbon nanotubes as a function of diameter. Figure from Ref. [13].

temperature measurements of the conductance (Fig. 3.15) which show Coulomb blockade oscillations with distinct peaks. The peak separation is well-described by considering the separation between the quantized energy levels of a nanotube of length 120 nm, which corresponds to the channel length in these experiments. Indeed, the Coulomb blockade energy is given by

$$E_{CB} = \frac{e^2}{C}, \tag{3.23}$$

where C is the capacitance between the nanotube and the backgate. Using Eq. (3.23) with the SiO_2 thickness of 500 nm and a nanotube diameter of 1.5 nm, we calculate the Coulomb blockade energy to be 44 meV, in reasonable agreement with the height of the Coulomb diamonds presented in Ref. [13]. The Coulomb oscillations are thus believed to originate from the weak coupling of the nanotube to the contacts as a consequence of tunnel barriers.

3.5 Metal/Oxide/Nanotube Contacts

While most of the research on the properties of nanotube/metal contacts has focused on fabricating ohmic contacts to obtain the lowest possible contact resistance, for electrical insulation and tunneling applications, it is important to develop contacts that have high resistance. Fig. 3.16 shows nanotubes between Pd electrodes, but where the Pd electrodes were exposed to O_2 plasma [19] in order to create an ultrathin PdO layer on their surface. As the solid symbols in Fig. 3.16 show, the current across such contacts is extremely small, and the properties of the contact are very stable up to moderate voltages. Auger spectroscopy has shown that the oxide is only about 2 nm thick. Thus, such

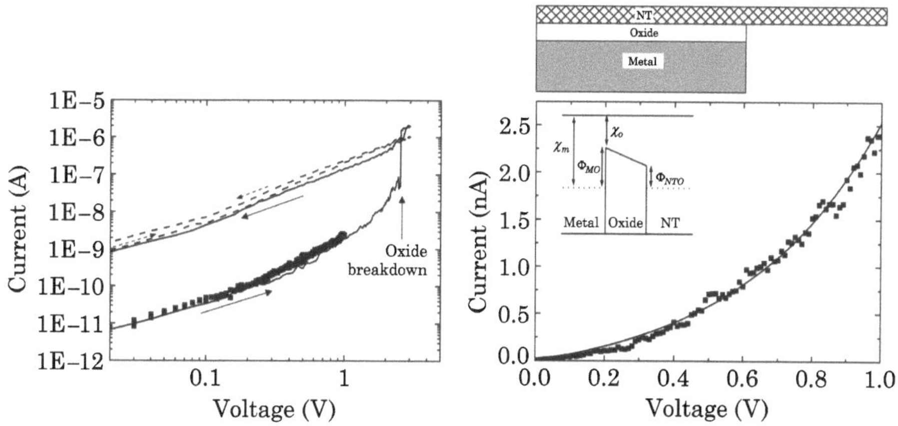

Figure 3.16 I/V curves for a device with Pd/PdO/nanotube contacts. The initial I/V curve shown with the black squares is entirely reversible and repeatable as long as the applied voltage is kept below 3 Volts. The solid line shows a ramp-up of the voltage up to 3 Volts, at which point the PdO breaks down, leading to a large increase in the device conductance. The new state of the contact is permanent as the voltage is decreased and increased again (solid and dashed lines). The panel on the right shows a sketch of the contact and of the expected band alignment. The main figure shows the exponential dependence of the current on voltage before the oxide is broken. Figure from Ref. [19].

ultrathin PdO oxides could be candidates for gate insulator materials. The current–voltage characteristics of these insulating contacts can be well described by tunneling across the oxide (illustration Fig. 3.16), with an expression of the form

$$I = I_0 \left(e^{V/V^*} - 1 \right) \tag{3.24}$$

where

$$V^* = \frac{\hbar}{ed}\sqrt{\frac{\phi_{MO} + \phi_{NT}}{m^*}}. \tag{3.25}$$

In this last equation, ϕ_{MO} is the height of the tunneling barrier at the metal-oxide interface, ϕ_{NTO} is the height of the tunneling barrier on the nanotube side, and m^* is the effective mass for tunneling. Excellent agreement with the experimental data is obtained if one assumes that ϕ_{MO} is equal to the Pd work function, ϕ_{NTO} is equal to the nanotube work function, and that the effective mass is equal to the free electron mass. Generally, it is expected that the tunnel barriers and the effective mass will be smaller than the assumed values. However, these uncertainties tend to cancel-out because of the ratio in Eq. (3.25). In addition, the uncertainty is reduced because of the square root in that equation. From the fit to the experimental data, the thickness of the oxide can be obtained using Eq. (3.25) and is found to equal 2 nm, in agreement with the Auger spectroscopy measurements.

A particularly intriguing feature of these Pd/PdO/nanotube contacts is an irreversible transition to a high conducting behavior when large voltages are applied. As Fig. 3.16 indicates, for voltages larger than about 2.5 V, the current across the contacts increases dramatically and irreversibly, eventually leading to low contact resistance. The origin of this effect is breakdown of the ultrathin oxide at the contact; interestingly, it has been found that exposure of these devices to air leads to complete recovery to the initial insulating state, presumably due to re-growth of the PdO in ambient. One may imagine using the oxide breakdown mechanism in a transistor fabrication method where the nanotube is laid across three Pd electrodes covered with PdO, and applying a large voltage across the source and drain to "form" the source and drain contacts while leaving the gate floating.

References

1. Y. Zhang, T. Ichihashi, E. Landree, F. Nihey and S. Iijima, "Heterostructures of single-walled carbon nanotubes and carbide nanorods", *Science*, Vol. 285, p. 1719, 1999.
2. J.J. Palacios, A.J. Pérez-Jiménez, E. Louis, E. SanFabián and J.A. Vergés, "First-principles phase-coherent transport in metallic nanotubes with realistic contacts", *Phys. Rev. Lett.*, Vol. 90, p. 106801, 2003.
3. M.P. Anantram and F. Léonard, "Physics of carbon nanotube devices", *Rep. Prog. Phys.*, Vol. 69, p. 507, 2006.
4. S.M. Sze, *Physics of semiconductor devices*, John Wiley & Sons, New York, 1981.
5. F. Léonard and J. Tersoff, "Role of Fermi-level pinning in nanotube Schottky diodes", *Phys. Rev. Lett.*, Vol. 84, p. 4693, 2000.
6. R. Martel, V. Derycke, C. Lavoie, J. Appenzeller, K.K. Chan, J. Tersoff and Ph. Avouris, "Ambipolar electrical transport in semiconducting single-wall carbon nanotubes", *Phys. Rev. Lett.*, Vol. 87, p. 256805, 2001.
7. J. Appenzeller, M. Radosavljevic, J. Knoch and Ph. Avouris, "Tunneling versus thermionic emission in one-dimensional semiconductors", *Phys. Rev. Lett.*, Vol. 92, p. 048301, 2004.
8. H. Ago, T. Kugler, F. Cacialli, W.R. Salaneck, M.S.P. Shaffer, A.H. Windle and R.H. Friend, "Work functions and surface functional groups of multiwall carbon nanotubes", *J. Phys. Chem.* B, Vol. 103, p. 8116, 1999.
9. S. Suzuki, C. Bower and Y. Watanabe, "Work functions and valence band states of pristine and Cs-intercalated single-walled carbon nanotube bundles", *Appl. Phys. Lett.*, Vol. 76, p. 4007, 2000.
10. A. Javey, J. Guo, Q. Wang, M. Lundstrom and H. Dai, "Ballistic carbon nanotube field-effect transistors", *Nature*, Vol. 424, p. 654, 2003.
11. Y. Yaish, J.-Y. Park, S. Rosenblatt, V. Sazonova, M. Brink and P.L. McEuen, "Electrical nanoprobing of semiconducting carbon nanotubes using an atomic force microscope", *Phys. Rev. Lett.*, Vol. 92, p. 046401, 2004.
12. Z. Chen, J. Appenzeller, J. Knoch, Y.-M. Lin and Ph. Avouris, "The role of metal–nanotube contact in the performance of carbon nanotube field-effect transistors", *Nano Lett.*, Vol. 5, p. 1497, 2005.
13. W. Kim, A. Javey, R. Tu, J. Cao, Q. Wang and H. Dai, "Electrical contacts to carbon nanotubes down to 1 nm in diameter", *Appl. Phys. Lett.*, Vol. 87, p. 173101, 2005.

14. Y. Nosho, Y. Ohno, S. Kishimoto and T. Mizutani, "Relation between conduction property and work function of contact metal in carbon nanotube field-effect transistor", *Nanotechnology*, Vol. 17, p. 3412, 2006.
15. F. Léonard and A.A. Talin, "Size-dependent effects on electrical contacts to nanotubes and nanowires", *Phys. Rev. Lett.*, Vol. 97, p. 026804, 2006.
16. J.W. Mintmire and C.T. White, "Universal density of states for carbon nanotubes", *Phys. Rev. Lett.*, Vol. 81, p. 2506, 1998.
17. Y. Xue and S. Datta, "Fermi-level alignment at metal–carbon nanotube interfaces: application to scanning tunneling spectroscopy", *Phys. Rev. Lett.*, Vol. 83, p. 4844, 1999.
18. N. Park and S. Hong, "Electronic structure calculations of metal–nanotube contacts with or without oxygen adsorption", *Phys. Rev.* B, Vol. 72, p. 045408, 2005.
19. F.E. Jones, A.A. Talin, F. Léonard, P.M. Dentinger and M.W. Clift, "Effect of electrode material on transport and chemical sensing characteristics of metal/carbon nanotube contacts", *J. Electron. Mater.*, Vol. 35, p. 1641, 2006.

4 Electronic Devices

4.1 Introduction

Given the semiconducting character of two thirds of carbon nanotubes, their high aspect ratio and structural robustness, it is natural to ask if semiconducting carbon nanotubes can be used as active elements in nanoscale electronic devices. Indeed, there has now been many demonstrations of such devices, ranging from two-terminal rectifiers to field-effect transistors. These demonstrations have spurred tremendous interest in the field of carbon nanotube electronics for several reasons. First, for a single nanotube device, the channel width is on the order of one nanometer, much smaller than state-of-the-art silicon transistors, promising higher device densities. Second, carbon nanotubes have low defect density, and electronic transport is expected to be less affected by defect scattering. Third, the carrier distribution is not as sensitive to temperature variations due to the van Hove singularities at the band edges. Finally, quantum confinement should be easier to achieve because of the small channel width, and thus single-electron devices should be more readily available. Balancing these advantages is the fact that the knowledge base, fundamental understanding, fabrication infrastructure, and device design are much less developed for carbon nanotubes compared to silicon. For example, while some of the fabricated carbon nanotube devices bear resemblance with traditional silicon devices, this chapter emphasizes the much different physics that governs the operation of the carbon nanotube devices. We have already seen in the previous chapter how one part of a carbon nanotube electronic device—the contacts—behaves much differently from that in conventional materials. The current chapter goes beyond these concepts to address the issue of the whole device, building from a discussion of simple two-terminal devices like *p–n* junctions to more complex devices such as transistors.

4.2 Rectifiers

Rectifiers are simple two-terminal devices that essentially allow current to flow for only one polarity of the applied voltage, the simplest examples being *p–n* junction diodes and Schottky diodes. While these are simple devices, they are used extensively by themselves and as part of more complex circuits; just as important however, is the fact that these systems serve as excellent testbeds to study and highlight the differences between nanotube-based devices and conventional ones. This section presents experimental and theoretical work aimed at realizing carbon nanotube rectifiers, and at understanding their basic operating principles. A central result of this section is that the electrostatics of carbon

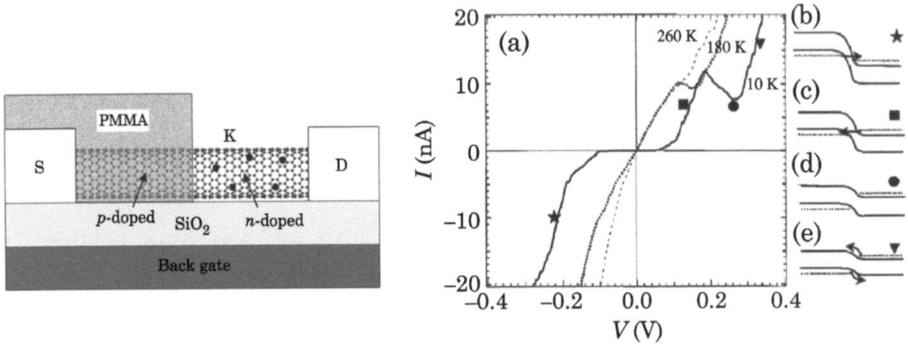

Figure 4.1 Schematic of a chemically doped carbon nanotube *p–n* junction and the associated *I–V* curve. The sketches labeled (b)–(e) show the band-bending along different points of the *I–V* curve, and the different transport regimes. Figures from, and after Ref. [1].

nanotube devices is much different than that of bulk devices because of the reduced dimensionality, and more specifically because of the weak electrostatic screening. We will see in Chapter 7 that this weak electrostatic screening also has an important impact on the optical properties of carbon nanotubes.

4.2.1 Experimental Realizations of Carbon Nanotube *p–n* Junctions

There are many possible strategies to achieve *p–n* junctions with carbon nanotubes. Examples include substitution of boron and nitrogen in the carbon lattice, doping by charge transfer from electrodes, atoms or molecules, or electrostatic control of the band-bending. Fig. 4.1 shows one of the strategies that has been implemented to fabricate such a device [1]. The method hinges on the fact that the synthesized carbon nanotubes assembled into devices in air are predominantly *p*-type (this is discussed at length in Chapter 8). Thus, it is only necessary to reverse the doping on one side of the nanotube to obtain a *p–n* junction. This can be accomplished by protecting half of the nanotube with the photoresist PMMA, and exposing the uncovered half to potassium, which is an electron donor. The associated current–voltage curve for such a device (Fig. 4.1) shows similarities with an Esaki diode, i.e. it shows negative differential resistance.

The role of dopants in *p–n* junctions is to create an electrostatic potential step at the junction. In traditional planar devices, doping is essentially the only way to generate such a potential step. In nanotubes however, one can take advantage of the quasi-one-dimensional geometry and use an external electrostatic potential to form the *p–n* junction. An example of this strategy [2] using a buried split-gate structure is shown in Fig. 4.2. The advantages of this technique are that no chemical doping of the nanotubes is required, and that

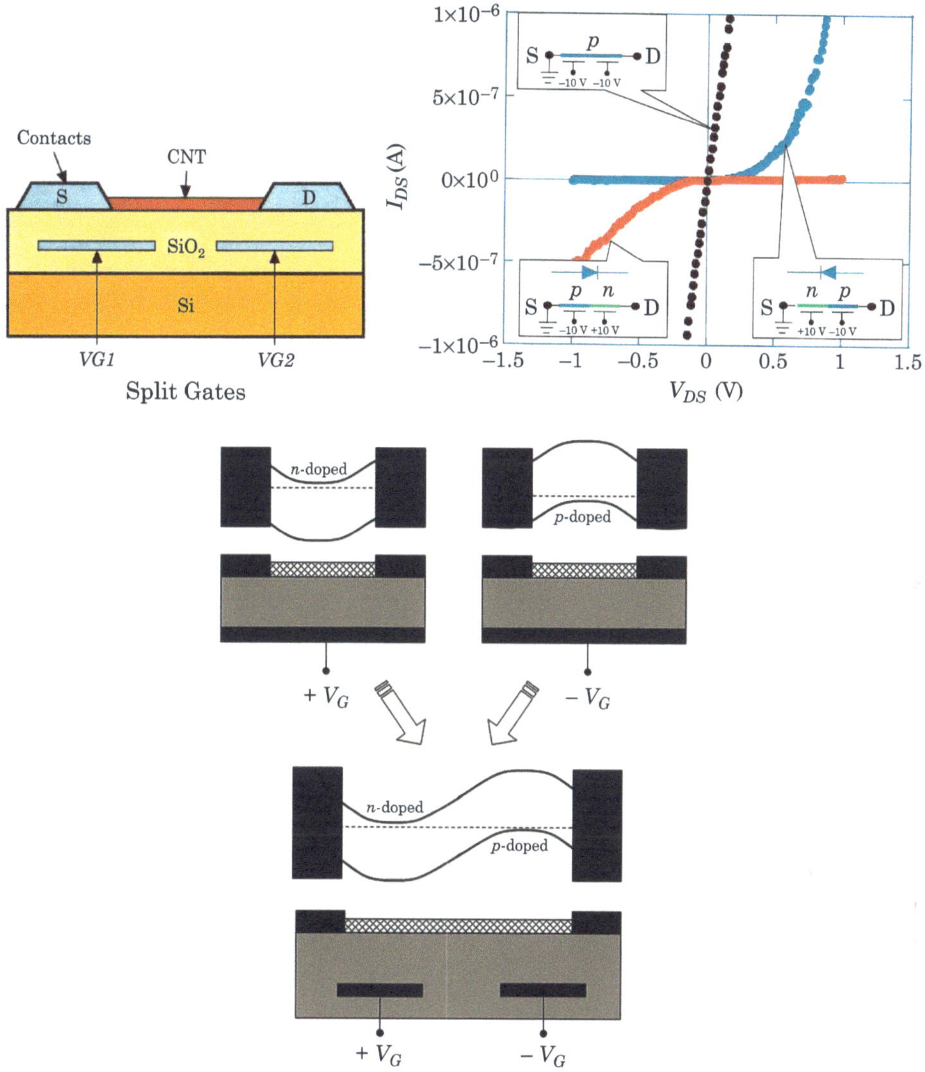

Figure 4.2 Top left: split-gate structure to create a carbon nanotube p–n junction without the need to chemically dope the nanotube. Top right: experimental current–voltage characteristics of such a device. Figures from Ref. [2]. Bottom: schematic illustrating the concept of the split gate architecture, including a sketch of the band-bending.

the device can be operated in several different modes in a controlled manner. Fig. 4.2 shows the I–V curve for this device for three regimes of operation, allowing the transformation of the device from a p–n diode, to a n–p diode, and to a transistor-like device.

To obtain an equation describing the current–voltage characteristics of carbon nanotube p–n junctions we consider the equation for the current through a ballistic conductor

$$I = \frac{4e^2}{h} \int T(E) \left[f_R \left(E - E_F^L \right) - f_L \left(E - E_F^R \right) \right] dE \tag{4.1}$$

where the subscripts R and L refer to the right and left leads, respectively. For nondegenerate doping of the carbon nanotube, the transport across the p–n junction is through thermionic emission over the potential step; to model this situation we assume that the transmission probability $T(E) = 1$ for electrons (holes) at energies above (below) the conduction (valence) band. The current is then

$$I = \frac{8e}{h} \int_{E_c^\infty}^{\infty} \left[f_R \left(E - E_F^L \right) - f_L \left(E - E_F^R \right) \right] dE \tag{4.2}$$

where E_c^∞ is the energy of the conduction band edge on the n-type side far from the junction. When the doping is not too large, $E_c^\infty - E_F^R \gg kT$, the Fermi functions can be approximated as

$$f = \frac{1}{1 + \exp\left(\frac{E - E_F}{kT}\right)} \approx \exp\left(\frac{E_F - E}{kT}\right) \quad \text{when } E - E_F \gg kT. \tag{4.3}$$

Substituting this approximation in the expression for the current gives

$$\begin{aligned} I &= \frac{8e}{h} \int_{E_c^\infty}^{\infty} \exp\left(\frac{E_F^R - E}{kT}\right) dE - \frac{8e}{h} \int_{E_c^\infty}^{\infty} \exp\left(\frac{E_F^L - E}{kT}\right) dE \\ &= \frac{8ekT}{h} \left[\exp\left(\frac{E_F^R - E_c^\infty}{kT}\right) - \exp\left(\frac{E_F^L - E_c^\infty}{kT}\right) \right]. \end{aligned} \tag{4.4}$$

Assuming that the band edge simply tracks the Fermi levels in the leads (i.e. far away from the junction), the difference $E_c^\infty - E_F^R$ is independent of applied voltage; furthermore, since we have $E_F^R - E_F^L = eV$ we obtain

$$I = I_0 \left[\exp\left(\frac{eV}{kT}\right) - 1 \right] \tag{4.5}$$

with

$$I_0 = \frac{8ekT}{h} \exp\left(\frac{E_F^R - E_c^\infty}{kT}\right). \tag{4.6}$$

Eq. (4.5) is the celebrated ideal diode equation describing rectifying behavior, except that here it was derived under the assumption of ballistic transport.

Further development of the buried-gate approach has shown that such carbon nanotube diodes can achieve this ideal rectifying behavior. Fig. 4.3 shows the measured I–V curves for a p–n diode, with the solid lines in the figure representing the modified diode equation

$$I = I_0 \left[\exp\left(\frac{eV}{nkT} \right) - 1 \right] \tag{4.7}$$

where n is the ideality factor. The right panel in Fig. 4.3 shows that at low bias, the experimental data satisfies this expression with an ideality factor $n = 1.2$, close to the ideal diode value $n = 1$. The inverse of the slope of the $d\ln(I)/dV$ curve

$$\alpha = \frac{dV}{d\ln(I)} = nkT/e \tag{4.8}$$

is plotted in the bottom inset in Fig. 4.3 as a function of temperature clearly indicating the linear dependence of α on temperature. The inset in the left panel indicates that at high bias, the measured values of α does not approach zero at small temperatures as Eq. (4.8) would suggest; the origin of this effect is unclear. Further analysis of the I–V curves shows that the transmission probability is about 0.2, indicating that there is some amount of scattering in the nanotube. The device behavior can thus be improved by reducing scattering; one source of scattering is due to the interaction of the carbon nanotube with the substrate, which can cause geometrical deformations of the nanotube and also distortions of the nanotube electronic structure. This source of scattering can be removed by suspending the nanotube over a trench in the channel region, as shown in Fig. 4.4. This device is fabricated by first growing a 400 nm thermal oxide on a heavily-doped Si substrate, on top of which two Mo split gates are fabricated using standard lithography, with a gate spacing between 0.5 micron and 1 micron. These split gates are then used as an etch mask to etch 250 nm of oxide. 150 nm of oxide is then deposited to form the gate dielectric for the split gates, and lithography is used to define source and drain electrodes. Carbon nanotubes were grown by chemical vapor deposition from catalyst particles on the electrodes. The current–voltage characteristics of this device follow the ideal diode equation with $n = 1$.

4.2.2 Theory of Carbon Nanotube p–n Junctions

While the experiments discussed in the previous section indicate that carbon nanotube p–$n$junctions can be fabricated and can behave as ideal diodes, such devices need to be examined more carefully to establish commonalities and

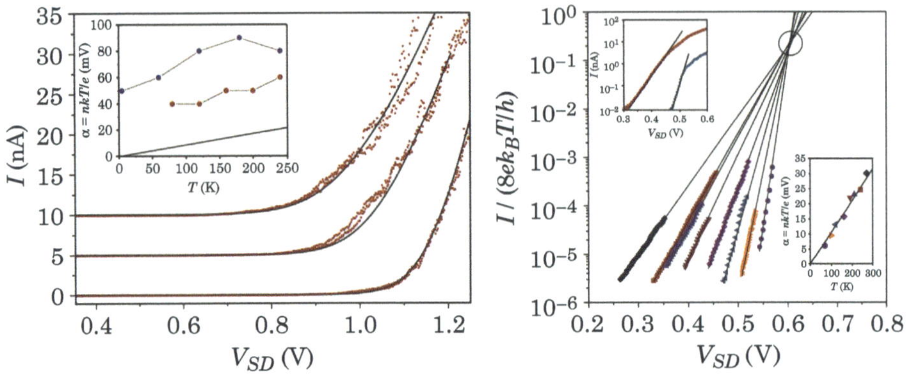

Figure 4.3 Measured current–voltage characteristics of carbon nanotube p–n junctions. The left panel shows fits of the ideal diode equation with a series resistor (1–6 MΩ) in the high forward bias regime at 80, 160, and 240 K (bottom to top, offset for clarity). Inset: α for two different devices. The solid line indicates the expected temperature dependence for an ideal diode. The right panel shows I–V curves in the low forward bias regime at various temperatures. Top inset: full I–V curves in forward bias at $T = 240$ K (filled squares) and 120 K (filled triangles). Bottom inset: temperature dependence of the inverse slope $\alpha = dV/d(\ln I)$. Figures from Ref. [3].

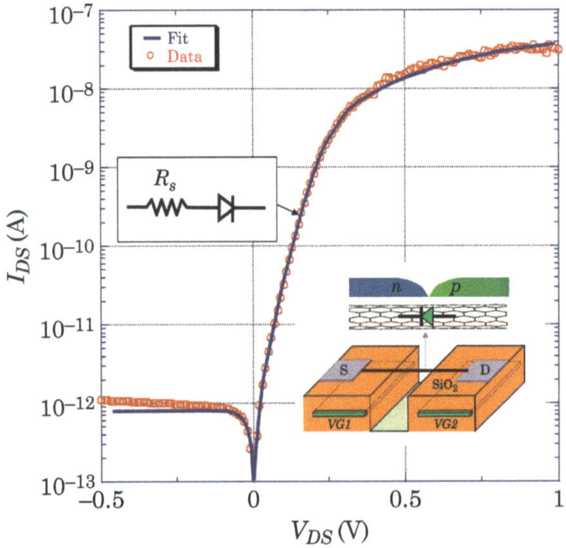

Figure 4.4 The inset shows a sketch of a buried-gate carbon nanotube p–n junction where the nanotube is suspended over a trench in the channel region. The main figure shows the measured current–voltage diode behavior (open symbols) as well as a fit to the ideal diode equation (solid line) including a contact resistance of 18 MΩ. Figure from Ref. [4].

differences with traditional devices. The behavior of nanotube p–n junctions can be understood by performing self-consistent calculations of the charge and electrostatic potential along the nanotube. Assuming azimuthal symmetry, the simplest model for the charge on the nanotube is

$$\sigma(z) = \frac{e}{\varepsilon} f - \frac{e}{\varepsilon} \int D(E, z) F(E)\, dE \tag{4.9}$$

where ε is the dielectric constant of the medium in which the nanotube is embedded, f is the doping fraction, $D(E, z)$ is the nanotube density of states at position z along the tube, and $F(E)$ is the Fermi function. The density of states can be expressed as

$$D(E, z) = \frac{a\sqrt{3}}{\pi^2 R V_0} \frac{|E + eV(z)|}{\sqrt{(E + eV(z))^2 - (E_g/2)^2}} \tag{4.10}$$

where $V(z)$ is the electrostatic potential along the nanotube. This expression for the spatial variation of the density of states is simply a rigid shift with the local electrostatic potential; while there are more sophisticated methods to calculate the actual local density of states and the occupation of the states that enters in the calculation of the charge, Eqs. (4.9) and (4.10) are sufficient to illustrate the general properties of nanotube p–n junctions.

The other equation necessary for the computations is the electrostatic potential $V(z)$ generated by the charge density $\sigma(z)$:

$$V(z) = \int K(z - z')\sigma(z')dz' \tag{4.11}$$

where σ is the charge per unit area on the nanotube and $K(z - z')$ is the free space electrostatic kernel for a hollow cylinder:

$$K(z - z') = \frac{R}{4\pi\varepsilon_0} \int_0^{2\pi} \frac{d\theta}{\sqrt{(z - z')^2 + 2R^2 - 2R^2 \cos\theta}}. \tag{4.12}$$

The procedure is therefore to solve self-consistently Eqs. (4.9) and (4.11) for a given doping on the nanotube. Fig. 4.5 shows results of such calculations for two doping fractions. Clearly, the band-bending in the nanotube is similar to what is observed in planar devices: a potential step at the junction and essentially flat bands away from the junction. The behavior is quite different, however, if one looks at the charge distribution. In a planar device, there is a region of constant charge near the junction, and no charge outside of that so-called depletion region. In contrast, for nanotubes, there is significant charging away from the junction. In fact, the charge decays only logarithmically away from the junction. This difference between planar and nanotube devices is again due to the different electrostatics of dipole sheets and dipole rings (see Chapter 3). In the planar device, having a dipole sheet at the junction is sufficient to ensure that the potential stays constant far away from the junction. For the dipole

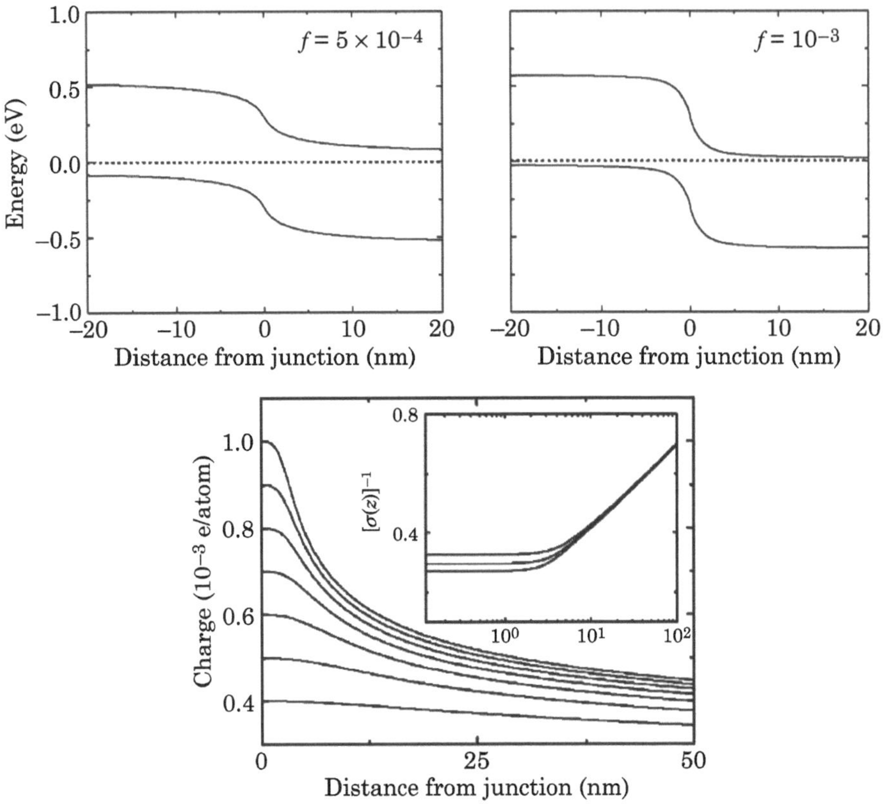

Figure 4.5 Calculated self-consistent band-bending and charge along a carbon nanotube p–n junction. The top two panels show the band-bending for two different doping fractions. The dotted line is the Fermi level. The bottom panel shows the charge distribution on the n-type side of the junction for different doping fractions, which can be read-off from the value of the vertical intercept. The inset in the bottom figure shows a scaling of all of the curves for the different doping fractions. Figure from Ref. [5].

ring however, the potential decays away from the junction. Since the potential must be constant far away from the junction, charge must continuously be added along the nanotube to keep the potential from falling. To illustrate this important effect in more detail, we consider mathematically the behavior of bulk and nanotube p–n junctions.

Bulk p–n junction

All derivations of the electrostatics of bulk p–n junctions essentially start with the assumption of a depletion region near the junction where the dopants are completely depleted and where the charge is a constant. It is usually assumed that this depletion region ends abruptly at the depletion width$W$ as illustrated in Fig. 4.6. For equal and opposite doping on the two sides of the p–n junction,

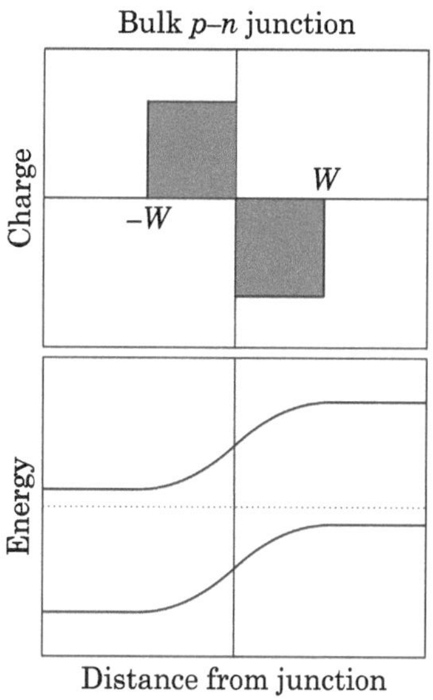

Figure 4.6 Charge distribution and band-bending for a symmetric bulk p–n junction. Dotted line is the Fermi level.

Poisson's equation for this charge distribution is

$$\frac{d^2V(z)}{dz^2} = \begin{cases} -\frac{\rho}{\varepsilon} & 0 < z < W \\ \frac{\rho}{\varepsilon} & -W < z < 0 \\ 0 & |z| > W \end{cases} \tag{4.13}$$

where ρ is the charge density. This second order differential equation is augmented by the boundary conditions that the potential reach a value at the depletion edge that gives charge neutrality in the semiconductor:

$$V(\pm W) = V_{\pm}. \tag{4.14}$$

The difference $V_+ - V_-$ gives the so-called built-in potential V_{bi}. Integrating Eq. (4.13) with this boundary condition leads to the solution

$$V(z) = \begin{cases} \frac{V_{bi}}{2} - \frac{\rho}{2\varepsilon}\left(z^2 + W^2\right) + A_+\left(z - W\right) & 0 < z < W \\ -\frac{V_{bi}}{2} + \frac{\rho}{2\varepsilon}\left(z^2 + W^2\right) + A_-\left(z + W\right) & -W < z < 0 \\ \frac{V_{bi}}{2} & z \geq W \\ -\frac{V_{bi}}{2} & z \leq -W. \end{cases} \tag{4.15}$$

The constants $A_\pm$ are determined from the continuity of the derivative of the potential at $z = \pm W$. Applying this condition gives the solution for the potential

$$V(z) = \begin{cases} \frac{V_{bi}}{2} - \frac{\rho}{2\varepsilon}(z-W)^2 & 0 < z < W \\ -\frac{V_{bi}}{2} + \frac{\rho}{2\varepsilon}(z+W)^2 & -W < z < 0 \\ \frac{V_{bi}}{2} & z \geq W \\ -\frac{V_{bi}}{2} & z \leq -W. \end{cases} \quad (4.16)$$

Continuity of the potential at $z = 0$ gives the expression for the depletion width

$$W_{\text{bulk}} = \sqrt{\frac{\varepsilon V_{bi}}{\rho}}. \quad (4.17)$$

The spatial variation of the potential is sketched in Fig. 4.6. The important point is that the assumption of a depletion region with constant charge naturally leads to a potential that is constant outside of the depletion region, and thus the semiconductor is charge-neutral even at long distances from the junction. This result turns out to be entirely fortuitous and is a special property of bulk, planar interfaces. Nanotubes, with their reduced dimensionality have a much different behavior, as we now discuss.

Carbon nanotube *p–n* junction

We first illustrate the problem that arises when the assumption of a constant depletion region is made for a carbon nanotube *p–n* junction. The critical difference with the bulk junction is that we no longer have an infinite dipole sheet, but rather have a nanometer-size dipole ring. The potential due to a dipole ring can be calculated as

$$V(z) = \frac{R}{4\pi\varepsilon_0} \int dz' \sigma(z') \int_0^{2\pi} \frac{d\theta}{\sqrt{(z-z')^2 + 2R^2 - 2R^2 \cos\theta}} \quad (4.18)$$

with

$$\sigma(z') = \begin{cases} \rho & 0 < z < W \\ -\rho & -W < z < 0 \\ 0 & |z| > W. \end{cases} \quad (4.19)$$

The potential due to the n-type side is obtained by direct integration

$$
\begin{aligned}
V_n(z) &= \frac{R\rho}{4\pi\varepsilon_0}\int_0^W dz' \int_0^{2\pi} \frac{d\theta}{\sqrt{(z-z')^2 + 2R^2 - 2R^2\cos\theta}} \\
&= \frac{R\rho}{4\pi\varepsilon_0}\int_0^{2\pi} d\theta \ln\left(\frac{z+\sqrt{2R^2(1-\cos\theta)+z^2}}{z-W+\sqrt{2R^2(1-\cos\theta)+(z-W)^2}}\right). \quad (4.20)
\end{aligned}
$$

Similarly, the potential due to the p-type region is

$$
V_p(z) = -\frac{R\rho}{4\pi\varepsilon_0}\int_0^{2\pi} d\theta \ln\left(\frac{-z+\sqrt{2R^2(1-\cos\theta)+z^2}}{-z-W+\sqrt{2R^2(1-\cos\theta)+(z+W)^2}}\right), \quad (4.21)
$$

so that the total potential is given by

$$
\begin{aligned}
V(z) &= V_n(z) + V_p(z) \\
&= \frac{R\rho}{4\pi\varepsilon_0}\int_0^{2\pi} d\theta \ln\left(\frac{z+\sqrt{2R^2(1-\cos\theta)+z^2}}{z-W+\sqrt{2R^2(1-\cos\theta)+(z-W)^2}}\right. \\
&\quad \left.\times \frac{z+W-\sqrt{2R^2(1-\cos\theta)+(z+W)^2}}{z-\sqrt{2R^2(1-\cos\theta)+z^2}}\right). \quad (4.22)
\end{aligned}
$$

For large z this expression becomes

$$
V(z) \approx \frac{R\rho}{\varepsilon_0}\frac{W^2}{z^2} \qquad |z| \gg R, W. \quad (4.23)
$$

Thus, under the assumption of a finite depletion region with constant charge, the potential decays far away from the junction. This is in contrast to the bulk planar junction where the potential is constant outside of the depletion region. The decay of the potential far from the junction implies that the bands would fall below the charge neutrality position and that the nanotube would be charged far from the junction. To prevent this unphysical situation from happening, charge must be added to the nanotube outside of the depletion region to prevent the potential from decaying. This is the basis of the long-range charging in nanotube junctions. To obtain an expression for the charge outside of the depletion region, we consider the charge distribution

$$\sigma(z) = \begin{cases} \rho \quad 0 < z < W \\ \alpha V(z) \quad z > W \\ -\rho \quad -W < z < 0 \\ \alpha V(z) \quad z < -W, \end{cases} \tag{4.24}$$

which is essentially a linear response model for the charge outside of the depletion region. The electrostatic potential is then

$$V(z) \approx V_0(z) + \alpha \int_{-\infty}^{-W} V(z')K(z-z')dz' + \alpha \int_{W}^{\infty} V(z')K(z-z')dz' \tag{4.25}$$

where $V_0(z)$ is the potential due to the finite width depletion regions. Since the potential is small in the depletion region, we extend the limits of integration in the integrals to get

$$V(z) \approx V_0(z) + \alpha \int_{-\infty}^{-\infty} V(z')K(z-z')dz'. \tag{4.26}$$

To obtain the asymptotic dependence of this integral equation, we transform the potential and the electrostatic kernel to Fourier space. The Fourier representation of the kernel is

$$K(q) = I_0(qR)\, K_0(qR) \tag{4.27}$$

where I_0 and K_0 are modified Bessel functions. The equation for the potential in Fourier space is

$$V(q) = V_0(q) + \alpha V(q)K(q) \tag{4.28}$$

giving the solution

$$V(q) = \frac{V_0(q)}{1-\alpha K(q)}. \tag{4.29}$$

For small q (long distances) the Bessel functions have asymptotic behavior $I_0(qR) \sim$ constant and $K_0(qR) \sim -\ln(qR)$ giving the potential

$$V(q) \sim \frac{V_0(q)}{\ln(qR)}. \tag{4.30}$$

The real space potential is given by the inverse transform

$$V(z) \sim \int_0^\infty V(q)\sin(qz)dq = \int_0^\infty \frac{V_0(q)\sin(qz)}{\ln(qR)}dq \sim \int_0^\infty \frac{\sin(qz)}{q\ln(qR)}dq. \quad (4.31)$$

For large z, and because the integrand is dominated by small q, the integral can be approximated as

$$V(z) \sim z\int_0^{q_c} \frac{1}{\ln(qR)}dq \quad (4.32)$$

where the cut-off wavenumber $q_c \sim 1/z$. The integral gives

$$V(z) \sim zEi\left[\ln\left(q_c R\right)\right] \sim \frac{ze^{\ln(q_c R)}}{\ln\left(q_c R\right)} \sim \frac{1}{\ln(R/z)}. \quad (4.33)$$

Thus, because the charge is linearly related to the potential through Eq. (4.24) the charge outside of the depletion region follows the asymptotic behavior

$$\sigma(z) \sim \frac{1}{\ln(z/R)} \quad \text{large } z; \quad (4.34)$$

this function varies extremely slowly with distance, leading to long range charging of the carbon nanotube. This expression is in excellent agreement with the numerical results presented in Fig. 4.5.

While there is long-range charging of the carbon nanotube, there is still a depletion-like region near the junction where the carriers are fully depleted and the charge is constant. It is interesting to obtain expressions for the width of this depletion region as a function of doping. To do so, we first obtain an equation for the built-in potential by assuming that the potential step at the junction is due solely to the charge in the depletion regions:

$$\begin{aligned} V_{bi} &= V_n(W) - V_p(-W) = 2V_n\left(W\right) \\ &= \frac{R\rho}{2\pi\varepsilon_0}\int_0^{2\pi} d\theta \ln\left(\frac{W + \sqrt{2R^2(1-\cos\theta) + W^2}}{\sqrt{2R^2(1-\cos\theta)}}\right) \\ &= \frac{R\rho}{2\pi\varepsilon_0}\int_0^{2\pi} d\theta \ln\left(\frac{\frac{W}{R} + \sqrt{2(1-\cos\theta) + \left(\frac{W}{R}\right)^2}}{\sqrt{2(1-\cos\theta)}}\right). \end{aligned} \quad (4.35)$$

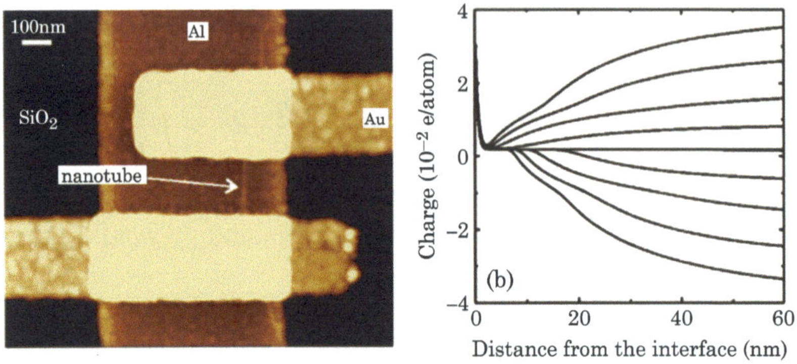

Figure 4.7 The left panel shows a scanning electron micrograph image of a nanotube between two electrodes. The right panel shows the charge along the nanotube, indicating long-distance charge transfer from the electrodes. Figure from Ref. [6].

For not too high doping, $W \gg R$ and Eq. (4.35) can be approximated as

$$V_{bi} \approx \frac{R\rho}{\varepsilon_0} \ln\left(\frac{2W}{R}\right), \tag{4.36}$$

giving the expression for the depletion width

$$W_{NT} \sim R \exp\left(\frac{\varepsilon_0 V_{bi}}{\rho R}\right). \tag{4.37}$$

As indicated by Eq. (1.47), the built-in potential depends logarithmically on the doping and is therefore a slowly varying function of doping, and the main dependence of the depletion width on the doping (Eq. (4.37)) comes from the $1/\rho$ factor in the exponential. As a consequence the depletion width is an extremely sensitive function of the doping, in contrast to the mild dependence on doping of bulk junctions, Eq. (4.17).

While we discussed the long distance charging in the context of p–n junctions, Schottky junctions between nanotubes and planar metals are also expected to show the same behavior, and this has been demonstrated experimentally. Fig. 4.7 shows a scanning electron micrograph image of a nanotube connecting two Au electrodes, and the associated charge distribution away from the contact. The long distance charging is observed, as predicted theoretically.

The much different charge distribution and electrostatics in nanotube junctions has dramatic impact on device design. For example, in traditional devices, the height of the potential step can be tailored by changing the doping. The depletion width in such devices depends weakly on the doping, thus allowing for precise control of the device properties. For nanotubes, however, the situation is quite different. Fig. 4.8 shows the calculated depletion width for the nanotube p–n junction as a function of doping.

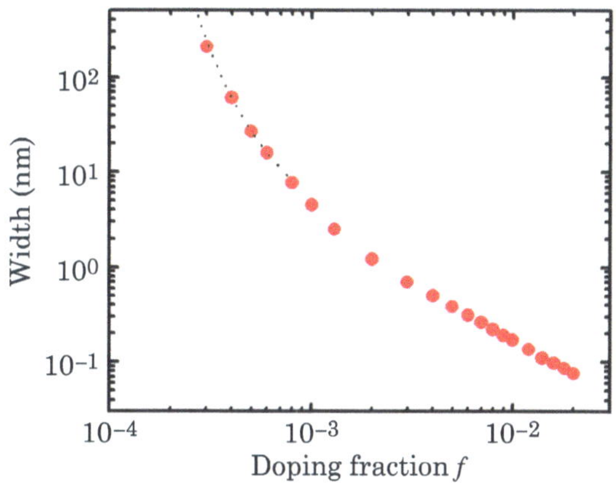

Figure 4.8 Calculated depletion width for a nanotube *p*–*n* junction as a function of doping. Figure from Ref. [5].

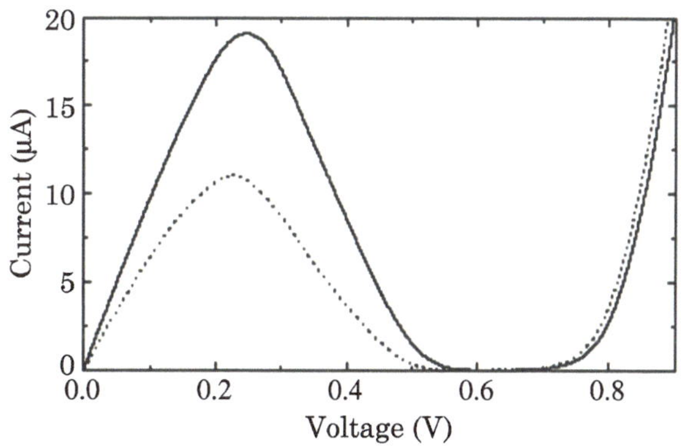

Figure 4.9 Calculated current–voltage curve for a nanotube *p*–*n* junction with high doping. The current–voltage curve shows negative differential resistance between 0.25 and 0.6 V. Figure from Ref. [7].

Clearly the depletion width is extremely sensitive to the doping, and thus fluctuations in dopant levels from device to device can significantly affect the device characteristics. Furthermore, at high doping, the depletion width is so small that tunneling across the potential step prevents the device from rectifying. This tunneling phenomenon is the basic operating principle behind negative differential resistance devices, and is observed in the experimental device of Fig. 4.1. It is thus interesting to model the properties of such devices. To do so requires computing the I–V curve, and this is done using the expression for the current in Eq. (4.1) where the transmission probability $T(E)$ for tunneling is calculated using the WKB approximation. Fig. 4.9 shows the results of such calculations, which indicate negative differential resistance, with a large ratio of maximum to minimum current.

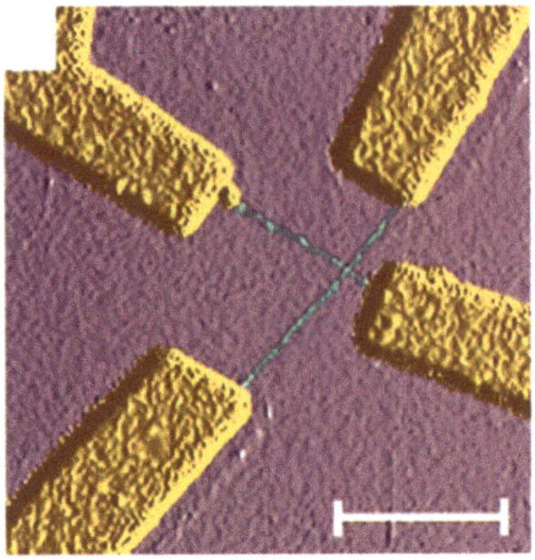

Figure 4.10 A nanotube device made of two crossing nanotubes. Figure from Ref. [8].

4.2.3 Metal–Semiconductor Junctions

Sections 4.2.1 and 4.2.2 described *intratube p–n* junctions, where rectification comes from modulation of the doping within a single nanotube. In addition, we discussed how contacts between nanotubes and metals can also act as Schottky diodes. In this section, we are concerned with metal–semiconductor rectifiers where both the metal *and* the semiconductor are carbon nanotubes. Such devices can be fabricated by combining two different nanotubes: Fig. 4.10 shows an experimental realization of one such device, consisting of two crossing nanotubes. Measurement of the individual conductance is used to determine the semiconducting or metallic character of each of the two nanotubes. Fig. 4.11 indicates that the current between the metallic and semiconducting nanotubes (curve labeled MS) shows rectification. This rectification behavior can be understood from the fact that the bandgap in a semiconducting nanotube arises from the opening of a symmetric gap around the Fermi points of a graphene sheet. Thus, the Fermi level is in the middle of the nanotube bandgap, and is at the same energy as the Fermi level in a metallic tube. This leads to the presence of a Schottky barrier at the crossing point between the two nanotubes equal to half the badgap, as illustrated in Fig. 4.11.

The same Schottky barrier concept can be used to create *intra*-tube metal–semiconductor junctions. Fig. 4.12 shows an image of a nanotube in a four probe measurement configuration, with a kink between the middle electrodes. Two-probe measurements show that one end of the nanotube is semiconducting, while the other end is metallic. Thus, the two segments of the nanotube correspond to different chiralities, and the angle at which they meet is determined by the presence of topological defects which allow a seamless junction. This type of junction shows strong rectification, and its working principle has been described theoretically in reference [9].

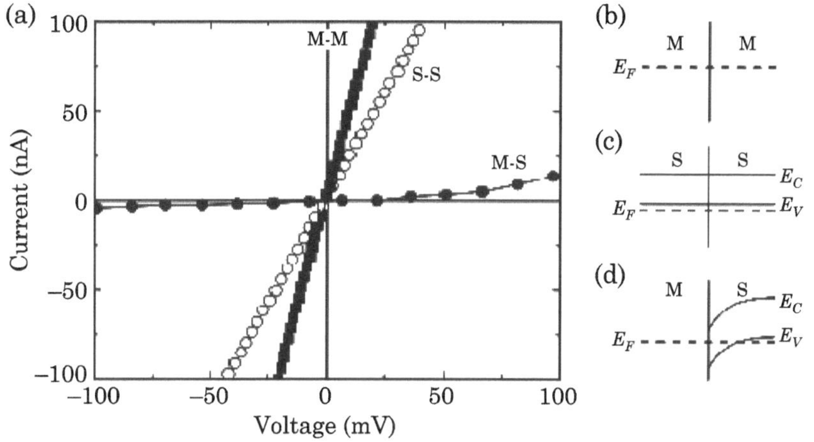

Figure 4.11 Measured *I*–*V* curves for devices like the one in Fig. 4.10. The metal–semiconductor (MS) junction shows rectifying behavior, due to a Schottky barrier at the junction, as illustrated in (d). Figure from Ref. [8].

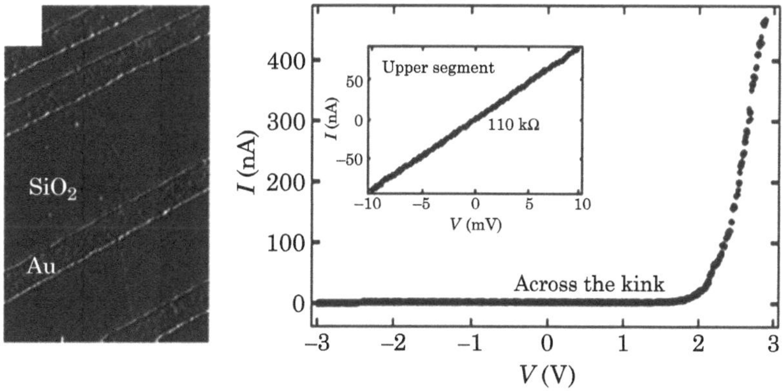

Figure 4.12 Intra-tube metal–semiconductor junction, and associated rectifying behavior. Figure from Ref. [10].

4.3 Field-Effect Transistors

Ever since its invention, the transistor has been the workhorse of the electronics industry, with field-effect transistors dominating the mass consumer market. Simple field-effect transistors are easier to realize, as compared to other types of transistors, such as bipolar junction transistors, in that no intricate doping control is necessary to demonstrate the field-effect. It is no surprise then that some of the initial devices made with nanotubes have been field-effect transistors [11,12]. Fig. 4.13 shows an atomic force micrograph image of one of the early nanotube transistors, which consists of a semiconducting nanotube bridging two Pt electrodes, and sitting on SiO_2 between the electrodes.

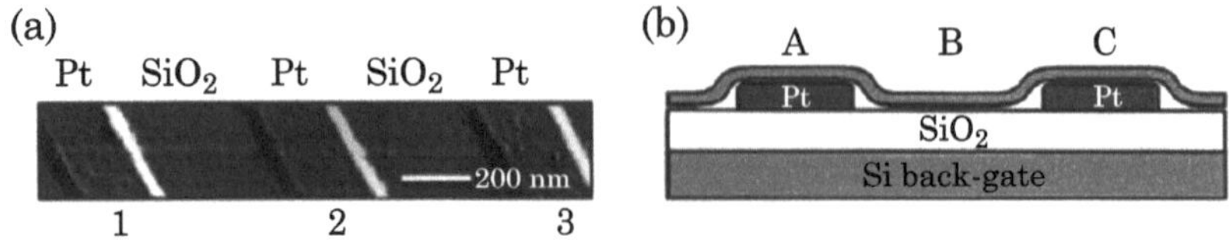

Figure 4.13 Atomic force micrograph image and sketch of original nanotube transistor. The transistor action is controlled by changing the voltage on the Si back gate. Figure from Ref. [11].

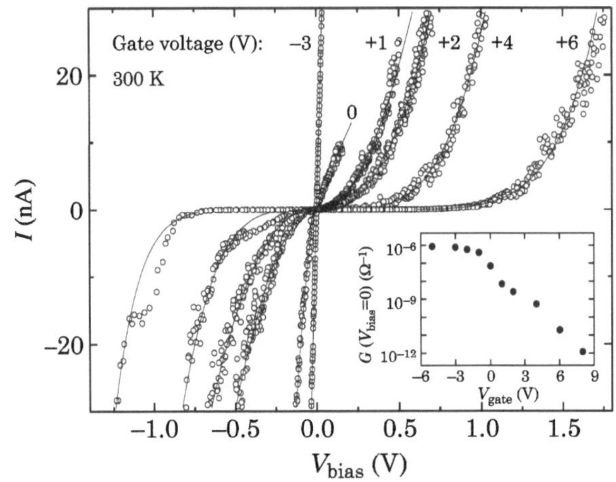

Figure 4.14 Current–voltage characteristics of an early nanotube transistor. Figure from Ref. [11].

A heavily doped Si substrate serves as a back gate, which controls the switching action of the transistor.

The drain current versus drain voltage characteristics of this transistor are shown in Fig. 4.14. In going from a gate voltage of −3 V to +6 V, the device changes from high to low conductance, which correspond to the ON and OFF states, and thus providing the switching action of the transistor.

Since this original device, there has been much experimental and theoretical progress in the understanding of the physics that governs the transistor action, and in improvement of the device performance. An important outcome of this work is the fact that the type of contact (ohmic or Schottky) has a profound influence on the device behavior.

4.3.1 Ohmic Contacts

As discussed in Chapter 3, ohmic contacts to nanotubes have been reported in the literature. Because of the ohmic contacts, the physics governing the transistor action is bending of the bands in the channel by the applied gate voltage. Theoretical work [13] to explain this behavior has been presented in the literature. Results of such work, based on quantum transport calculations, are presented in Fig. 4.15. As illustrated in this figure, the device consists of a single-wall, semiconducting carbon nanotube, embedded in metal contacts on either side, defining the source and drain. Between the source and drain electrodes, an insulating dielectric surrounds the nanotube up to a radius of 10 nm. A cylindrical gate of radius 10 nm wraps the dielectric and serves to control the device behavior. In the calculations, the nanotube and the metals are separated by a van der Waals distance of 0.3 nm. (The insulator has dielectric constant of 3.9, as for SiO2, and it is also separated from the tube by 0.3 nm.) A zigzag nanotube of index (17,0) is considered, which has a radius of 0.66 nm and band gap 0.55 eV. The metal Fermi level is chosen to be 1 eV below the nanotube midgap before self-consistency. (For the nanotube midgap 4.5 eV below the vacuum level, this corresponds to a metal work function of 5.5 eV.) Panel (a) in the figure shows the calculated zero-bias conductance as a function of the gate voltage. The device shows three regimes: in regime I the conductance is high, corresponding to the ON state of the transistor. In this regime, the bands are essentially flat (Fig. 4.15 (b)) so there is little scattering of electrons at the Fermi level. (Even though the transport is "ballistic", a spatially varying potential can cause scattering of electrons.) Since the conduction band has a degeneracy of two, the conductance in this regime saturates to a value close to two quanta of conductance. As the gate voltage is increased, the conductance decreases sharply, and the transistor enters the OFF regime. This regime is characterized by a large barrier in the middle of the nanotube that blocks the electrons (there is a small leakage current due to source–drain tunneling). As the gate voltage is further increased, the channel is driven into inversion. While for micron-sized channels this inversion would lead to a permanent turn-on of the conductance, for nanometer-sized channels the situation is quite different. In this case, the band-bending creates an electrostatic quantum dot in the middle of the nanotube, leading to the appearance of localized energy levels. Thus, the inversion regime in nanotube transistors with nanoscale channels consists of resonant tunneling through these discrete levels, leading to a peak in the conductance in regime III. This regime is expected to have intriguing behavior such as high frequency response and has yet to be explored experimentally.

As we discussed in Chapter 3, ohmic contacts to carbon nanotubes have been realized experimentally using palladium as the contact metal. Nanotube field-effect transistors with palladium contacts (Fig. 4.16) and channel lengths of a few hundred of nanometers have achieved ON state conductances close to two quanta of conductance [14]. However, increasing the channel length to a few microns severely decreases the ON state conductance, consistent with the electron–phonon scattering mean-free path of 1.5 microns. Thus for ohmically-contacted nanotubes, the ON state conductance is limited by scattering in the

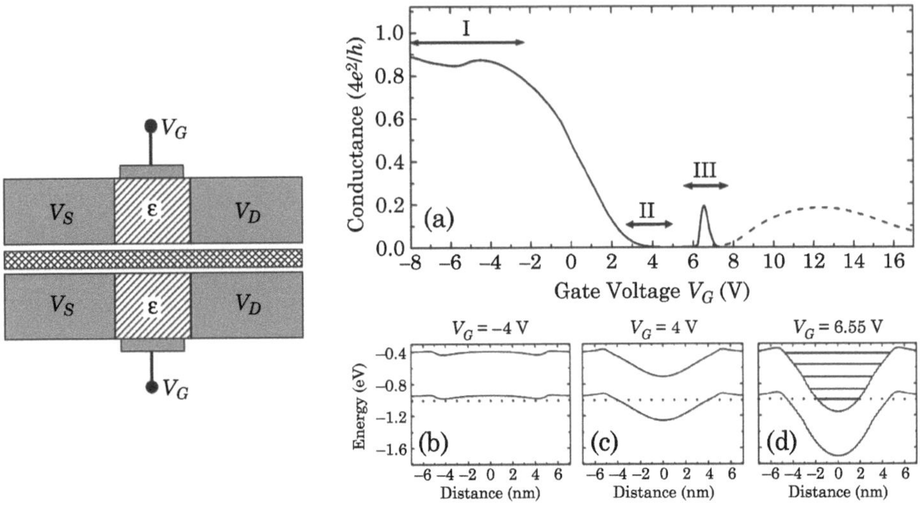

Figure 4.15 Left: schematic cross section of a carbon nanotube field-effect transistor with a cylindrical gate. The gray areas are the gate and the metallic source and drain contacts to the nanotube. The hatched areas represent the dielectric that surrounds the nanotube, and the cross-hatched area is the nanotube. The source–drain separation is 10 nm; the cylindrical gate has a radius of 10 nm. (a) The calculated zero-bias conductance of the nanotube field-effect transistor. (b–d) Band-bending associated with regimes I, II and III of (a). Dotted line is the Fermi level. Figure from Ref. [13].

channel, and channel lengths less than the electron–phonon scattering mean-free path are essentially ballistic.

The temperature dependence of the ON state conductance also provides further evidence for the presence of ohmic contacts. Indeed, using Eq. (2.11) and assuming perfect transmission through the contacts and the nanotube, we can obtain the temperature-dependent ON state conductance as [15]

$$G = \frac{4e^2}{h} \frac{e^{\Delta/kT}}{1 + e^{\Delta/kT}} \tag{4.38}$$

where $\Delta = E_v - E_F$ represents the position of the Fermi level in the valence band. The monotonically decreases with increasing temperature in agreement with the data presented in Fig. 3.8. Thus, even in carbon nanotube field-effect transistors without electron–phonon scattering it is expected that the conductance will decrease with increasing temperature, and can be reduced by as much as a factor of two at room temperature compared to its low-temperature value.

The main conclusion of this section is that the behavior of ohmic nanotube transistors is determined by changes in the band-bending and scattering in the channel region. As we will see in the next section, nanotube transistors with Schottky contacts behave much differently.

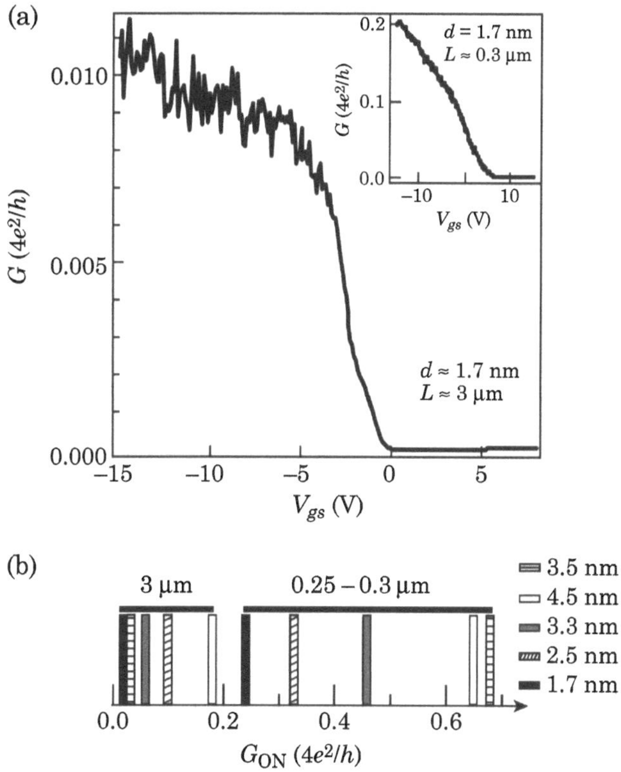

Figure 4.16 (a) Transfer characteristics of carbon nanotube field-effect transistors with ohmic palladium contacts. The main panel shows that a device with a channel of 3 microns reaches only 1/100th of the maximum possible ON conductance. However, when the channel length is reduced to 0.3 microns, the ON state conductance is within a factor of 5 of the maximum possible value. (b) Distribution of ON state conductance for a number of devices illustrates the role of the channel length. Figure from Ref. [14].

4.3.2 Schottky Contacts

As we have discussed in Chapter 3, electrical contacts to carbon nanotubes are often characterized by the presence of Schottky barriers. Normally, the current across such contacts is dominated by thermionic emission, where electrons must be thermally excited over the Schottky barrier. However, if the band-bending near the contact is very sharp, electrons can tunnel through the barrier, leading to a much increased current. This is precisely the effect that governs the operation of Schottky barrier nanotube transistors, as illustrated in Fig. 4.17.

In part (a) of this figure, the band-bending is sketched for the OFF state of the transistor. At this gate voltage, the band-bending near the contact is small, the tunneling length is long, and consequently the tunneling current is small. Increasing the gate voltage as in panel (b) raises the bands in the middle of the nanotube, leading to a much sharper band-bending at the contacts. This

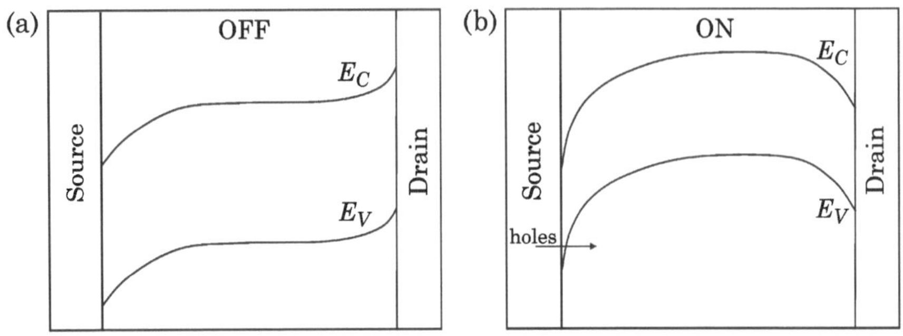

Figure 4.17 Band-bending in a Schottky barrier nanotube transistor, for two values of the gate voltage. Figure after Ref. [16].

reduces the tunneling distance at the contacts and leads to larger current. The device operation is thus controlled by modulation of tunneling at the contacts, a mechanism that is entirely different form conventional transistors and ohmic contact nanotube field-effect transistors. This effect of the gate voltage on the contact behavior has been demonstrated using computer simulations [17]. Fig. 4.18 shows a cross-section of the nanotube transistor considered for the calculations. The calculations proceed as described earlier, with the charge and the potential on the nanotube calculated self-consistently; the transmission due to tunneling through the band-bending at the contact is obtained from the WKB approximation, and the conductance is calculated from the transmission using the Landauer formula, Eq. (4.1). For a calculated potential variation $V(z)$ along the length of the carbon nanotube, the transmission probability within the WKB approximation is given by

$$T(E) = \exp\left(-2\int_{z_1}^{z_2} k(E,z)dz\right) \tag{4.39}$$

where $k(E,z)$ is the imaginary part of the wavevector as a function of position along the carbon nanotube and z_1 and z_2 are the classical turning points. Tunneling through the contact band-bending is illustrated in Fig. 4.19, indicating that the imaginary part of the wavevector depends on distance along the nanotube because of the band-bending. To obtain an expression for the imaginary part of the wavevector as a function of energy and distance, we begin by deriving a simple expression for its dependence on energy. As discussed in Chapter 1, near the Fermi points the band structure is given by

$$E = \frac{3a\gamma}{2}\sqrt{\left(\frac{E_g}{3a\gamma}\right)^2 + k^2} \tag{4.40}$$

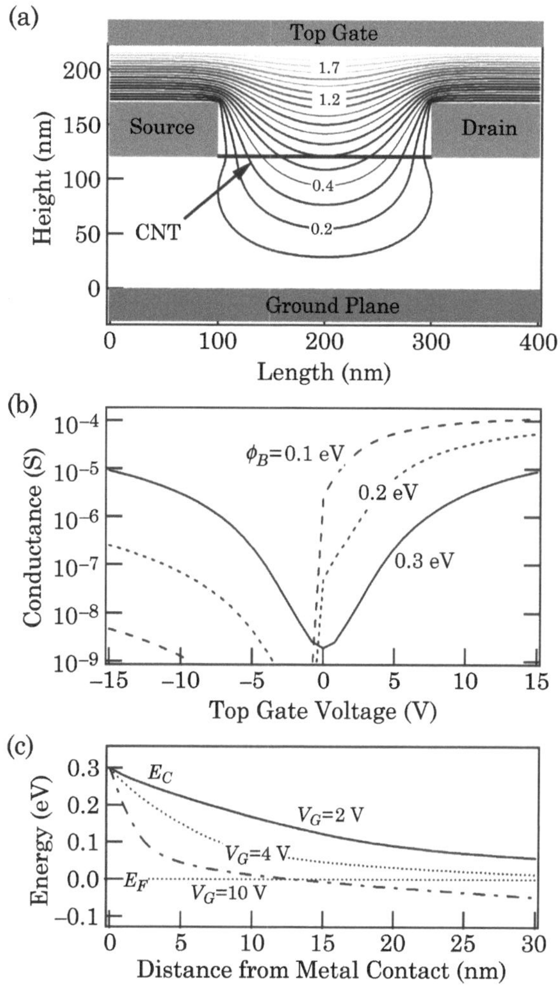

Figure 4.18 (a) Sketch of a carbon nanotube transistor with Schottky barriers at source and drain. Solid lines are lines of constant electrostatic potential for a gate voltage of 2 V. (b) Room temperature conductance as a function of gate voltage for different values of the electron Schottky barrier height. (c) Band-bending near the contact for three values of the gate voltage. Figure from S. Heinze.

and expressing k in terms of E from this equation gives

$$k(E) = \frac{2}{3a\gamma}\sqrt{E^2 - (E_g/2)^2}. \tag{4.41}$$

For values of $|E| < E_g/2$ the wavevector is imaginary; this imaginary value is zero at the band edges and reaches a maximum at midgap. The dependence of Im(k) on energy is shown in Fig. 4.19 for a nanotube with 0.6 eV bandgap. For this nanotube, the maximum value of the imaginary part of k is about 0.5 nm^{-1} corresponding to a decay length of about 2 nm; one might therefore

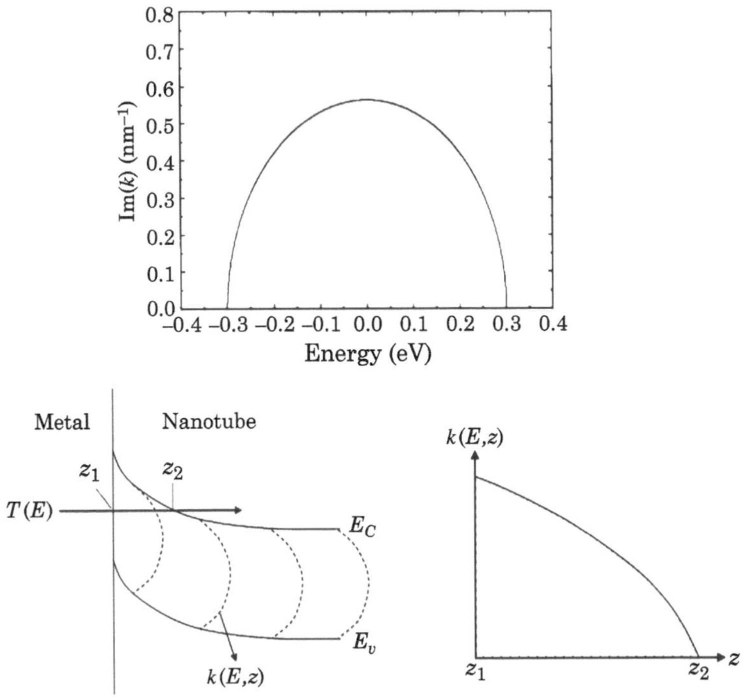

Figure 4.19 Top: imaginary part of the electron wavevector for a carbon nanotube with a bandgap of 0.6 eV. The imaginary part of the wavevector is zero at the band edges and reaches a maximum at midgap. Bottom left: illustration of tunneling through band-bending at a nanotube/metal contact, with the transmission probability $T(E)$, the imaginary part of the wavevector $k(E, z)$ and the two classical turning points z_1 and z_2. Bottom right: sketch of the imaginary part of the wavevector as a function of distance along the nanotube. The area under the curve enters the calculation of the transmission probability.

expect that tunneling will become important when the tunneling length is less than about 10 nm.

To calculate the tunneling probability from Eq. (4.39) the spatial dependence of k is also needed. We assume that the electrostatic potential $V(z)$ is known and simply shift the nanotube band structure locally by this electrostatic potential. Therefore, the energy and spatial dependence of the imaginary part of the wavevector is

$$k(E, z) = \frac{2}{3a\gamma}\sqrt{(E_g/2)^2 - [E + eV(z)]^2} \tag{4.42}$$

and the transmission probability is

$$T(E) = \exp\left\{-\frac{4}{3a\gamma}\int_{z_1}^{z_2}\left(\frac{E_g^2}{4} - [E + eV(z)]^2\right)^{1/2} dz\right\}. \tag{4.43}$$

Once the transmission probability is known for the relevant range of energies, the conductance is calculated from

$$G = \frac{4e^2}{h} \int_{-\infty}^{\infty} T(E) \left(-\frac{\partial f}{\partial E} \right) dE. \tag{4.44}$$

Fig. 4.18 (b) shows the calculated conductance as a function of gate voltage for different values of the electron Schottky barrier (the barrier for holes is simply the nanotube bandgap minus the electron barrier). The main curve in this figure corresponds to a situation where the metal Fermi level is exactly in the middle of the nanotube bandgap. The barriers for electrons and holes are equal and the device turns on at both positive (electron conduction) and negative (hole conduction) values of the gate voltage. For unequal electron and hole barriers, the conductance versus gate voltage curve is asymmetric, with much larger current for the side with the smallest barrier (in the case of Fig. 4.18 (b) the smallest barrier is for electrons). The origin of the conductance modulation by the gate voltage is illustrated in Fig. 4.18 (c). There it is shown that an increase of the gate voltage leads to a sharper band-bending in the vicinity of the contact, substantially reducing the tunneling length at the Fermi level (from 40 nm to 10 nm in this particular case). Since tunneling probabilities depend exponentially on tunneling length, the device conductance increases very rapidly with increase in gate voltage.

4.3.3 Subthreshold Swing

A key performance parameter of field-effect transistors is the subthreshold swing S defined as

$$S = \left| \frac{d \log_{10} G}{dV_g} \right|^{-1} \tag{4.45}$$

and is a measure of the effectiveness of the gate in changing the channel conductance. In a traditional channel-controlled transistor, the smallest subthreshold swing is achieved when a change in the gate voltage leads to the same shift of the bands, i.e. the gate fields are unscreened (Fig. 4.20). This corresponds to a situation where

$$\frac{dE_{c,v}}{d\,(eV_g)} = -1. \tag{4.46}$$

The conductance can be calculated from the expression in Eq. (4.44), assuming that the transmission probability is equal to 1 for energies below the

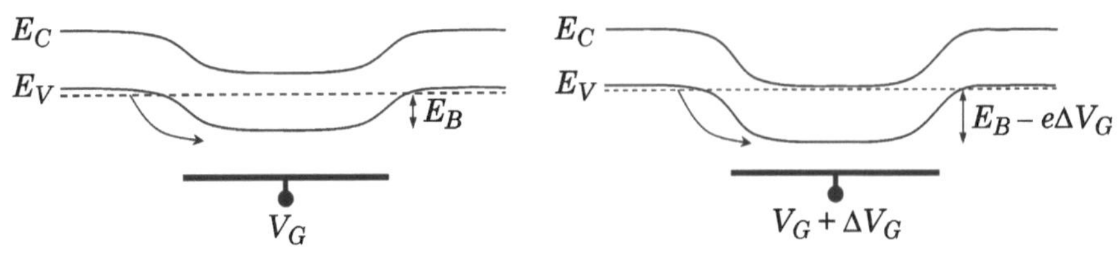

Figure 4.20 Illustration of the ideal subthreshold swing in a p-type field-effect transistor. At finite temperature, the small bias conductance is governed by thermal excitation of holes over a potential barrier E_B in the middle of the channel. The ideal device behavior occurs when a change ΔV_G in the gate voltage is unscreened and leads to a change $-e\Delta V_G$ in the position of the band edge in the middle of the channel.

minimum of the valence band. Under this assumption, the conductance reads

$$G = 2G_0 \exp\left(\frac{E_v - E_F}{kT}\right) \tag{4.47}$$

and the subthreshold swing is

$$S = \left|\frac{d\log_{10} G}{dV_g}\right|^{-1} = \left(\frac{d\log_{10} G}{d\left(E_v/e\right)}\right)^{-1} = \frac{kT}{e}\ln 10 \tag{4.48}$$

which is equal to 60 meV/decade at room temperature.

This traditional derivation of the subthreshold swing assumes that the device characteristics are controlled by band-bending in the channel and that conduction is determined by thermal excitation of carriers; it applies to carbon nanotube field-effect transistors with ohmic contacts. In practice, the gate oxide screens the gate fields, and the coupling between the gate and the channel is not perfect, which causes the subthreshold swing to be larger than the ideal value given by Eq. (4.48). The typical strategy to reduce the subthreshold swing is to decrease the oxide thickness to improve the coupling, and this will be discussed in the next section. However, because the switching mechanism is fundamentally different in Schottky barrier nanotube transistors, the geometrical requirements for improving device performance are also quite different. Indeed, in the case of the Schottky barrier nanotube transistor the band-bending at the contact is the key issue that affects device performance, and enhancing the effect of the gate on this band-bending is the key challenge in device design. For example, it is clear from the equipotential lines in Fig. 4.18 (a) that the contacts significantly screen the gate-induced electric fields. Two ways have been proposed to improve the impact of the gate voltage on the band-bending at the contact [17]. The first approach follows that of conventional transistors and consists of decreasing the gate oxide thickness. Fig. 4.21 shows the calculated conductance versus gate voltage curve for the device of Fig. 4.18 (a), for a fixed contact thickness of 50 nm, but with oxide thicknesses ranging from 60 to 120 nm. As expected,

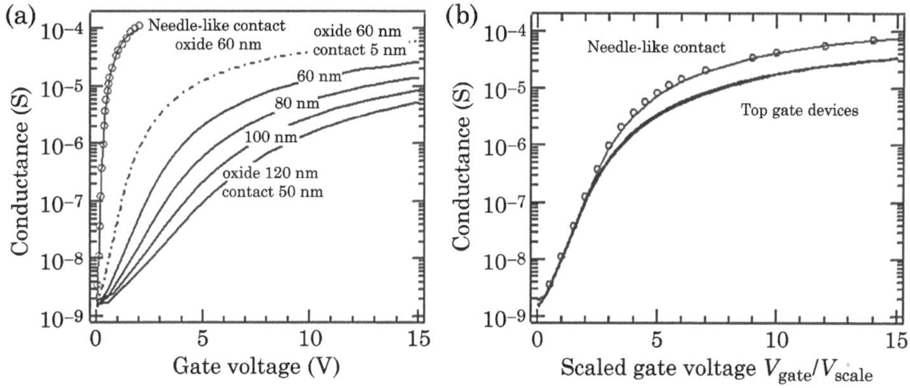

Figure 4.21 (a) Influence of transistor geometry on the device characteristics of Schottky barrier nanotube transistors. The four right-most curves correspond to different thicknesses of the oxide above the nanotube, as labeled, with all other parameters as in Fig. 4.18. (The curve for 100 nm is the same as that in Fig. 4.18 (b) for a Schottky barrier height of 0.3 eV.) The dot–dashed curve shows the conductance when the contact thickness is reduced to 5 nm. The curve at the left corresponds to a needle-like metal electrode and cylindrical gate. Open circles are calculated as for the other curves, solid curve uses the exact electrostatic kernel. All calculations are at room temperature. (b) The graph shows the same curves, but for each curve the gate voltage is rescaled by the voltage at which the conductance is 10^{-8} S. (The five rightmost curves cannot be distinguished on this scale, forming a single line). Figure from S. Heinze.

this improves the device performance, with a larger ON conductance and lower threshold voltage. More importantly, the dash-dotted line in Fig. 4.21 shows that reducing the thickness of the contact from 50 nm to 5 nm leads to further performance improvement, a mechanism that does not occur in traditional channel-controlled devices. From this insight, it is clear that the ultimate behavior is obtained for a needle-like contact and a cylindrical gate, as the open circles in Fig. 4.21 indicate.

The much different physics behind the operation of Schottky barrier nanotube transistors has important implications on the scaling of various performance parameters with device dimensions. As discussed above, it was predicted that reducing the thickness of the gate insulator improves the subthreshold swing because it allows the gate to more effectively modulate the band-bending at the contact. Such a behavior has been verified experimentally by fabricating nanotube transistors with gate oxide thickness between 2 and 20 nm. The current–voltage characteristics of these transistors shown in Fig. 4.22 clearly indicate that the devices with the smaller gate oxides have smaller subthreshold swings. This behavior is quantitatively plotted in the right panel of Fig. 4.22 which shows that the subthreshold swing decreases rapidly as the gate oxide thickness is reduced, but that the improvement is smaller for reduction of the gate oxide thickness below 5 nm. This can be understood by considering the unscreened band-bending at the contact for a double-gate geometry with a

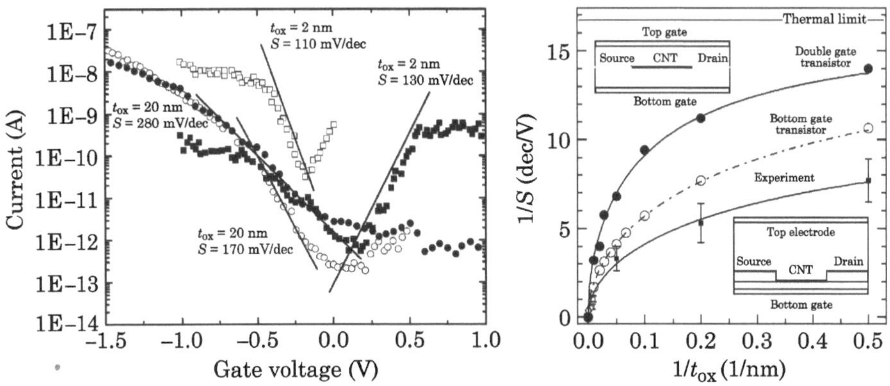

Figure 4.22 The left panel shows the current vs gate voltage characteristics of Schottky barrier carbon nanotube transistors for different gate oxide thicknesses. The right panel shows the measured and calculated (see text for detail) subthreshold swing as a function of the inverse gate oxide thickness. Figures from Ref. [18] and from S. Heinze.

vanishingly thin contact [18], as illustrated in the top inset in the right panel of Fig. 4.22. (By unscreened, we mean that the charge on the carbon nanotube is not considered in the calculation. This is a good approximation as long as the Fermi level is more than kT away from the band edge.) For this device geometry, the potential near the contact is [18]

$$V(z) = \frac{2V_g}{\sqrt{\pi}} \left(\frac{z}{t_{\text{ox}}} \right)^{1/2} \tag{4.49}$$

where t_{ox} is the gate oxide thickness. Substituting this expression in Eq. (4.43) gives

$$\begin{aligned} T(E) &= \exp\left\{ -\frac{4}{3a\gamma} \int_{z_1}^{z_2} \left(\frac{E_g^2}{4} - \left[E + \frac{2eV_g}{\sqrt{\pi}} \left(\frac{z}{t_{\text{ox}}} \right)^{1/2} \right]^2 \right)^{1/2} dz \right\} \\ &= \exp\left\{ -\left(\frac{\tilde{V}}{V_g} \right)^2 h(s) \right\} \end{aligned} \tag{4.50}$$

where

$$h(s) = \int_{\min(0,-1-s)}^{1-s} t \left(1 - [s+t]^2 \right)^{1/2} dt \tag{4.51}$$

and

$$\tilde{V} = \left(\frac{\pi E_g^3}{12a\gamma} \right)^{1/2} t_{\text{ox}}^{1/2}. \tag{4.52}$$

Since the transmission depends on the ratio $\tilde{V}/V_g$, $\tilde{V}$ plays the role of a rescaling voltage. The conductance is calculated from

$$G = \frac{4e^2}{h} \int_{-\infty}^{\infty} T(E) \left(-\frac{\partial f}{\partial E} \right) dE = \frac{2e^2 E_g}{hkT} H(x, y) \tag{4.53}$$

with

$$H(x, y) = \int_{-\infty}^{\infty} e^{-h(s)/x^2} \left(-\frac{e^{sy}}{(1 + e^{sy})^2} \right) ds \tag{4.54}$$

and

$$x = \frac{V_g}{\tilde{V}}, \quad y = \frac{E_g}{2kT}. \tag{4.55}$$

These expressions can be simplified by assuming that $y = E_g/2kT \gg 1$ to obtain

$$G = \frac{4e^2}{h} \exp \left[-\frac{1}{3} \left(\frac{\tilde{V}}{V_g} \right)^2 \right]. \tag{4.56}$$

The conductance increases exponentially with gate voltage and scales with $t_{\text{ox}}^{1/2}$ since $\tilde{V}$ is proportional to $t_{\text{ox}}^{1/2}$. The threshold voltage can be defined as the gate voltage at which the conductance decreases by a factor of 1000 from its saturation value:

$$V_{\text{th}} = \frac{1}{\sqrt{9 \ln 10}} \tilde{V} \propto t_{\text{ox}}^{1/2}. \tag{4.57}$$

Thus the threshold voltage is also z proportional to $t_{\text{ox}}^{1/2}$. The subthreshold swing at the threshold voltage is then

$$S = \left(\frac{d\log G}{dV_g} \right)^{-1} \Bigg|_{V_{\text{th}}} \propto \tilde{V} \propto t_{\text{ox}}^{1/2} \tag{4.58}$$

and is also proportional to $t_{\mathrm{ox}}^{1/2}$. The important point is that this scaling behavior of the subthreshold swing with gate oxide thickness is different from that in conventional MOSFETs because the switching is controlled by Schottky barriers at the contacts.

As the thickness of the oxide becomes quite small, the above analysis breaks down since the fields at the contact no longer follow the simple expression used for the calculations. But in the limit of very thin oxides, the subthreshold swing must approach the theoretical limit of $(kT/e)\ln 10$; a simple empirical expression for S is

$$S = \left[\alpha t_{\mathrm{ox}} + \left(\frac{kT}{e}\ln 10\right)^2\right]^{1/2}. \tag{4.59}$$

The solid lines in the right panel of Fig. 4.22 show that such an expression describes well the experimental and computational results. It is important to note that the scaling with $t_{\mathrm{ox}}^{1/2}$ is a consequence of the particular form of the electrostatic potential at the contact. For a different device geometry this functional form may be different giving another scaling form. In particular, for an infinitely thick contact [18], the electrostatic potential varies as $z^{2/3}$ leading to a scaling $t_{\mathrm{ox}}^{2/3}$.

4.3.4 High-κ Dielectrics

High-κ dielectrics for gate insulators have been the subject of intense research in the semiconductor industry. The driving force behind this activity is that reducing the channel length in transistors also requires a reduction of the gate oxide thickness to maintain the gate efficiency. For SiO_2, current technologies utilize a thickness of only a few atomic layers, and further reduction of this thickness leads to channel-gate leakage current because of tunneling. For this reason, alternative materials to SiO_2 are being explored, with the requirement that they have the same or larger gate capacitance and low leakage current. Because capacitance is essentially the ratio of the dielectric constant to the gate insulator thickness, high-κ dielectrics allow the gate oxide thickness to be increased (to limit leakage) while maintaining or even increasing the capacitance. At the time of writing of this manuscript, major semiconductor manufacturers are announcing that chips using HfO_2 ($\kappa = 25$) as the gate insulator will soon be available.

Integrating new high-κ dielectrics in existing silicon technology is a complex process which is very much dependent on the ability to grow high-quality thin oxide layers and the ability to form good interfaces with the channel material. The same issues are faced in carbon nanotube transistors. While it has been demonstrated that SiO_2 is a relatively good substrate for carbon nanotubes,

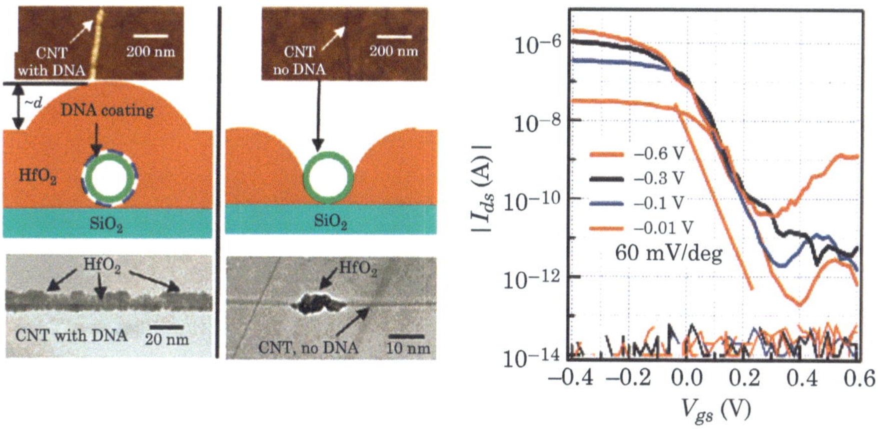

Figure 4.23 Left image: growth of HfO_2 by atomic layer deposition on carbon nanotubes with and without a DNA coating. Top images in both panels are atomic force microscope images of carbon nanotubes with about 5 nm of HfO_2. Middle panels are sketches of the expected wetting behavior. The bottom panels show transmission electron microscopy images of a 5 nm layer of HfO_2 on the carbon nanotubes. Good wetting is seen in the case of DNA-functionalized nanotubes. Right image: transfer characteristics of a carbon nanotube transistor with 3 nm of HfO_2 as the gate dielectric. The subthreshold swing approaches the theoretical room-temperature limit of 60 mV/decade. Figures from Ref. [19].

it is by no means obvious that high-κ dielectrics will form good interfaces with the nanotubes. An example of this is shown in Fig. 4.23 [19]. There the goal is to coat a carbon nanotube with a thin layer of HfO_2 as a gate insulator. Unfortunately, the HfO_2 has poor wetting properties on bare carbon nanotubes, and it is impossible to form a continuous, uniform layer directly on the nanotubes. However, Fig. 4.23 also shows that coating the nanotubes with DNA allows the formation of a good oxide layer. Transistors fabricated with such DNA-coated nanotubes and a 3 nm HfO_2 gate insulator can achieve low subthreshold swings because of the large gate capacitance.

One high-κ material that has been found to form good interfaces with nanotubes when grown using atomic layer deposition is ZrO_2 [20]. Fig. 4.24 shows that the deposition of ZrO_2 over a carbon nanotube sitting on SiO_2 gives a highly conforming interface between the ZrO_2 and the nanotube. This property allows the fabrication of top-gated field-effect transistors with thin layers of high-κ dielectric, which show low subthreshold swing and low leakage current (Fig. 4.25).

4.3.5 Logic Circuits

The discovery of transistor behavior in carbon nanotubes was critical in demonstrating their potential for future electronic devices. The next step in this progress is to demonstrate that combining several devices can lead to logic

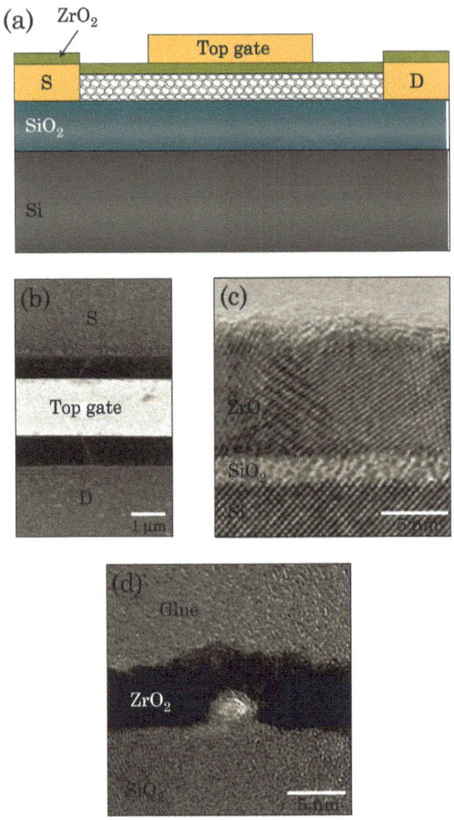

Figure 4.24 ZrO_2 as a high-κ dielectric for carbon nanotube field-effect transistors. (a) shows a sketch of the device with the top gate geometry. (b) Scanning electron microscopy image of a carbon nanotube transistor viewed from the top. (c) Transmission electron microscopy (TEM) image of the interface between ZrO_2 and SiO_2. (d) TEM image of the ZrO_2/SiO_2 interface including a carbon nanotube (bright circular spot in the image). Figure from Ref. [20].

function. While this may appear relatively simple at first glance, it requires the ability to control and combine the properties of p-type and n-type devices, and carbon nanotube devices are typically p-type. One strategy to form n-type nanotubes is to utilize the strong effects of gases on the nanotube electronic properties (this will be discussed in more detail in Chapter 8). For example, Fig. 4.26 shows that vacuum annealing or exposure to potassium of initially p-type carbon nanotube transistors converts them to n-type devices. For the vacuum annealed nanotubes, this conversion to n-type behavior can be reversed by exposing the nanotube to oxygen, as indicated in Fig. 4.27.

Therefore, one strategy to make both p-type and n-type devices for logic circuits is to first convert the nanotubes to n-type by vacuum annealing. The nanotubes that are to stay n-type are protected by PMMA, while the others are exposed to oxygen and converted to p-type. The two types of transistors

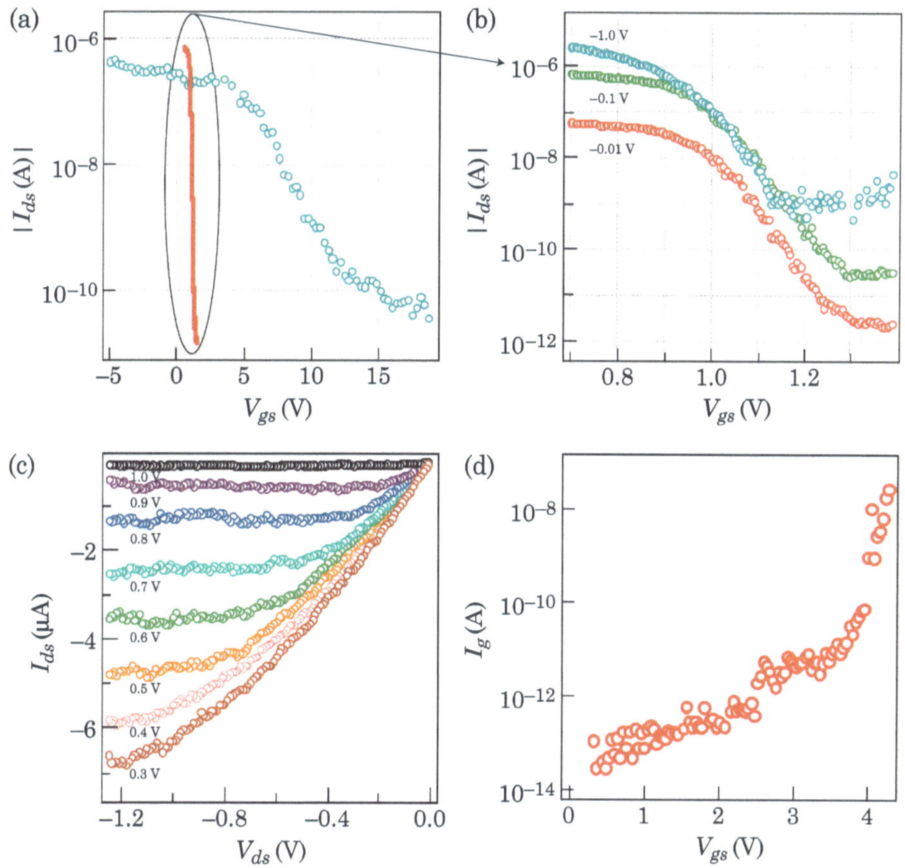

Figure 4.25 Characteristics of a *p*-type nanotube field-effect transistor with high-κ gate insulator. (a) I_{ds}–V_{gs} curves recorded under $V_{ds} = -100$ mV with bottom Si/SiO2 gate (circles) and top-gate/ZrO2 (solid curve) respectively.The bottom-gate was grounded during top-gate operation. (b) I_{ds}–V_{gs} curves recorded with top-gate/ZrO2 at bias voltages of $V_{ds} = -10$ mV, -0.1 V and -1 V, respectively. (c) I_{ds}–V_{ds} curves of the transistor recorded for various top-gate voltages at 0.1 V steps. (d) Gate leakage current versus top-gate voltage. The leakage current is negligible (at the pA level) until $V_{gs} > 3$ V. Figure from Ref. [20].

are then electrically connected to form an inverter circuit (illustration Fig. 4.27 (c)). An example of the measured V_{in} versus V_{out} characteristics of such devices is shown in panel (d) of Fig. 4.27, indicating that a positive V_{in} gives a negative V_{out}, and vice-versa; the device thus inverts the input voltage.

More complicated logic circuits using a larger number of individual nanotube transistors have also been demonstrated. Fig. 4.28 shows the operation of NOR, OR, NAND, and AND gates, as well as an oscillator, utilizing up to six individual nanotube transistors. While the frequency of the ring oscillator is rather low, this is a consequence of the simple connection of the transistors through coaxial cables, and the performance can in principle be significantly improved by direct interconnects on the chip.

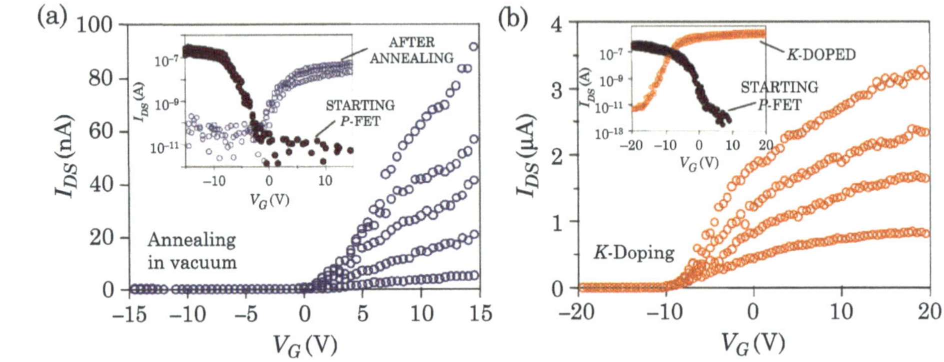

Figure 4.26 Conversion of p-type carbon nanotube transistors to n-type by (a) vacuum annealing at 700 K for 10 min, (b) exposure to potassium. Figure from Ref. [21].

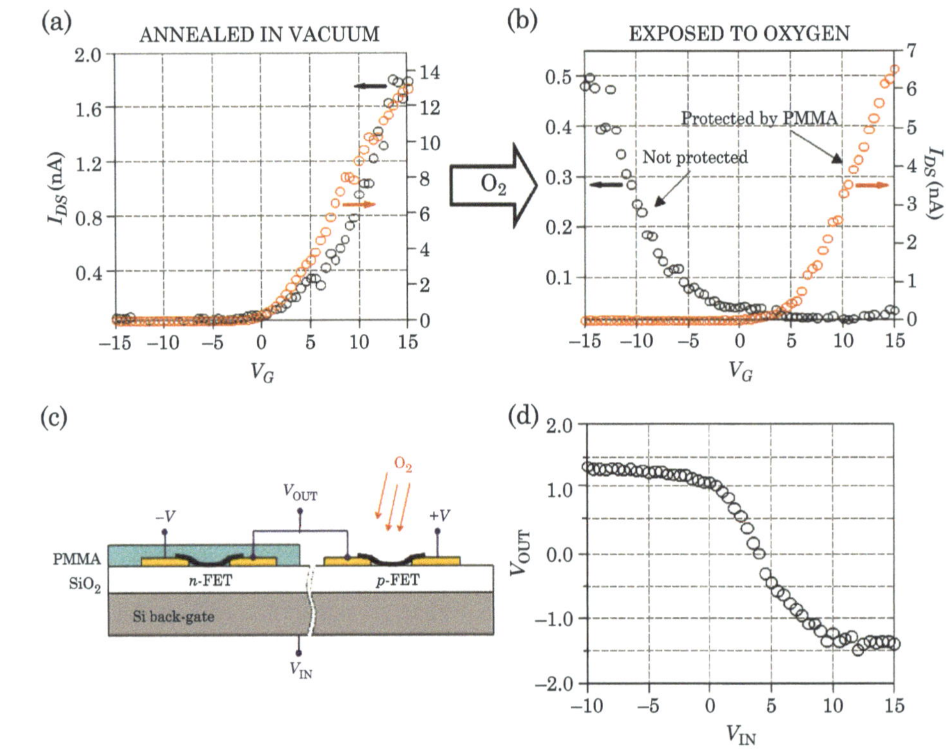

Figure 4.27 Fabrication of a voltage inverter ("NOT" logic gate) using two nanotube field-effect transistors. Initially the two transistors are p-type. One of them is protected by PMMA, the other is not. (a) After vacuum annealing both transistors are converted to n-type. (b) The two devices are exposed to oxygen (10^{-3} Torr of oxygen for 3 min). The unprotected transistor converts back to the original p-type, while the protected nanotube remains n-type. (c) The two complementary transistors are wired as shown in the schematic. (d) Characteristics of the resulting intermolecular inverter are shown. Figure from Ref. [21].

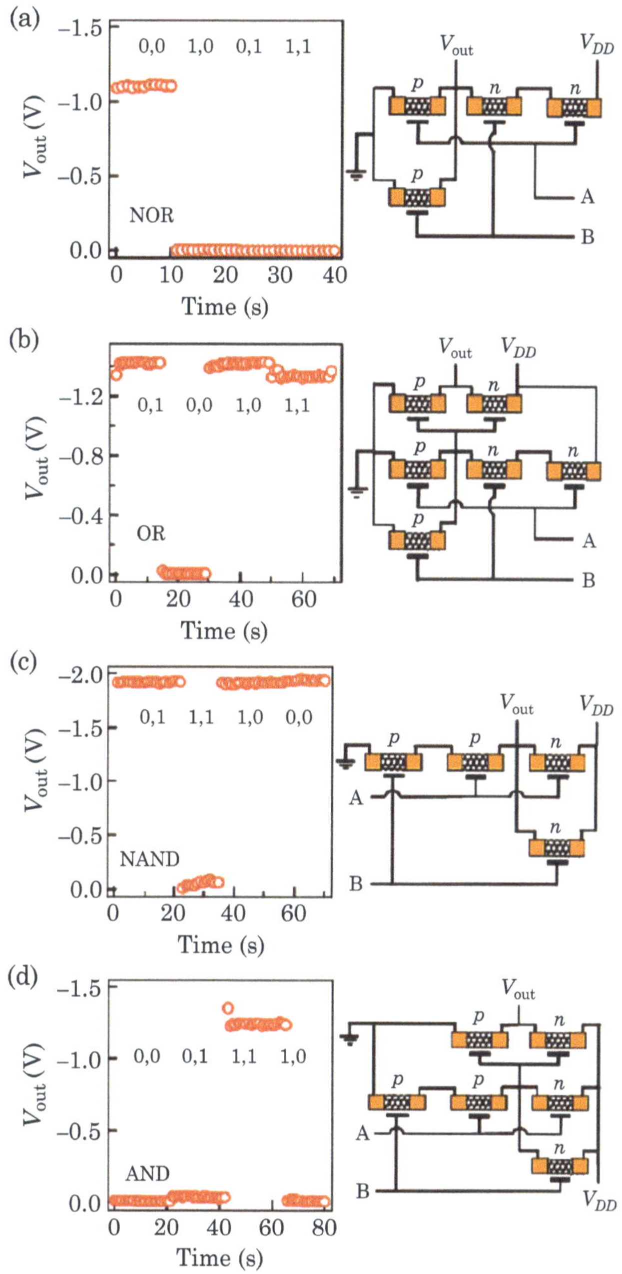

Figure 4.28 Several types of logic circuits using several carbon nanotube transistors. (a) NOR; (b) OR; (c) NAND; (d) AND gates. The bottom figure shows a ring oscillator. Figure from Ref. [22].

Recent progress in carbon nanotube logic circuits is the fabrication of these circuits on a *single* carbon nanotube. Fig. 4.29 shows a single bundle of carbon nanotubes making contact to three electrodes, producing two *p*-type transistors

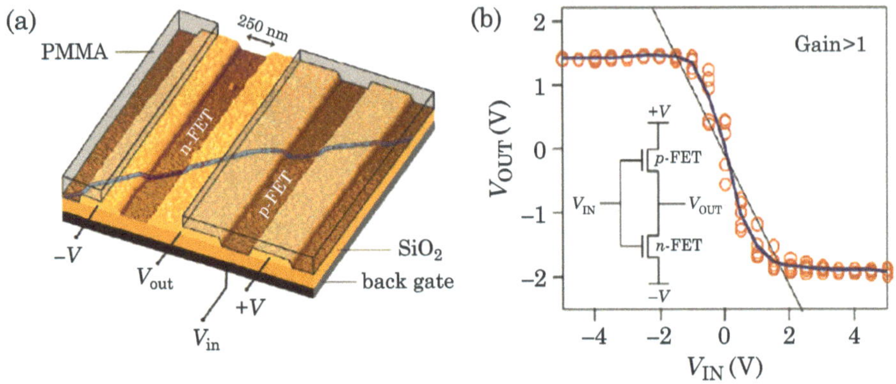

Figure 4.29 (a) Atomic force microscope image showing the geometry of a carbon nanotube inverter. A single nanotube bundle is positioned over the gold electrodes to produce two p-type nanotube field-effect transistors in series. The device is covered by PMMA and a window is opened by e-beam lithography to expose part of the nanotube. Potassium is then evaporated through this window to produce a n-type field-effect transistor, while the other nanotube remains p-type. (b) Characteristics of the resulting intramolecular voltage inverter. Open circles are raw data for five different measurements on the same device ($V = \pm 2$ V). The line is the average of these five measurements. The thin straight line corresponds to an output/input gain of one. Figure from Ref. [21].

in series. This nanotube is then covered with PMMA, and a window is opened between two of the electrodes. Exposure of this channel to potassium causes a transformation from p-type to n-type, and a common back gate to the two channels is used to operate the device as an inverter.

Extension of this idea of making a circuit entirely on a single carbon nanotube has recently shown that multistage ring oscillators can be fabricated using long carbon nanotubes. Fig. 4.30 shows a scanning electron microscope image of an 18 micron carbon nanotube with multiple source, drain and gates. The arrangement actually consists of 10 nanotube transistors (5 p-type, 5 n-type) arranged in groups of two (1 p-type, 1 n-type) to form 5 inverters in series. Interestingly, the approach to make the p-type and n-type transistors is different from what we have already discussed. Here, two different metals (Pd and Al) are used as the gate material to make the two types of transistors. The idea is that the large work function difference between Pd and Al (about 1 eV) effectively acts like an extra gate voltage, shifting the I–V curve on the V_g axis. Thus, one takes advantage of the ambipolar nature of the transport, and brings the positive V_g turn-on in the window of the negative V_g turn-off: as one device turns-off the other turns on giving the inverter characteristics. This is illustrated in Fig. 4.30 (b) which shows about a 1 V shift of the Pd I–V curve to the left when Pd is replaced with Al.

The frequency dependence of the signal is shown in Fig. 4.30 (d) for different values of V_{dd} showing resonance frequencies in the range of 13 to 52 MHz. The magnitude of the resonant frequency and dependence on V_{dd} can be understood

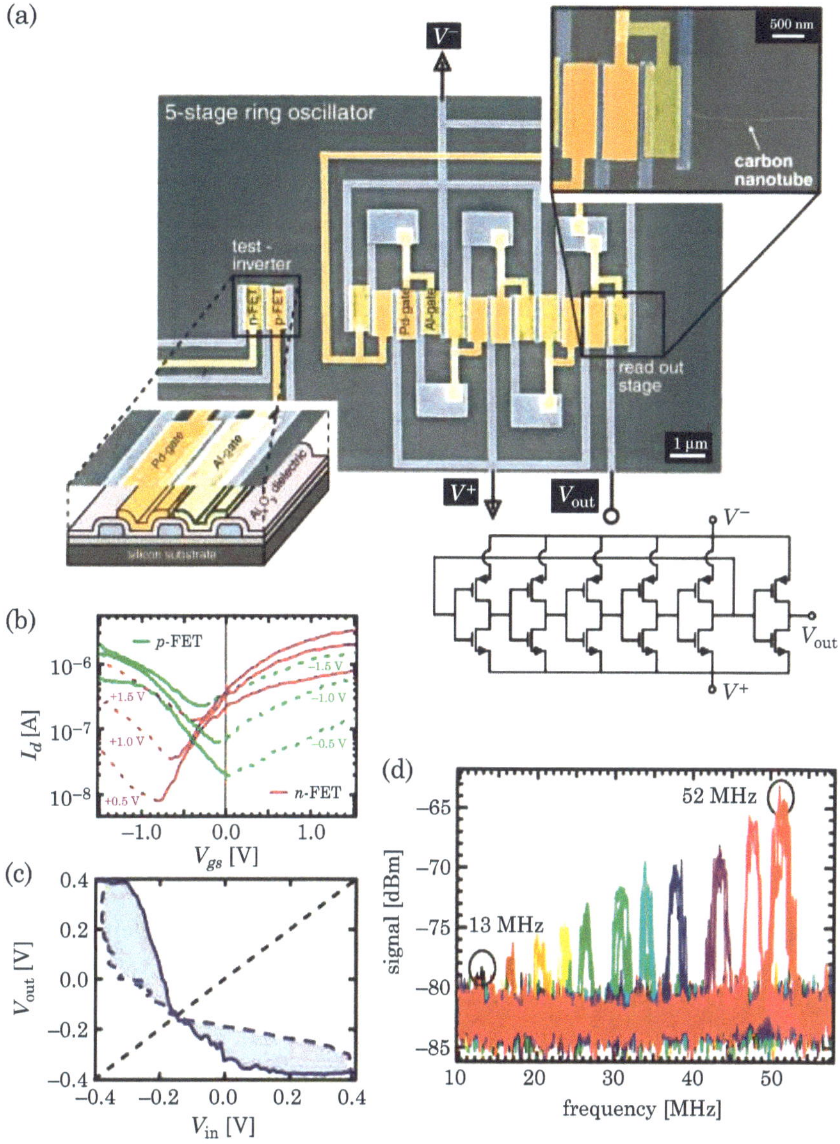

Figure 4.30 Multistage ring oscillator fabricated on a single carbon nanotube. (a) Scanning electron microscope image of the five-stage ring oscillator, as well as test inverter. (b) Current–voltage characteristics of the p-type transistors with Pd gate, and the n-type transistors with Al gate. (c) Characteristics of the test inverter. (d) Frequency response of the ring oscillator. The different colored curves correspond to different values of V_{dd} equal to 0.5 V and 0.56 V to 0.92 V (in 0.04 V increments) from left to right. Figure from Ref. [23].

as follows: the expected frequency for a N-stage ring oscillator is [23]

$$f = \frac{1}{2\tau N} \tag{4.60}$$

where τ is the stage delay time given by the RC constant

$$\tau = RC = \frac{VC}{I}. \tag{4.61}$$

Therefore, the resonance frequency can be expressed as

$$f = \frac{I}{2CVN}. \tag{4.62}$$

The capacitance per stage in the oscillator is estimated to be 1.8 fF, while for $V_{dd} = 0.92$ V the DC current in a single transistor is about 1 µA. For a 5 stage oscillator this gives a frequency $f = 59$ MHz in good agreement with the experiment. The point is that this measured resonance frequency is due to the parasitics and not to the intrinsic properties of the nanotube. The increase of the resonance frequency with increasing V_{dd} is a consequence of the increase in the current through a single transistor as V_{dd} is increased (i.e. the current is not saturated with V_{gs}). Another consequence of this effect seen in Fig. 4.30 (d) is the increase of the signal power with increase in V_{dd}. The signal attenuation is due to the impedance mismatch between the output of the ring oscillator and the spectrum analyzer.

4.3.6 Mobility

While carbon nanotubes are lauded for their promise as ballistic conductors, for long enough channel lengths, scattering with defects, phonons, etc. eventually leads to diffusive behavior, and the introduction of a mobility μ.

The mobility of the charge carriers is defined through the relation

$$\mu = \sigma / n \tag{4.63}$$

where σ is the conductivity and n is the carrier concentration. For diffusive transport, the conductivity is related to the conductance$G$ as

$$\sigma = GL \tag{4.64}$$

where L is the length of the nanotube. For a p-type channel near threshold, the charge density is proportional to deviations of the gate voltage from the threshold voltage:

$$n = C\left(V_{\text{th}} - V_g\right) \tag{4.65}$$

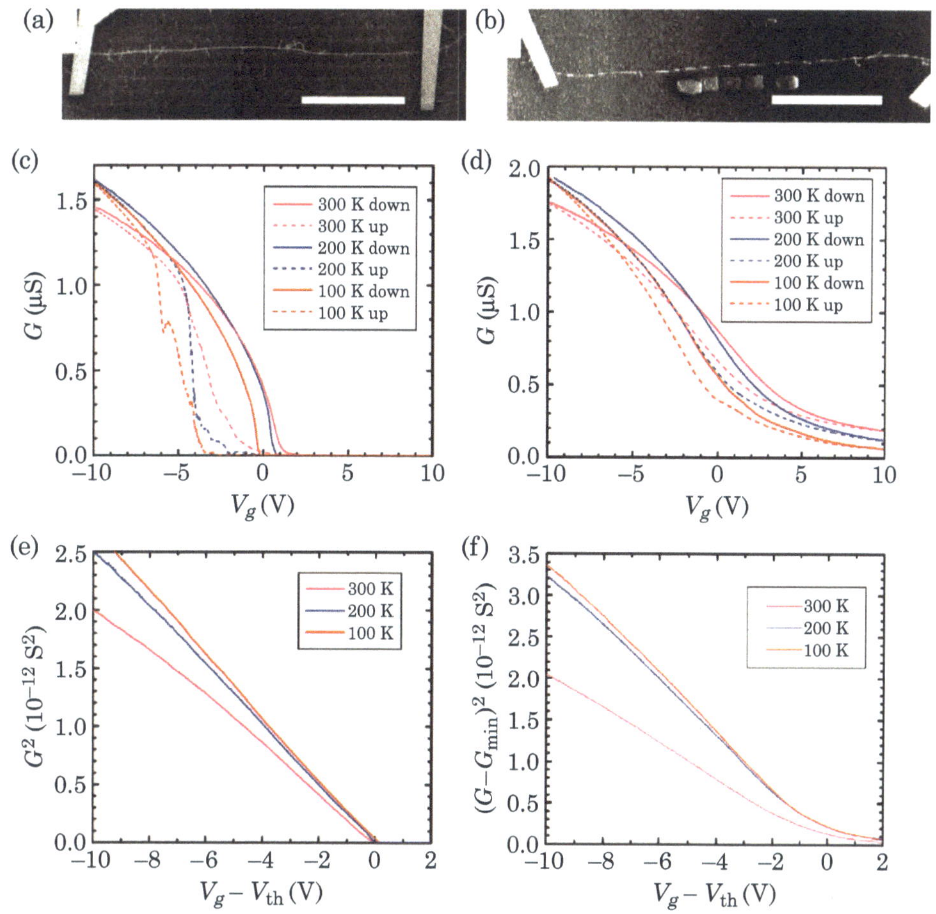

Figure 4.31 Behavior of long-channel carbon nanotube transistors, indicating extremely large field-effect and intrinsic mobilities. The scale bar in (a) is 100 μm long. Figure from Ref. [24].

and the mobility is therefore

$$\mu = \frac{L}{C}\frac{G}{V_{\text{th}} - V_g}. \tag{4.66}$$

Another quantity of interest when evaluating device performance is the field-effect mobility defined as

$$\mu_{\text{FE}} = \frac{L}{C}\frac{\partial G}{\partial V_g}. \tag{4.67}$$

Fig. 4.31 shows the measured field-effect mobility, which attains a value of 79 000 cm^2/Vs at room temperature, almost two orders of magnitude larger

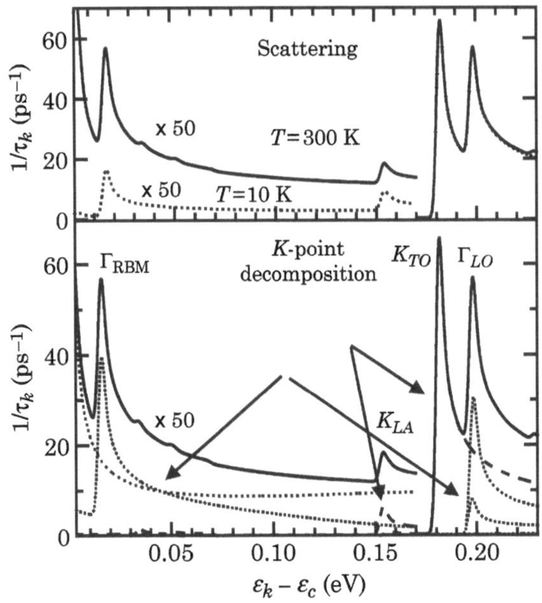

Figure 4.32 Electron-phonon scattering time vs electron energy relative to conduction band edge, for a (25,0) tube. Solid curves are for $T = 300$ K, dotted curves are for $T = 10$ K. Figure from V. Perebeinos.

than that typically achieved in Si MOSFETs. The intrinsic mobility exceeds the highest reported mobility in bulk materials, 77 000 cm^2/Vs, the Hall mobility of InSb. (It should be noted that subsequent experiments [25] have obtained field-effect mobilities in the range 1500–20 000 cm^2/Vs and have suggested that the very large mobilities could originate from a nonuniform response of the nanotube to the gate field.)

Without a detailed atomistic study of the particular nanotube used in the experiments, it is difficult to ascertain the role of defects in causing the diffusive behavior. However, phonons undoubtedly cause electron scattering, leading to diffusive behavior. Theoretical work has addressed this issue by considering electron–phonon scattering in semiconducting carbon nanotubes [26]. The approach is similar to what was presented in Chapter 2 for electron–phonon scattering in metallic carbon nanotubes: the scattering time and scattering length are calculated using Fermi's golden rule for the electron–phonon interaction and the Boltzmann equation is solved in the presence of an electric field to obtain the carrier velocity and mobility as a function of the electric field from the nonequilibrium carrier distribution. The essential difference here is the much different band structure of semiconducting and metallic nanotubes.

The results of such calculations are presented in Fig. 4.32 for a (25,0) semiconducting carbon nanotube. We first note that the binding energy, representing the distortions of the band structure due to the electron–phonon interaction, tends to be small and relatively independent of energy. This effect

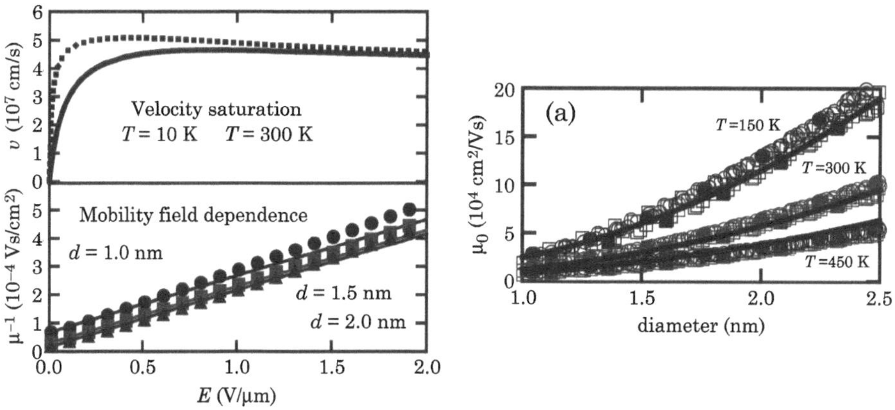

Figure 4.33 Left (top): drift velocity vs electric field at $T = 300$ K (solid curve) and $T = 10$ K (dotted curve) in a (25,0) nanotube. Left (bottom): inverse mobility at $T = 300$ K vs electric field for a (25,0) nanotube (triangles); (19,0) nanotube (squares); and (13,0) nanotube (circles). Right: zero-field mobility vs tube diameter, for tubes of many different chiralities. Solid lines are from analytical model, Eq. (4.69). Figures from V. Perebeinos.

was discussed in Chapter 1 in the context of the bandgap renormalization by phonons, and will not be discussed further here. Instead, we focus on the scattering time as a function of energy in the conduction band, as shown in the panels of Fig. 4.32. It is clear from this picture that the scattering time is sensitive to both energy and temperature. The scattering time can be as small as a few hundredths of picoseconds, but can also reach picoseconds for energies not too far from the conduction band edge. Thus, for devices where injection of carriers is a fraction of an eV in the valence or conduction bands, ballistic transport should be possible in defect-free nanotubes over long distances. It turns out that almost all of the scattering is from phonons near the Γ and K points of the graphene Brillouin zone. The lowest energy phonon band gives negligible scattering. The second lowest energy phonon mode is an acoustic mode that gives large scattering near the band edge, and correspondingly small scattering time. For the (25,0) nanotube, the third lowest energy phonon mode is the radial breathing mode with energy of 15 meV, leading to a dip in the scattering time at about 0.02 eV above the conduction band edge. The next phonon mode with important electron–phonon scattering is a longitudinal acoustic mode which gives a scattering length of about 5 microns (the scattering length is simply the scattering time times the Fermi velocity). Longitudinal optical phonons at the Γ and K points give the strongest scattering, significantly reducing the scattering length to the 20–40 nm range.

For device applications and comparison with experiment, what matters is the total effect of all of the phonon modes on the conductivity, as captured by the mobility. Fig. 4.33 shows the calculated drift velocity and mobility as a function of the electric field. The maximum drift velocity for this particular nanotube is 5×10^7 cm/s which is about a factor of five larger than silicon. The mobility obtained from the saturation velocity is plotted in Fig. 4.33 (b), which attains values of 5×10^3 cm^2/Vs at fields of 1 V/μm. More importantly, the graph indicates that a simple relation exists between the mobility and the electric field:

$$\mu^{-1} = \mu_0^{-1} + v_s^{-1} E \tag{4.68}$$

where v_s is the saturation velocity and E is the electric field. The zero-field mobility μ_0 takes a wide range of values depending on the diameter of the nanotube: from 15 000 cm^2/Vs for a (13,0) nanotube to 65 000 cm^2/Vs for a (25,0) nanotube. Indeed this dependence on diameter is plotted in Fig. 4.33 for different temperatures. The behavior can be captured by the phenomenological relation

$$\mu_0\,(t,d) = \mu_1 \frac{300\ \text{K}}{T} \left(\frac{d}{1\ \text{nm}} \right)^{\alpha} \tag{4.69}$$

with $\mu_1 = 12\,000$ cm^2/V s and $\alpha = 2.26$.

Since the mobility due to electron–phonon scattering is sensitive to the presence or absence of available states, carrier concentration can have a strong impact on the mobility because states in the conduction or valence bands become occupied by the carriers. These carriers may originate from doping, or from capacitive coupling to an external gate (see for example Eq. (4.65) and the experimental results of Fig. 4.31). To explore this effect, theoretical calculations have used a Boltzmann transport approach with electron–phonon interactions to calculate the mobility as a function of the carrier density [27]. Results of such calculations at low fields are presented in Fig. 4.34. The mobility initially increases with an increase in the carrier density, reaches a maximum and decreases at high carrier concentrations. To understand the origin of this behavior, we consider the classical relationship between the mobility, the scattering time and the average effective mass

$$\mu = \frac{e\tau}{m_{\text{av}}}. \tag{4.70}$$

The introduction of an average effective mass is because states in the bands are occupied over an energy range that reflects the carrier density, so this leads in principle to an energy-dependent effective mass. Introducing an energy-averaged effective mass allows for a simple analysis using Eq. (4.70). This average effective mass is obtained from the equation

$$\frac{1}{m_{\rm av}} = \frac{\int_{E_g/2}^{E_F} m^{-1}(E)dE}{\int_{E_g/2}^{E_F} dE} = \frac{1}{E_F - E_g/2} \int_{E_g/2}^{E_F} \frac{1}{\hbar^2}\frac{\partial^2 E}{\partial k^2}\, dE \qquad (4.71)$$

where E_F is the position of the Fermi level for a given doping level. At zero temperature, it is obtained from the expression

$$\begin{aligned} f &= \int_{E_g/2}^{E_F} D(E)dE \\ &= D_0 \int_{E_g/2}^{E_F} \frac{E}{\sqrt{E^2 - (E_g/2)^2}} dE \\ &= D_0\sqrt{E_F^2 - (E_g/2)^2} \end{aligned} \qquad (4.72)$$

or

$$E_F = \sqrt{(f/D_0)^2 + (E_g/2)^2}. \qquad (4.73)$$

In these equations, f is the doping fraction in electrons/atom and $D_0 = a\sqrt{3}/\pi^2 R\gamma$. To calculate the average effective mass we use the expression for the subband energy

$$E(k) = \frac{E_g}{2}\sqrt{1 + 9R^2k^2} \qquad (4.74)$$

to get

$$\frac{\partial^2 E}{\partial k^2} = \frac{9E_g^4 R^2}{8E^3} \qquad (4.75)$$

and the average effective mass is

$$m_{\rm av} = \frac{4\,\hbar^2}{18 D_0 E_g^2 R^2} \frac{4f^2 + (D_0E_g)^2}{\sqrt{4f^2 + (D_0E_g)^2} + D_0E_g}. \qquad (4.76)$$

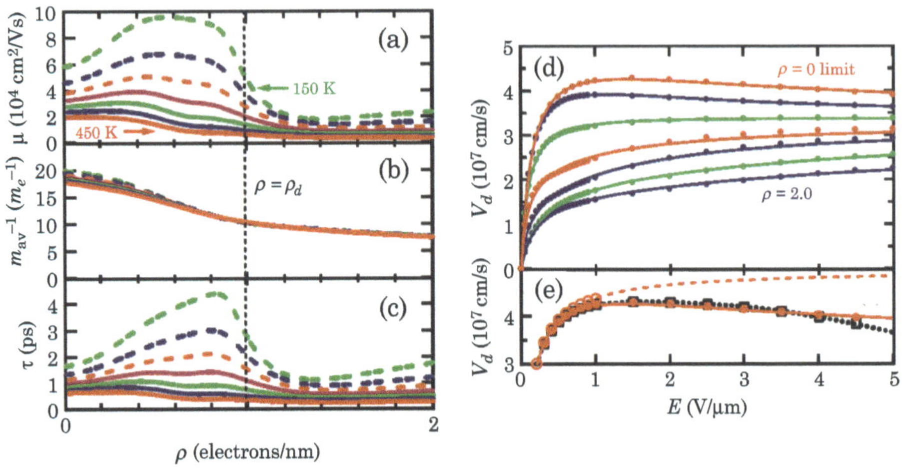

Figure 4.34 (a) Electron mobility at low fields for a (19,0) carbon nanotube as a function of the carrier density. (b) Contribution to the mobility of the inverse effective mass. (c) Contribution to the mobility of the scattering time. Temperatures from top to bottom in each panel are 150, 200, 250, 300, 350, 400 and 450 K. The vertical dashed line indicates the doping density at which the second subband becomes occupied. (d) Drift velocity at room temperature as a function of electric field for different carrier densities for a (19,0) nanotube. Curves from top to bottom correspond to carrier densities of 0, 0.25, 0.5, 0.75, 1.0, 1.5 and 2.0 electrons/nm. (e) Results for $\rho = 0$ are compared with single band models: single hyperbolic band (dotted line) and single parabolic band (dashed line). Figures from Ref. [27].

For typical nanotubes, the product $D_0 E_g$ is small, and the above equation simplifies to

$$m_{\rm av} = m_0 \sqrt{1 + \frac{4f^2}{D_o^2 E_g^2}} \tag{4.77}$$

where m_0 is the effective mass at the band edge. The important point here is that the average effective mass monotonically increases with the carrier concentration (see Fig. 4.34 (b)), and is only weakly dependent on temperature, so that the nonmonotonic behavior of the mobility with carrier density and its temperature dependence arise from the scattering time. Indeed, from the mobility calculated using the Boltzmann transport approach and average effective mass from Eq. (4.77), the scattering time can be obtained from Eq. (4.70). Fig. 4.34 (c) shows the scattering time as a function of the carrier density. The scattering time initially increases with increasing carrier density since the number of available states for scattering decreases as the Fermi level is higher in the conduction band. At a certain carrier concentration f, the Fermi level reaches the second subband, opening up additional channels for scattering; this leads to a decrease of the scattering time. Additional calculations [27] show that the scattering time is essentially proportional to the diameter of the carbon nanotube.

The high field behavior is studied by calculating the drift velocity v_d from the occupation function in the Boltzmann equation [27]. Fig. 4.34 (d) shows the field-dependence of v_d for different values of the carrier concentration. In the absence of carriers, the drift velocity reaches a maximum at a field of 1 V/μm and then decreases at larger fields. In contrast, at high carrier concentrations the drift velocity increases continuously. It is tempting to invoke the occupation of the second subband as the reason for the nonmonotonic field dependence of the drift velocity. However, comparisons of the results with those of a single-band model (Fig. 4.34 (e)) indicate that when the single band is described by a hyperbolic band profile, the Boltzmann results are well reproduced. A parabolic model for the single band does not match the full numerical calculations, and fails to give a maximum in v_d. Therefore, the nonparabolicity of the first subband is responsible for the nonmonotonic behavior.

4.3.7 Short-Channel Effects

Short channel effects in transistors can lead to serious degradation of the device characteristics as the different length scales (channel length, gate insulator thickness, etc.) are reduced to smaller and smaller dimensions. In conventional transistors made with silicon or other bulk semiconductor materials, short channel effects have been extensively studied, and the basic physics is well understood. In contrast, short channel effects in carbon nanotube transistors has received relatively little attention, despite the obvious importance for applications. The key issue here is that, because of the reduced dimensionality of carbon nanotubes, the Coulomb interaction has different implications than for bulk devices. In this section we discuss modeling work aimed at studying short channel effects in carbon nanotube field-effect transistors with ohmic contacts[15].

Fig. 4.35 shows a sketch of the device under consideration: it consists of a semiconducting carbon nanotube embedded in source and drain electrodes made of palladium. In the channel region, the nanotube is surrounded by SiO_2, and the SiO_2 is itself wrapped by a cylindrical gate of radius R_G. The simulations use a (17,0) zigzag nanotube, which has a bandgap of 0.55 eV in the tight binding model used for the calculations. The calculations proceed by obtaining self-consistently the charge and the potential on the carbon nanotube. The potential is calculated by solving Poisson's equation in three-dimensions, with the charge on the nanotube and the boundary conditions at the source and drain contacts, the gate, and at the dielectric/air interface. The charge and ultimately the current are calculated using the nonequilibrium Green's function technique, allowing for a quantum description of the system in the ballistic transport regime. The metal Fermi level in the contacts is taken to be 1 eV below the nanotube midgap before self-consistency. Results of these calculations for the current as a function of gate voltage are shown in Fig. 4.36, for a gate radius of 3 nm. For a channel length of 10 nm, the ON/OFF ratio is about 1000,

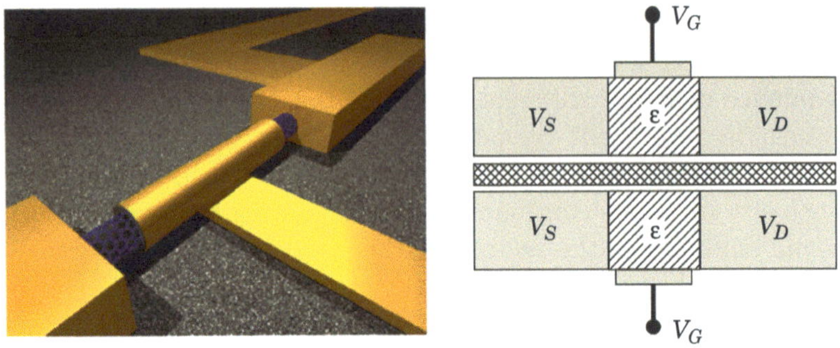

Figure 4.35 Sketch of the nanotube transistor used for device simulations. The nanotube is embedded in metals at its two ends, and in a dielectric in the channel region. The dielectric is wrapped by a cylindrical gate. In the left panel, the separation between the contacts and the central dielectric region is to illustrate the structure in the channel; in the calculations, the contacts touch the dielectric as shown in the right panel. Figures from Refs [15] and [13].

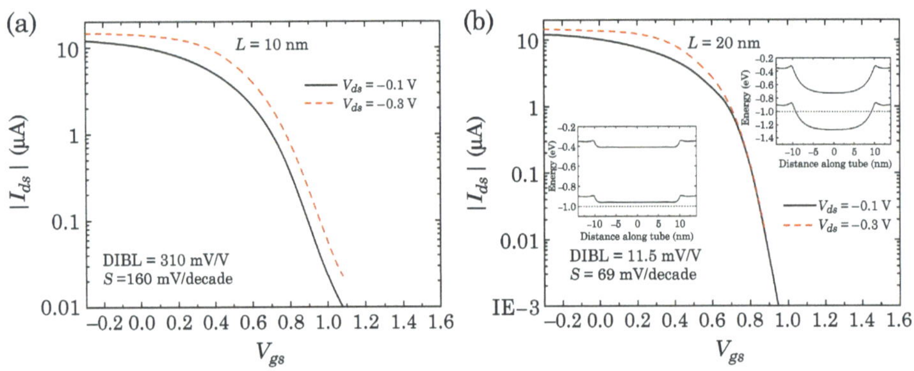

Figure 4.36 Current as a function of gate–source voltage for channel lengths of (a) 10 nm and (b) 20 nm. The gate radius is 3 nm. The insets in (b) show the band-bending for $Vgs = 0$ (left) and $Vgs = 1$ V (right). Figure from Ref. [15].

and the subthreshold swing is 160 mV/decade. An increase of the source–drain voltage from −0.1 V to −0.3 V causes a shift of the threshold voltage by 310 mV, a signature of drain-induced barrier lowering (DIBL), as will be discussed below. Increasing the channel length to 20 nm significantly improves both the subthreshold swing and the DIBL, which become 69 mV/decade and 11.5 mV/decade, respectively. It is worth mentioning that at $V_g = 0$, the nanotube is significantly hole-doped due to the long-range charge transfer from the contacts (left inset in Fig. 4.36).

The short channel effects are also seen in the current versus source–drain voltage, as indicated in Fig. 4.37. The 20 nm channel device shows saturation of the current for large values of the source–drain voltage, but the 10 nm device shows no saturation at all. The origin of this effect is DIBL, and plots of the band-bending along the carbon nanotube confirm this behavior (Fig. 4.38).

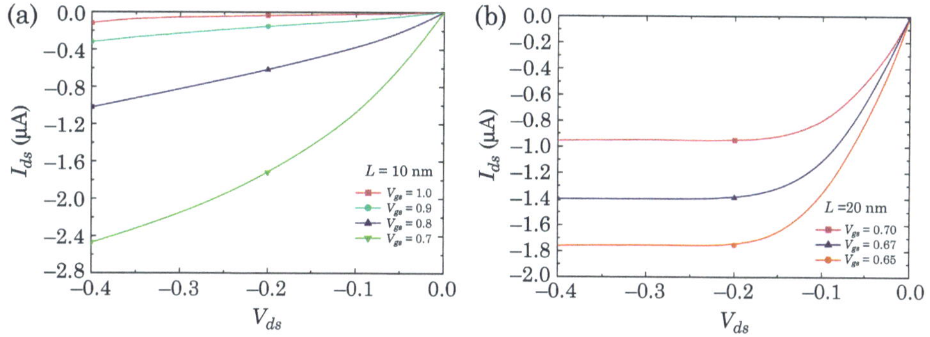

Figure 4.37 Current as a function of drain–source voltage for the carbon nanotube transistor of Fig. 4.35 for (a) 10 nm channel length and (b) 20 nm channel length. In both panels the gate radius is 3 nm. Figure from Ref. [15].

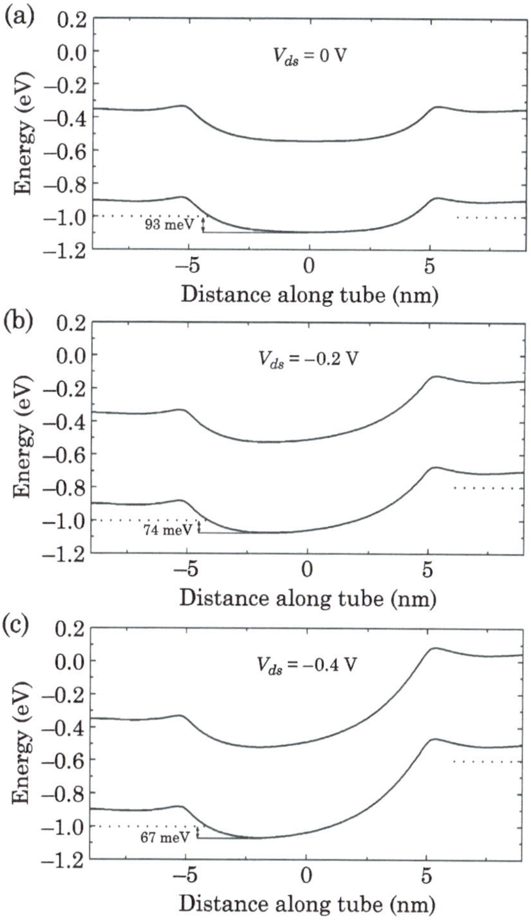

Figure 4.38 Calculated self-consistent band-bending for the 10 nm channel device with a 3 nm gate radius, for a gate–source voltage of 0.8 V. The solid lines are the valence and conduction band edges; horizontal dotted lines are the metal Fermi levels in the contacts. Figure from Ref. [15].

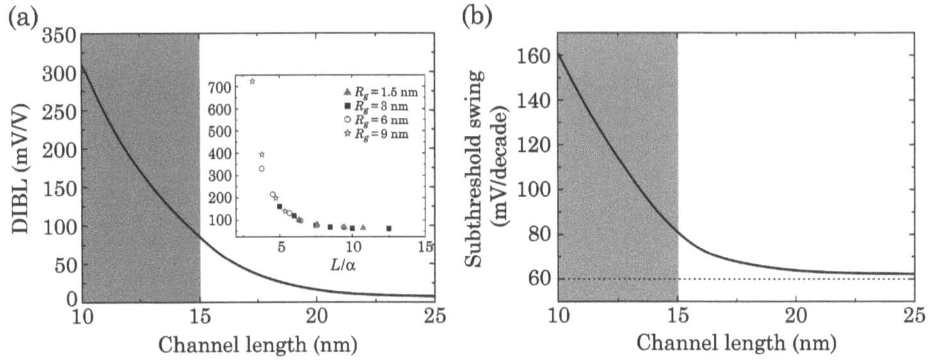

Figure 4.39 (a) and (b) show the variation of DIBL and subthreshold swing on channel length, respectively. Shaded areas are regions where the short channel effects are larger than typical device requirements. The inset in (a) shows collapse of the data for several devices of different channel lengths and gate radii upon scaling of the channel length. The horizontal dotted line in (b) is the theoretical limit for the subthreshold swing. Figure from Ref. [15].

At $V_g = 0.8$ V and $V_{ds} = 0$ there is a barrier of 93 meV that significantly impairs hole conduction; as the source–drain voltage is increased, the barrier is reduced, causing an increase in the current and a lack of current saturation at large source–drain voltages. This voltage-dependent barrier is also at the origin of the DIBL in Fig. 4.36.

One can expect that for longer channel lengths the source and drain contacts will have less impact on the barrier in the middle of the channel, and that the gate will be more effective at modulating the height of that barrier. This is indeed the case, as Fig. 4.39 indicates. There it is shown that both the DIBL and the subthreshold swing decrease rapidly with an increase of the channel length. Typical device requirements are that DIBL be less than 100 mV/V and for the subthreshold swing to be less than 80 mV/decade. For the cylindrical nanotube device with a 3 nm radius SiO_2 gate insulator and a (17,0) nanotube, this requires the channel to be larger than 15 nm. This constraint depends on the thickness of the gate insulator. The inset in Fig. 4.39 shows that scaling of the DIBL is obtained if the channel length is scaled by the factor

$$\alpha = \sqrt{R_g + l}. \tag{4.78}$$

For the particular device geometry and nanotube discussed here, it is found that $l = 1$ nm. It is important to point out that electrostatic analyses for cylindrical gate transistors [28,29] predict a scaling quantity proportional to the oxide thickness. This type of scaling relation does not lead to collapse of the data onto a single curve. It is thought that the unusual dielectric response of the nanotube [13], strong charge transfer from the contacts and the actual device geometry render the conventional analyses inapplicable; more work is needed to address these issues.

4.3.8 Crosstalk

Ever increasing computing power not only requires devices with short channels, but also a large density of these devices on a single chip. At high density, interactions between neighbor devices (crosstalk) can become significant, even more so for carbon nanotubes since all of the carriers are constrained to a surface shell and can be impacted by electrical fields. Thus, it is important to assess the role of interactions between nearby carbon nanotube devices, a topic that is also important for multi-nanotube devices that are becoming more common (an example is shown in Fig. 4.40). Computer simulations [30] are addressing this issue by considering the impact of nanotube density on the characteristics of carbon nanotube transistors, including both semiconducting–semiconducting and semiconducting–metallic interactions. Fig. 4.40 shows sketches of planar carbon nanotube transistors with multiple parallel nanotubes in the channel. The key question is how the spacing d between nanotubes influences the ON and OFF states of the transistor, as well as the properties of the contacts.

Fig. 4.41 shows the results of self-consistent quantum transport calculations for the conductance of a single semiconducting nanotube in the array. The results indicate that the overall conductance decreases as the separation between nanotubes decreases. This behavior has two origins: the first is that in the ON state, there is significant charge on the nanotubes, which interacts with the nearby charge due to other nanotubes. This is energetically costly, and to lower its energy, the system lowers the charge on the nanotubes, effectively decreasing the conductance. The second cause is that because of the charge repulsion, the band alignment in the contact is modified such that the charge on the nanotubes is decreased; this leads to the metal Fermi level being in the bandgap, and the resulting Schottky barrier decreases the current in the ON state.

To quantify the impact of the nanotube density on the conductance, Fig. 4.42 shows the conductance at $V_{gs} = 0$ as a function of the nanotube spacing for different channel lengths. Irrespective of the channel length, the conductance follows a dependence $A\,[1 - \exp(-d/\beta)]$ indicating that it is exponentially sensitive to the nanotube separation. From the figure, it is clear that the parameter β depends on the length of the channel, with a larger value of β for larger channel lengths. Indeed, if one defines a 10% reduction in the conductance as the onset of crosstalk, then plotting the separation d at which this reduction happens as a function of the channel length gives the diagram of Fig. 4.42. Clearly, very short channel devices can have very high packing densities, while long channel devices are limited to tube separations of 15 nm. A notable aspect of the results of Fig. 4.42 is that the value of d is independent of the channel length for large L. At first glance, one would expect that a larger channel length leads to larger total charge on the nanotubes and thus larger interaction energy. However, screening of the Coulomb interaction by the planar gate leads to a different behavior, as we now discuss.

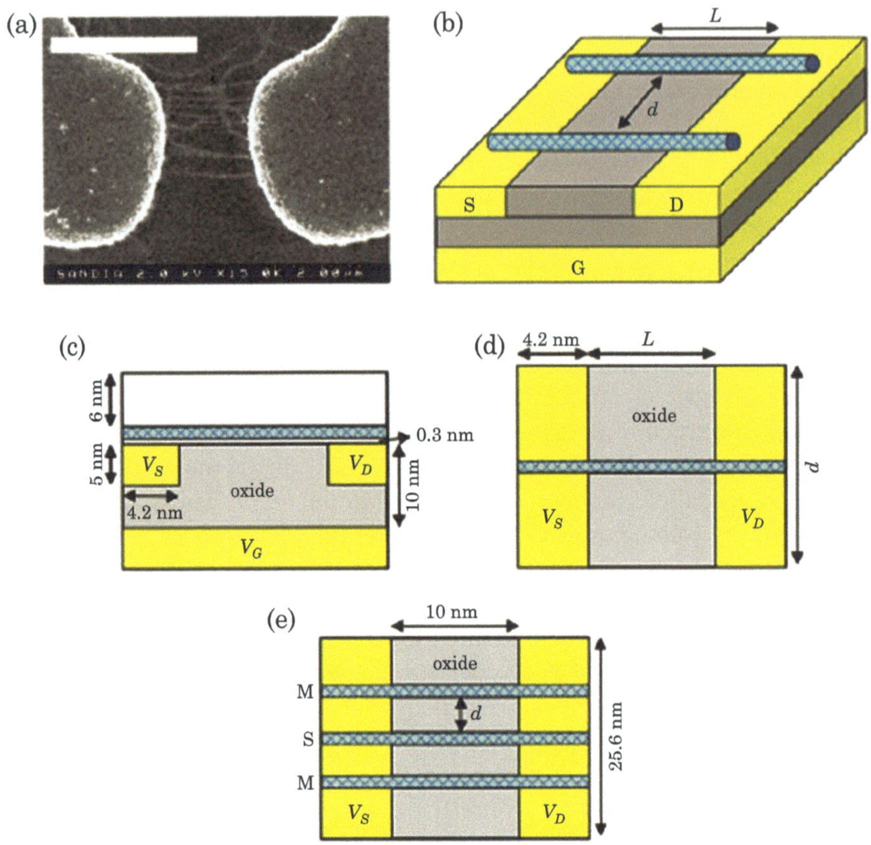

Figure 4.40 (a) shows a scanning electron micrograph image of a carbon nanotube device containing multiple nanotubes. (b) shows a sketch of the device used in the calculations, with the distance L indicating the channel length and d the nanotube separation. (c) shows a side view of the device, while (d) and (e) show top views of the unit cells used to study semiconducting–semiconducting and semiconducting–metallic interactions, respectively. Figure from Ref. [30].

To understand the role of interactions between semiconducting carbon nanotubes, we consider the simplified geometry of Fig. 4.43. There, two uniformly charged semiconducting nanotubes of length L are separated from each other by a distance d, and both are at a distance L_G from an infinite metallic plane held at potential V_G.

The electrostatic potential on one of the nanotubes due to the other can be calculated using an image potential construction (Fig. 4.43 (a)) to be

$$V = V_g + \frac{\lambda}{4\pi\varepsilon} \ln \left(\frac{\left(L + \sqrt{L^2 + 4d^2}\right)\left(-L + \sqrt{L^2 + 4d^2 + 16L_g^2}\right)}{\left(-L + \sqrt{L^2 + 4d^2}\right)\left(L + \sqrt{L^2 + 4d^2 + 16L_g^2}\right)} \right) \quad (4.79)$$

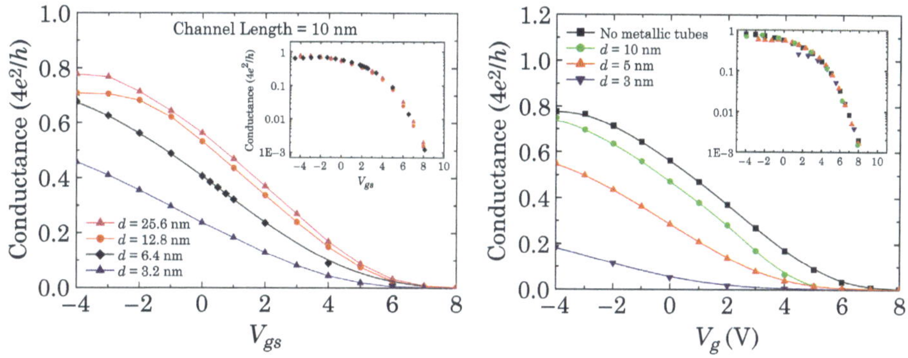

Figure 4.41 Conductance of a (17,0) nanotube transistor as a function of gate–source voltage, for different separations of the nanotube from other semiconducting nanotubes (left) or metallic nanotubes (right). In each figure, the inset shows the collapse of the data after rescaling and shifting of the gate voltage. Figure from Ref. [30].

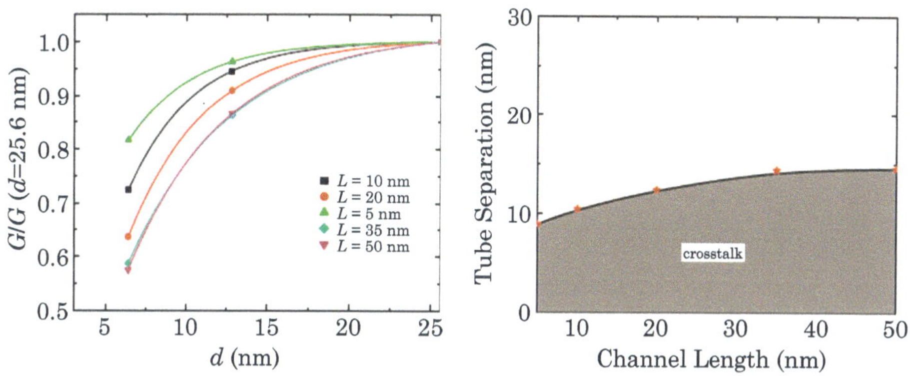

Figure 4.42 The left figure shows the conductance at $V_{gs} = 0$ versus nanotube separation, for different channel lengths. Solid lines are fits of the form $A[1 - \exp(-d/\beta)]$. The figure on the right shows the nanotube separation below which crosstalk becomes important. Figures from Ref. [30].

where λ is the charge per unit length on the nanotubes. In the long channel limit $L \gg d$ the last equation becomes

$$V = V_g + \frac{\lambda}{4\pi\varepsilon} \ln \frac{\sqrt{d^2 + 4L_g^2}}{d}. \tag{4.80}$$

Thus in this limit the potential is independent of the channel length, and the length scale that competes with the nanotube separation is the gate insulator thickness, due to screening of the fields by the gate. Taking a potential of ΔV as a criterion for the importance of intertube effects gives the nanotube separation

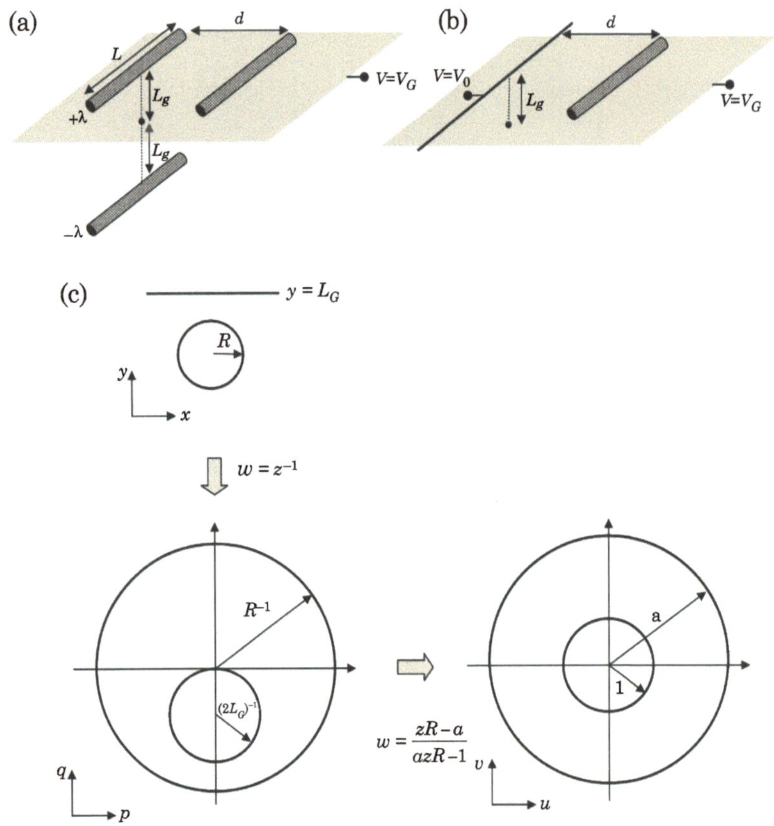

Figure 4.43 Sketch of situations considered for analytical calculation of interactions between (a) semiconducting nanotubes, (b) semiconducting and metallic nanotubes. Figures from Ref. [30]. (c) illustrates the conformal mapping procedure to solve the electrostatic problem associated with (b), see text for details.

below which crosstalk becomes important:

$$d^* = \frac{2L_g}{\sqrt{\exp\left(\frac{8\pi\kappa\varepsilon_0\Delta V}{\lambda}\right) - 1}}. \tag{4.81}$$

Thus the gate insulator thickness L_G sets a length scale for d^*, but is exponentially reduced due to the screening by the dielectric insulator. Since d^* is only 15 nm for uncovered nanotubes on SiO_2, very high device densities should be possible by embedding the nanotubes in the insulator and using high-κ dielectrics.

The general behavior of interactions between semiconducting nanotubes can be understood from further analysis of Eq. (4.80). Near threshold, the charge on the nanotube can be written as $\lambda = \lambda_0 (V_g - V_{\text{th}})$, where V_{th} is the threshold voltage. Substitution of this expression in Eq. (4.80) gives

$$V = \left(1 + \frac{\lambda_0}{2\pi\varepsilon} \ln \frac{\sqrt{d^2 + 4L_g^2}}{d}\right) V_g - \frac{\lambda_0 V_{\mathrm{th}}}{2\pi\varepsilon} \ln \frac{\sqrt{d^2 + 4L_g^2}}{d}. \tag{4.82}$$

Hence, the gate voltage is rescaled and shifted by the interaction between nanotubes. This behavior is indicated in the inset of Fig. 4.41, showing good collapse of the data after appropriate rescaling and shifting of V_g.

As mentioned above, interactions between semiconducting and metallic nanotubes are also important for multi-nanotube devices, and may also be important for all-nanotube devices where metallic nanotubes are used as interconnects. Fig. 4.41 plots the conductance of a semiconducting carbon nanotube in close proximity to metallic nanotubes. The same qualitative behavior observed for semiconducting–semiconducting interactions is seen here as well. This behavior can be understood by considering the situation of Fig. 4.43 (b): a semiconducting nanotube carrying charge density λ is at a distance d from a metallic nanotube. Both nanotubes are at distance L_G from an infinite metallic plane held at potential V_G. The metallic nanotube plays a different role than the semiconducting nanotube in the case of semiconducting–semiconducting interactions in that the metallic nanotube has plenty of charge to screen the electric fields; thus, the metallic nanotube can be approximated has a line of constant electrostatic potential, equal to V_0. To calculate the electrostatic potential on the semiconducting nanotube due to the metallic nanotube we solve Laplace's equation

$$\nabla^2 V = 0 \tag{4.83}$$

with the boundary conditions

$$V = \begin{cases} V_0 & \text{for } x^2 + y^2 = R^2 \\ V_G & \text{for y} = L_g. \end{cases} \tag{4.84}$$

The solution to this problem can be obtained using conformal mapping. The general idea behind conformal mapping is to take advantage of the properties of complex functions to transform a complicated differential equation problem to a simpler problem in a new coordinate system. In the particular case of Eqs. (4.83) and (4.89), we will show that this problem can be transformed to one consisting of coaxial cylinders held at constant potential, for which the Laplace equation has a simple solution. Once that solution is obtained, it can be converted back to the original coordinate system using coordinate transformations.

We consider the complex function $w_1 = p(x, y) + iq(x, y)$ which satisfies the Laplace equation by virtue of the Cauchy–Rieman relations between the real and imaginary part. We first perform a conformal mapping

$$w_1 = f_1(z) = \frac{1}{z} \tag{4.85}$$

where $z = x + iy$. This transforms the straight line $y = L_g$ into a circle of diameter L_G^{-1} touching the origin in the (p, q) plane, and transforms the circle $x^2 + y^2 = R^2$ into the circle $p^2 + q^2 = R^{-2}$ as illustrated in Fig. 4.43. From Eq. (4.85) we can obtain

$$\begin{aligned} p &= -\frac{y}{x^2 + y^2} \\ q &= \frac{x}{x^2 + y^2}. \end{aligned} \tag{4.86}$$

Furthermore, a second mapping $w_2 = u(p, q) + iv(p, q)$

$$w_2 = f_2(z) = \frac{zR - a}{aRz - 1} \tag{4.87}$$

with $z = p + iq$ and

$$a = \frac{L_G}{R}\left(1 + \sqrt{1 - (R/L_G)^2}\right) \approx \frac{L_G}{R} \tag{4.88}$$

shifts the center of the circle obtained from the straight line $y = L_g$ to the origin, and rescales its radius to 1. This mapping leaves the circle centered at the origin undisplaced, but rescales its radius to the value a. The relationships between (u, v) and (p, q) are

$$\begin{aligned} u &= \frac{aR^2\left(p^2 + q^2\right) - \left(a^2 + 1\right)Rp + a}{a^2R^2\left(p^2 + q^2\right) - 2aRp + 1} \\ v &= \frac{\left(a^2 - 1\right)Rq}{a^2R^2\left(p^2 + q^2\right) - 2aRp + 1}. \end{aligned} \tag{4.89}$$

Note that the sequence of mappings has created a boundary value problem of two coaxial circles, and Laplace's equation can be easily solved for this problem. The solution is

$$V(u, v) = V_0 + \frac{V_G - V_0}{\ln\left(a^2\right)} \ln\left(u^2 + v^2\right). \tag{4.90}$$

The potential in the original (x, y) space is obtained with the use of Eqs. (4.86) and (4.89). Since we are interested in the potential caused on a nanotube parallel to the nanotube held at potential V_0 and at the same distance from the gate, we focus on the potential $V(x,0)$. Then we have $(p,q) = \left(0, x^{-1}\right)$ and

$$
\begin{aligned}
u &= \frac{aR^2 + ax^2}{a^2R^2 + x^2} \approx \frac{L_G}{R}\frac{x^2}{L_G^2 + x^2} \\
v &= \frac{\left(a^2-1\right)Rx}{a^2R^2 + x^2} \approx \frac{L_G^2}{R}\frac{x}{L_G^2 + x^2}.
\end{aligned}
\tag{4.91}
$$

This gives the solution at a distance d

$$
\begin{aligned}
V(d,0) = {} & \frac{V_G}{2\ln(L_G/R)} \ln\left(\frac{d^2L_G^2}{R^2\left(L_G^2 + d^2\right)}\right) \\
& + V_0\left[1 - \frac{1}{2\ln(L_G/R)}\ln\left(\frac{d^2L_G^2}{R^2\left(L_G^2 + d^2\right)}\right)\right].
\end{aligned}
\tag{4.92}
$$

Much like the situation for semiconducting–semiconducting interactions, the interaction of semiconducting nanotubes with metallic nanotubes leads to a rescaling and shift of the gate voltage as shown in the inset of Fig. 4.41.

4.3.9 Noise

Noise in electronic devices can have many sources, including $1/f$ noise, thermal noise, shot noise, random-telegraph-signal (RTS) noise, etc. Of these, $1/f$ noise has raised the most scientific interest because it is ubiquitous in many systems but yet its origin is unclear. The name "$1/f$ noise" originates from the dependence of the current power spectrum $P(f)$ on the frequency f

$$
P(f) = \int \left(\langle I(t)I(0)\rangle - \langle I\rangle^2\right) e^{ift} dt = A\frac{I^2}{f}. \tag{4.93}
$$

In traditional semiconductor electronic devices, two origins for this behavior have been proposed: (1) fluctuations in carrier density due to fluctuations in charge trapping at surface states [31]; and (2) mobility fluctuations [32]. In the case of mobility fluctuations, the power spectrum takes the form

$$
P(f) = \frac{\alpha_H I^2}{N_c f} \tag{4.94}
$$

where α_H is known as Hooge's constant and N_c is the total number of carriers in the system. This expression is also utilized as an empirical equation describing $1/f$ noise in general. Central to this equation is the assumption that fluctuations in the number of carriers are responsible for the current fluctuations; as we now

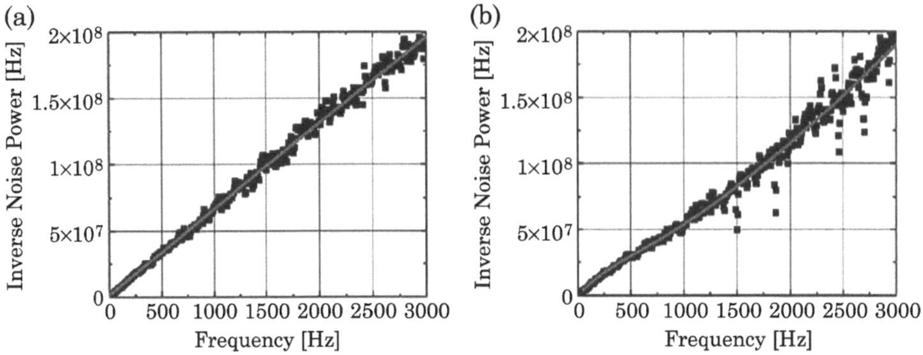

Figure 4.44 Noise behavior in single nanotube transistors. (a) shows a $1/f$noise behavior. (b) shows that in addition to the $1/f$ behavior, an additional noise component is seen in some devices. Figure from Ref. [33].

discuss, this hypothesis is valid for long carbon nanotubes where the transport is diffusive but breaks down for ballistic carbon nanotube transistors.

Fig. 4.44 shows the inverse of the noise power in carbon nanotube transistors with long channel lengths such that the transport is diffusive. It is clear from the power spectrum that the main component of the noise is a $1/f$ behavior, with some additional noise component observed in some devices (Fig. 4.44 (b)). This additional noise component is well described by adding a small contribution from RTS noise to the total noise:

$$P(f) = A\frac{I^2}{f} + B\frac{I^2}{1+(f/f_0)^2}. \tag{4.95}$$

More importantly however is the dependence of the noise power on the number of carriers in the nanotube. To extract this dependence, note that the number of carriers in a field-effect transistor is equal to

$$N_c = C_g L \left| V_g - V_{\text{th}} \right| / e \tag{4.96}$$

and the noise power is given by

$$P(f) = A\frac{I^2}{f} = \frac{\alpha_H e I^2}{C_g L \left| V_g - V_{\text{th}} \right| f}. \tag{4.97}$$

Fig. 4.45 shows the measured value of $1/A$ as a function of $|V_g - V_{\text{th}}|$ indicating a linear relationship; this behavior is consistent with an inverse dependence on the number of carriers. Eq. (4.97) is further supported by measurements on devices with different channel lengths, which show a linear dependence of the noise power spectrum on the channel length (Fig. 4.45).

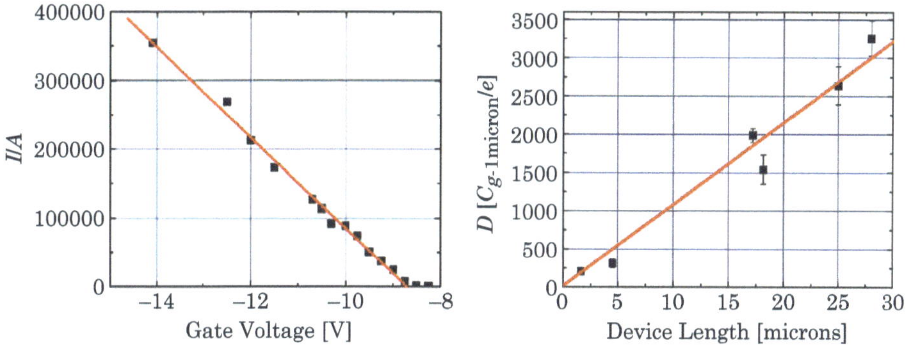

Figure 4.45 The left panel shows the dependence of the coefficient of the noise spectrum on the gate voltage. The right panel is the dependence on the length of the carbon nanotube channel. Figure from Ref. [33].

The above discussion focused on long carbon nanotubes where the transport is diffusive. For such diffusive transport, a body of work has established the fundamental aspects of the $1/f$ noise in terms of mobility fluctuations. For carbon nanotubes shorter than the scattering mean-free path, the transport is ballistic, and not much is known about noise in ballistic systems. To address this gap, recent experiments and theory have begun to look at this situation for both Schottky barrier and channel-controlled carbon nanotube transistors [34].

Fig. 4.46 shows the measured current–gate voltage characteristics of a Schottky barrier carbon nanotube transistor with a 600 nm channel as well as the amplitude of the power spectrum A from Eq. (4.93). Because the channel length is less than the electron–phonon scattering mean-free path, this device is believed to operate in the ballistic regime. Clearly, the amplitude of the power spectrum is much larger near the threshold voltage. To explain the origin of this behavior we consider the impact of fluctuating charge traps in the gate oxide on the current. It is well known that such charge fluctuations lead to $1/f$ noise because the charge state is thermally activated [35].

For a Schottky barrier device with ballistic transport, the current is a function of the electric field at the contact which in turn depends on the gate voltage and the device geometry. The geometrical factors can be captured in a parameter S_g and the electric field at the contact is then equal to V_g/S_g. The current is then a function of V_g/S_g:

$$I = I\left(V_g/S_g\right). \tag{4.98}$$

Fluctuating charge traps lead to fluctuating electric fields in the vicinity of the carbon nanotube, including near the contact. These fluctuating electric fields can be modeled as $\gamma F(t)$ where γ is the strength of the fluctuations and $F(t)$ is a dimensionless function of time with $1/f$ power spectrum. The total electric

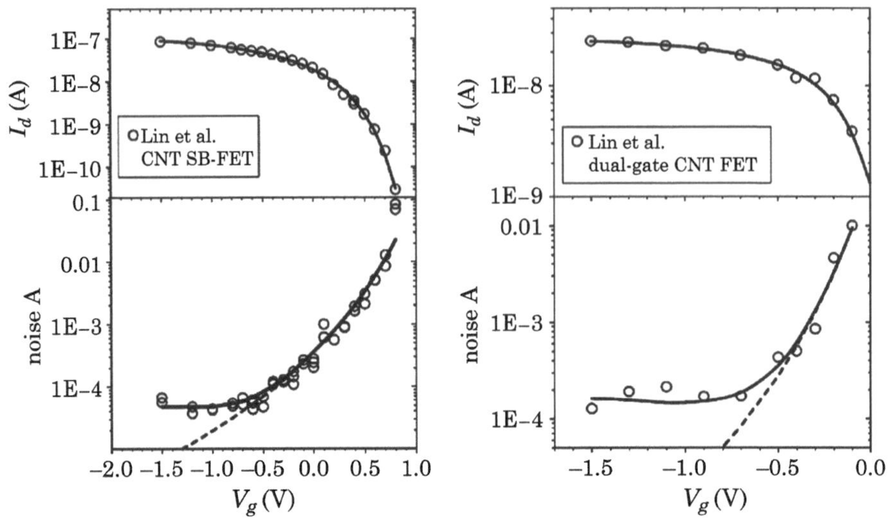

Figure 4.46 Measured and calculated noise in a Schottky barrier (left) and a channel-controlled (right) carbon nanotube transistor. Solid and dashed lines are theoretical fits, see text for details. Figure from Ref. [34].

field at the contact is $V_g/S_g + \gamma F(t)$ and the current will be

$$I = I\left(V_g/S_g + \gamma F(t)\right) \approx I_0\left(V_g\right) + \gamma S_g \frac{dI}{dV_g} F(t) \tag{4.99}$$

where the last expansion assumes a small electric field due to the noise. From this expression, we can calculate the current power spectrum as

$$\begin{aligned} P(f) &= \int \left(\langle I(t)I(0)\rangle - \langle I\rangle^2\right) e^{ift} dt \\ &= \gamma^2 S_g^2 \left(\frac{dI}{dV_g}\right)^2 \int \langle F(t)F(0)\rangle\, e^{ift} dt \\ &= \gamma^2 S_g^2 \left(\frac{dI}{dV_g}\right)^2 \frac{1}{f}. \end{aligned} \tag{4.100}$$

Therefore, the amplitude of the power spectrum is

$$A = \gamma^2 S_g^2 \left(\frac{d\ln I}{dV_g}\right)^2. \tag{4.101}$$

This immediately shows that the amplitude of the noise is related to the term $d\ln I/dV_g$ which is largest in the subthreshold regime, explaining the general behavior of the experimental measurements of Fig. 4.46. By fitting

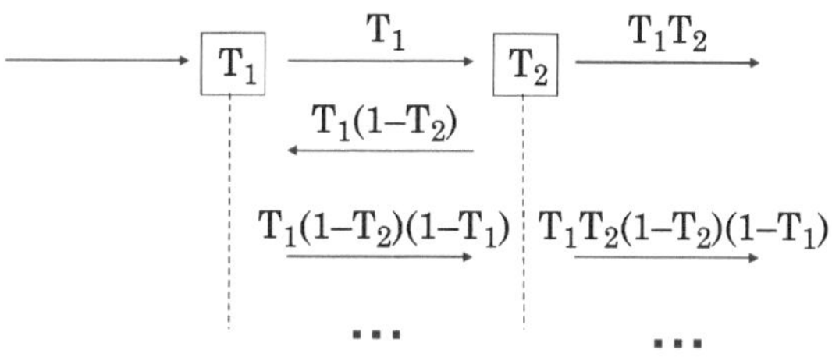

Figure 4.47 Illustration of the procedure to calculate the total transmission probability across two scatterers in series.

the experimentally measured current with a smooth function, the noise can be calculated from Eq. (4.100); the dashed line in Fig. 4.46 shows very good quantitative agreement with experiment near the threshold voltage.

Deviations from the theoretical model arise when the current is large, corresponding to a larger transmission probability through the contact Schottky barrier. In that case, the fluctuations in the channel become important and can no longer be ignored. To model this situation, we consider the transmission probability through two scattering regions in series, corresponding to the Schottky barrier (transmission T_{SB}) and the channel (transmission T_{ch}). The procedure for calculating the total transmission probability for scatterers in series is illustrated in Fig. 4.47.

The total transmission is given by summing each of the scattering paths

$$T = T_1T_2 + T_1T_2\,(1-T_1)\,(1-T_2) + T_1T_2\,(1-T_1)^2\,(1-T_2)^2 + \cdots$$
$$= \frac{T_1T_2}{1-(1-T_1)\,(1-T_2)}. \tag{4.102}$$

In terms of the transmission probabilities for the Schottky barrier and the channel, we have

$$T = \frac{T_{\mathrm{SB}}T_{\mathrm{ch}}}{T_{\mathrm{SB}} + T_{\mathrm{ch}} - T_{\mathrm{SB}}T_{\mathrm{ch}}}. \tag{4.103}$$

Assuming that both the transmission probabilities are small this equation simplifies to

$$T = \frac{T_{\mathrm{SB}}T_{\mathrm{ch}}}{T_{\mathrm{SB}} + T_{\mathrm{ch}}} \tag{4.104}$$

and from the Landauer formula the total transmission translates into the conductance

$$G = G_0\frac{T_{\mathrm{SB}}T_{\mathrm{ch}}}{T_{\mathrm{SB}} + T_{\mathrm{ch}}} \tag{4.105}$$

and the resistance

$$R_{\text{tot}} = \frac{1}{G_0}\frac{T_{\text{SB}} + T_{\text{ch}}}{T_{\text{SB}}T_{\text{ch}}} = \frac{1}{G_0 T_{\text{SB}}} + \frac{1}{G_0 T_{\text{ch}}} = R_{\text{SB}} + R_{\text{ch}}. \tag{4.106}$$

Thus the transmissivities in series lead to a total resistance that consists of two series resistances due to the Schottky barrier and the channel. For series resistors with fluctuations δR_{ch} and δR_{SB}, the total noise amplitude is given by

$$A = \frac{(\delta R_{\text{SB}})^2 + (\delta R_{\text{ch}})^2}{(R_{\text{SB}} + R_{\text{ch}})^2}. \tag{4.107}$$

Re-arranging this equation leads to a noise amplitude that is the sum of the Schottky barrier and channel fluctuations

$$A = A_{\text{SB}} + A_{\text{ch}}\left(\frac{R_{\text{ch}}}{R_{\text{tot}}}\right)^2 = \gamma^2 S_g^2 \left(\frac{d\ln I}{dV_g}\right)^2 + \alpha_{\text{ch}} I^2 \tag{4.108}$$

where $\alpha_{\text{ch}} = A_{\text{ch}}\left(R_{\text{ch}}/V_{sd}\right)^2$ and $A_{\text{ch}} = \left(\delta R_{\text{ch}}\right)^2 / R_c^2$ is the noise amplitude of the channel resistance. The important point is that the additional series resistance due to the channel increases the noise, with an amplitude that is proportional to I^2. In regions where the current changes little with the gate voltage (as in the ON state) this contribution to the noise dominates. Indeed, a numerical fit of Eq. (4.108) to the experimental data shows an improved agreement in the large current regime (solid line in Fig. 4.46).

For carbon nanotube transistors with ohmic contacts, the current is controlled by the height of the barrier in the middle of the channel. Still in that case the current can be written as $I = I\left(V_g/S_g\right)$ where the parameter S_g now describes the effectiveness of the gate at changing the barrier height in the middle of the channel. Charge trap fluctuations cause fluctuations in the height of the barrier which can again be written in the form $\gamma F(t)$. Therefore the analysis presented above for the Schottky barrier transistor carries directly over to the case of channel-controlled devices. Such channel-controlled devices can be achieved by fabricating ohmic contacts or by using a double-gate approach: the back gate is used to modulate the contacts in a Schottky barrier transistor to increase their transparency while a top gate is used to modulate the channel conductance. Fig. 4.46 shows the measured current as a function of the top gate voltage for such a device. The noise is well described by the ballistic noise theory including both the near and far threshold behavior.

References

1. C. Zhou, J. Kong, E. Yenilmez and H. Dai, "Modulated chemical doping of individual carbon nanotubes", *Science*, Vol. 290, p. 1552, 2000.

2. J.U. Lee, P.P. Gipp and C.M. Heller, "Carbon nanotube *p–n* junction diodes", *Appl. Phys. Lett.*, Vol. 85, p. 145, 2004.
3. K. Bosnick, N. Gabor and P. McEuen, "Transport in carbon nanotube *p–i–n* diodes", *Appl. Phys. Lett.*, Vol. 89, p. 163121, 2006.
4. J.U. Lee, "Photovoltaic effect in ideal carbon nanotube diodes", *Appl. Phys. Lett.*, Vol. 87, p. 073101, 2005.
5. F. Léonard and J. Tersoff, "Novel length scales in nanotube devices", *Phys. Rev. Lett.*, Vol. 83, p. 5174, 1999.
6. A. Bachtold, P. Hadley, T. Nakanishi and C. Dekker, "Logic circuits with carbon nanotube transistors", *Science*, Vol. 294, p. 1317, 2001.
7. F. Léonard and J. Tersoff, "Negative differential resistance in nanotube devices", *Phys. Rev. Lett.*, Vol. 85, p. 4767, 2000.
8. M.S. Fuhrer, J. Nygård, L. Shih, M. Forero, Y-G. Yoon, M.S.C. Mazzoni, H.J. Choi, J. Ihm, S.G. Louie, A. Zettl and P.L. McEuen, "Crossed nanotube junctions", *Science*, Vol. 288, p. 494, 2000.
9. A. Odintsov, "Schottky barriers in carbon nanotube heterojunctions", *Phys. Rev. Lett.*, Vol. 85, p. 150, 2000.
10. Z. Yao, H.W.Ch. Postma, L. Balents and C. Dekker, "Carbon nanotube intramolecular junctions", *Nature*, Vol. 402, p. 273, 1999.
11. S.J. Tans, A.R.M. Verschueren and C. Dekker, "Room-temperature transistor based on a single carbon nanotube", *Nature*, Vol. 393, p. 49, 1998.
12. R. Martel, T. Schmidt, H.R. Shea, T. Hertel and Ph. Avouris, "Single- and multi-wall carbon nanotube field-effect transistors", *Appl. Phys. Lett.*, Vol. 73, p. 2447, 1998.
13. F. Léonard and J. Tersoff, "Multiple functionality in nanotube transistors", *Phys. Rev. Lett.*, Vol. 88, p. 258302, 2002.
14. A. Javey, J. Guo, Q. Wang, M. Lundstrom and H. Dai, "Ballistic carbon nanotube field-effect transistors", *Nature*, Vol. 424, p. 654, 2003.
15. F. Léonard and D.A. Stewart, "Properties of short channel ballistic carbon nanotube transistors with ohmic contacts", *Nanotechnology*, Vol. 17, p. 4699, 2006.
16. J. Appenzeller, J. Knoch, V. Derycke, R. Martel, S. Wind and Ph. Avouris, "Field-modulated carrier transport in carbon nanotube transistors", *Phys. Rev. Lett.*, Vol. 89, p. 126801, 2002.
17. S. Heinze, J. Tersoff, R. Martel, V. Derycke, J. Appenzeller and Ph. Avouris, "Carbon nanotubes as Schottky barrier transistors", *Phys. Rev. Lett.*, Vol. 89, p. 106801, 2002.
18. S. Heinze, M. Radosavljevic, J. Tersoff and Ph. Avouris, "Unexpected scaling of the performance of carbon nanotube transistors", *Phys. Rev.* B, Vol. 68, p. 235418, 2003.
19. Y. Lu, S. Bangsaruntip, X. Wang, L. Zhang, Y. Nishi and H. Dai, "DNA functionalization of carbon nanotubes for ultrathin atomic layer deposition of high κ dielectrics for nanotube transistors with 60 mV/decade switching", *J. Am. Chem. Soc.*, Vol. 128, p. 3158, 2006.
20. A. Javey, H. Kim, M. Brink, Q. Wang, A. Ural, J. Guo, P. Mcintyre, P. Mceuen, M. Lundstrom and H. Dai, "High-κ dielectrics for advanced carbon nanotube transistors and logic gates", *Nature Mater.*, Vol. 1, p. 241, 2002.
21. V. Derycke, R. Martel, J. Appenzeller and Ph. Avouris, "Carbon nanotube inter- and intramolecular logic gates", *Nano Lett.*, Vol. 292, p. 453, 2001.
22. A. Javey, Q. Wang, A. Ural, Y. Li and H. Dai, "Carbon nanotube transistor arrays for multistage complimentary logic and ring oscillators", *Nano Lett.*, Vol. 2, p. 929, 2002.
23. Z. Chen, J. Appenzeller, Y.-M. Lin, J. Sippel-Oakley, A.G. Rinzler, J. Tang, S.J. Wind, P.M. Solomon and P. Avouris, "An integrated logic circuit assembled on a single carbon nanotube", *Science*, Vol. 311, p. 1735, 2006.
24. T. Dürkop, S.A. Getty, E. Cobas and M.S. Fuhrer, "Extraordinary mobility in semiconducting carbon nanotubes", *Nano Lett.*, Vol. 4, p. 35, 2004.

25. X. Zhou, J.-Y. Park, S. Huang, J. Liu and P.L. McEuen, "Band structure, phonon scattering, and the performance limit of single-walled carbon nanotube transistors", *Phys. Rev. Lett.*, Vol. 95, p. 146805, 2005.
26. V. Perebeinos, J. Tersoff and Ph. Avouris, "Electron–phonon interaction and transport in semiconducting carbon nanotubes", *Phys. Rev. Lett.*, Vol. 94, p. 086802, 2005.
27. V. Perebeinos, J. Tersoff and Ph. Avouris, "Mobility in semiconducting carbon nanotubes at finite carrier density", *Nano Lett.*, Vol. 6, p. 205, 2006.
28. S.-H. Oh, D. Monroe and J.M. Hergenrother, "Analytic description of short-channel effects in fully-depleted double-gate and cylindrical, surrounding-gate MOSFETs", *IEEE Electron Device Lett.*, Vol. 21, p. 445, 2000.
29. D.L. John, C. Castro and D.L. Pulfrey, "Electrostatics of coaxial Schottky-barrier field-effect transistors", *IEEE Trans. Nano*, Vol. 2, p. 175, 2003.
30. F. Léonard, "Crosstalk between nanotube devices: contact and channel effects", *Nanotechnology*, Vol. 17, p. 2381, 2006.
31. A.L. McWhorter, R. H. Kingston, Ed., *Semiconductor surface physics*, University of Pennsylvania Press, Philadelphia, 1957.
32. F.N. Hooge, "$1/f$ noise is no surface effect", *Phys. Lett.* A, Vol. 29, p. 139, 1969.
33. M. Ishigami, J.H. Chen, E.D. Williams, D. Tobias, Y.F. Chen and M.S. Fuhrer, "Hooge's constant for carbon nanotube field effect transistors", *Appl. Phys. Lett.*, Vol. 88, p. 203116, 2006.
34. J. Tersoff, "Low-frequency noise in nanoscale ballistic transistors", *Nano Lett.*, Vol. 7, p. 194, 2007.
35. P. Dutta and P.M. Horn, "Low frequency fluctuations in solids: $1/f$ noise", *Rev. Mod. Phys.*, Vol. 53, p. 497, 1981.

5 Electromechanical Devices

The impact of mechanical deformations on the electronic structure and electronic transport properties of materials is of both fundamental and applied interest. One example of this is the recent development of strained silicon by the semiconductor industry, which leads to increased carrier mobility and improved transistor performance. Another example is the intense research and development work in the area of Micro ElectroMechanical Systems (MEMS). Carbon nanotubes are particularly promising for Nano ElectroMechanical Systems (NEMS) because of their high structural strength, small size, high aspect ratio, and small weight. As we will see in this chapter, device demonstrations in this area include actuators, oscillators and electromechanical memories. Further development of these devices requires a detailed understanding of the impact of mechanical deformations on the electronic structure of carbon nanotubes. Thus, before embarking on a discussion of nanotube electromechanical devices, we first present a discussion of the impact of mechanical deformations on nanotube electronic properties, which arises because of coupling between σ and π orbitals, and by changing bond lengths and bond angles. The behavior is rich, and depends in exquisite detail on the chirality of the carbon nanotubes and on the type of mechanical deformation. In Section 5.1, we discuss the impact of bending on the electronic properties of carbon nanotubes. In Sections 5.2 and 5.3, we discuss the change in bandgap under uniaxial strain and radial deformation, respectively. Recent examples of devices based on electromechanical response are discussed in Section 5.4.

5.1 Bending

Usually, nanotubes are not perfectly straight in electronic transport experiments or in devices because strong van der Waals interactions with the substrate and metal contacts often cause the nanotube to bend, as illustrated in Fig. 5.1. Meandering of nanotubes on substrates is readily observed, and originates from deformations of the nanotube introduced during processing, and that are "frozen-in" by the interaction with the surface. One questions is: How does bending affect the electronic properties and the conductance of the nanotubes? In addition to changing bond lengths and bond angles, bending can also cause a deformation of the nanotube cross-section, and even cause buckling. For strong enough bending, significant σ–π interactions can come into play. Given these possible effects, it is quite surprising that nanotubes are very resistant to bending, even for large bending angles that significantly distort the nanotube cross-section.

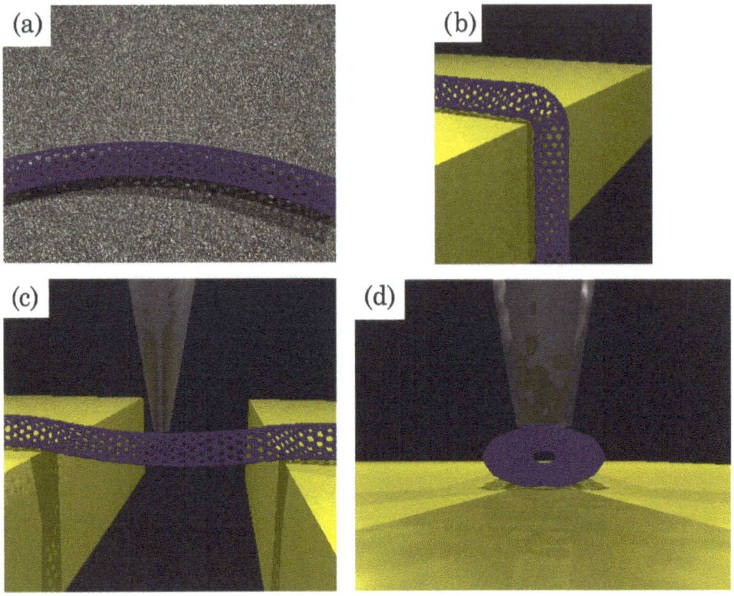

Figure 5.1 Examples of nanotube deformations. (a) A nanotube meanders on a substrate. (b) A nanotube bent around edges, as a result of strong van der Waals interactions. (c) A nanotube is suspended between supports, with a tip applying force. (d) The cross-section of a nanotube lying on a hard substrate is deformed using a tip.

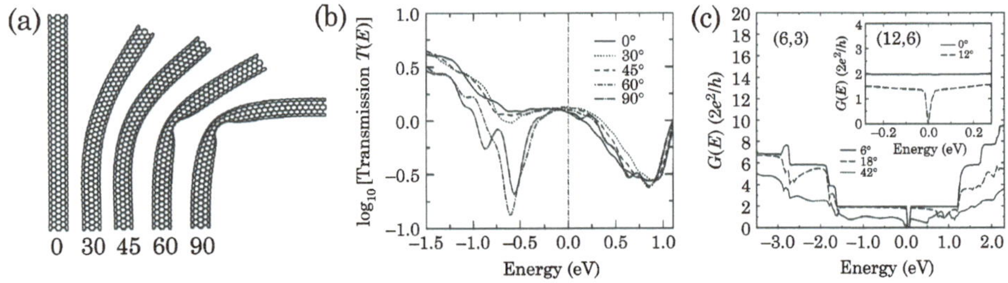

Figure 5.2 Impact of bending on nanotube conductance. (a) A (6,6) armchair nanotube at various bending angles. (b) The transmission versus energy for various bending angles corresponding to (a). (c) Transmission versus energy of (6,3) and (12,6) chiral nanotubes at various bending angles. In comparison to armchair and zigzag nanotubes, chiral nanotubes show a larger change in transmission at the Fermi energy even at small bending angles. Figures from Refs [1] and [2].

The evidence for this robustness comes in part from atomistic modeling on specific nanotubes. Fig. 5.2 (a) shows a (6,6) nanotube bent by up to 90 degrees [1]. It is clear from this picture that buckling of the nanotube occurs for bending angles larger than 45 degrees. What is remarkable is that despite this strong distortion of the nanotube structure, the transmission probability for electrons

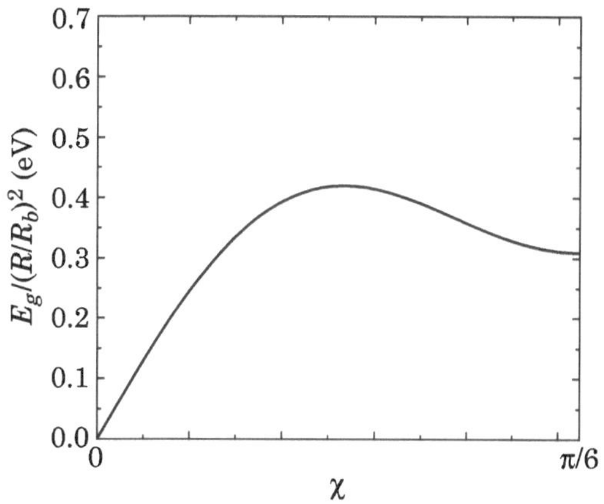

Figure 5.3 Bend-induced bandgap for a metallic nanotube as a function of the nanotube chirality. Figure after Ref. [6].

at the Fermi level (and hence the conductance) is essentially unchanged from that of the straight nanotube. The resilience of the conductance to bending has also been observed in computational modeling of (17,0) nanotubes [3], (6,6) nanotubes [4,5], and (10,10) nanotubes [2]. It is important to point out that exceptions to this rule have been observed. For example, it has been found that a (12,6) nanotube can open a gap even under bend angles as small as 12 degrees (Fig. 5.2 (c)). While these atomistic simulations on individual nanotubes are essential, it is difficult to provide a general picture of the effect of bending on nanotube electronic structure from these case-by-case studies. Instead, a general analytical picture has been provided by considering the impact of *weak* bending on the nanotube electronic properties, by considering perturbations of bond lengths and bond angles in a tight-binding model [6]. The results of these calculations indicate that bending breaks the azimuthal symmetry and introduces interactions between subbands with different angular momenta, modifying the nanotube electronic structure. In particular, for metallic nanotubes, a bandgap opens, with a magnitude that depends quadratically on the ratio of the nanotube radius and the bend radius. This result is illustrated graphically in Fig. 5.3.

As an example of the magnitude of the bend-induced bandgap, we consider a (18,0) zigzag nanotube, which has a radius of 0.7 nm. For a bending radius of 10 nm (which is already quite small), a bandgap of 1.5 meV opens. Thus, the magnitude of the bend-induced bandgap in the absence of strong lattice distortions is relatively small. The implication for electromechanical and electronic devices is that bending of nanotubes is not expected to strongly affect the electronic properties, unless the bending is so large that strong deformations of the nanotube cross-section arise; and even in that case, certain symmetries preserve the high conductance of certain nanotubes, as we saw in Fig. 5.2.

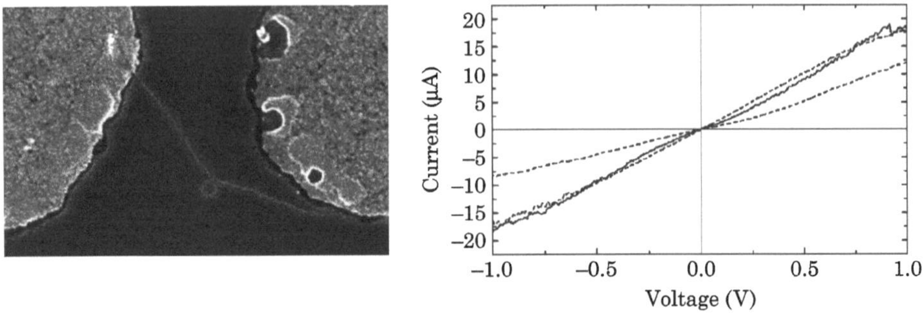

Figure 5.4 Left: scanning electron micrograph of a carbon nanotube device assembled using solution-based ac dielectrophoresis. The nanotube loops over itself in the middle of the channel, and contacts Pd electrodes at its two ends. Right: measured I–V curve of this device (solid line) compared with similar devices without a loop (dashed lines). Figure from Ref. [3].

5.1.1 Impact of Bending on Electronic Transport

Our discussion in this chapter has focused so far on the impact of bending on the electronic structure of carbon nanotubes. However, for device applications, what often matters is the *conductance* of nanotubes containing bends. To discuss this issue, we present in this section a case study: that of a carbon nanotube loop. The experimental situation is shown in Fig. 5.4 [3]. There, a carbon nanotube is connected to two Pd electrodes, and loops over itself in the middle of the channel. Such topological deformations arise here because the nanotube device was assembled using ac dielectrophoresis. In this method, a solution containing the nanotubes is deposited over the electrodes, and an ac voltage applied to the electrodes polarizes the nanotubes, which become attracted to the high electric field region between the electrodes. It is easy to imagine that during this solution-based assembly, nanotubes may "loop" over themselves as they move in the solution. The question is whether such loops can dramatically affect the conductance of the device.

The measured current–voltage characteristics of this device is shown in Fig. 5.3 as a solid line, and is compared to two similar devices where there is no loop. Clearly, the conductance is not much affected by a loop of this size (radius about 60 nm), being as large as the most conducting device without a loop. The impact of the loop on the conductance can be understood by performing quantum transport calculations through looped nanotubes [3]. For zigzag nanotubes, the wavefunction for the mth ring is written as

$$\phi_m^J(\vec{r}) = \frac{1}{\sqrt{n}} \sum_{l=1}^{n} e^{iJ\theta_{ml}} X(\vec{r} - \vec{r}_{ml}) \tag{5.1}$$

where the index l labels each of the n atoms around the nanotube circumference, and the angle θ_{ml} denotes the angular position of each atom around the circumference. J is the angular momentum labeling each of the n bands originating from quantization of the circumferential wavevector, and as before $X(\vec{r})$ is the wavefunction for the p_z orbital. We assume that the main effect of bending the nanotube is to change the bond length, and neglect corrections due to the change in orientation of the p_z orbitals. The Hamiltonian matrix elements between neighboring rings under these assumptions is

$$\begin{aligned} H_{2m,2m+1} &= \frac{1}{n}\sum_{l=1}^{n} e^{i(L-J)\theta_{2ml}}\gamma\left(a_{2ml}\right) \\ H_{2m,2m-1} &= \frac{1}{n}\sum_{l=1}^{n} e^{i(L-J)\theta_{2ml}}\left[e^{i\pi J/n}\gamma\left(a^{+}_{2ml}\right)+e^{-i\pi J/n}\gamma\left(a^{-}_{2ml}\right)\right] \end{aligned} \tag{5.2}$$

with $H_{2m,2m\pm1} = \left\langle \phi^{L}_{2m}\right| H \left|\phi^{J}_{2m\pm1}\right\rangle = \left\langle \phi^{L}_{2m\pm1}\right| H \left|\phi^{J}_{2m}\right\rangle$. Here, γ is the overlap integral, and a_{ml}, a^{+}_{ml} and a^{-}_{ml} are the three bond lengths for atom l of ring m, with a_{ml} the bond length with the nearest-neighbor atom at azimuthal angle θ_{ml}, and $a^{\pm}_{ml}$ the bond length with the nearest-neighbor atom at azimuthal angle $\theta_{ml} \pm \pi/n$. Note that bending of the nanotube introduces an angle-dependent overlap integral, breaking the azimuthal symmetry and mixing the bands. We consider a circular ring of radius R_b containing an integer number of unit cells $n_u = 2\pi R_b/3a$. The angle θ between two rings is equal to $\delta = a/R_b$ for bonds along the nanotube axis and $\delta/2$ otherwise; choosing the middle of two rings with the bond along the nanotube axis as the reference, the angle for the mth ring can be written as

$$\varphi_m = \begin{cases} \dfrac{\delta}{4}(3m-1); & m \text{ odd} \\ \dfrac{\delta}{4}(3m-2); & m \text{ even.} \end{cases} \tag{5.3}$$

Using the coordinate system of Fig. 5.5, the length of a bond between atoms il and jk can be calculated from

$$a_{il,jk} = \sqrt{r^2_{il} + r^2_{jk} - 2r_{il}r_{jk}\cos\left(\varphi_{il}-\varphi_{jk}\right) + \left(\xi_{il}-\xi_{jk}\right)^2} \tag{5.4}$$

where $\xi_{ml} = (d/2)\sin\theta_{ml}$ is the distance between atom i and a plane through the center of the nanotube. The radial distance from the bending axis is $r_{ml} = R_b - (d/2)\cos\theta_{ml}$. With these prescriptions, the perturbed bond lengths are given by

$$\begin{aligned} a_{ml} &= \sqrt{2}r_{ml}\sqrt{1-\cos\delta} \\ a^{\pm}_{ml} &= \sqrt{\left(2r^2_{ml} + 2r_{ml}\Delta r_{l\pm}\right)\left(1-\cos\delta/2\right) + \Delta r^2_{l\pm} + \Delta\xi^2_{l\pm}} \end{aligned} \tag{5.5}$$

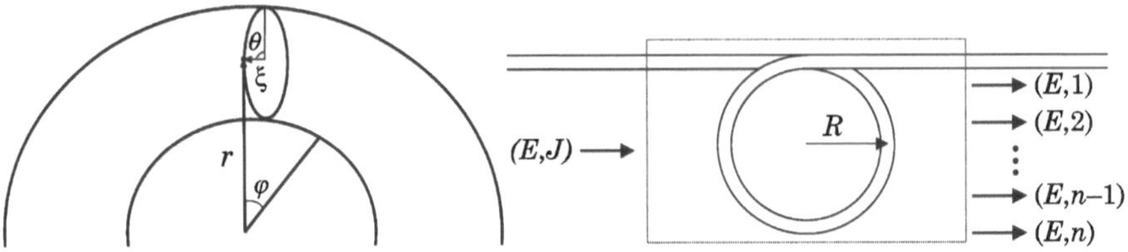

Figure 5.5 The left panel shows the coordinate system used to calculate the bond lengths between carbon atoms when the nanotube is bent. θ is the angle along the nanotube circumference, φ is the angle around the bending axis, and ξ is the coordinate perpendicular to the plane of the figure. The right panel shows the scattering region in a quantum transport calculation of the conductance of a looped nanotube. The figure shows that symmetry breaking due to bending leads to mixing of the bands and scattering into multiple channels. An electron at energy E in band J can be scattered to any other band at energy E. Figure from Ref. [3].

with

$$\begin{aligned}\Delta r_{i\pm} &= \frac{d}{2}\left[\cos\theta_i - \cos\left(\theta_i \pm \pi/n\right)\right] \\ \Delta \xi_{i\pm} &= \frac{d}{2}\left[\sin\theta_i - \sin\left(\theta_i \pm \pi/n\right)\right].\end{aligned} \tag{5.6}$$

The qualitative behavior is seen by expanding a_{ml} for small δ: $a_{ml} \approx a\left[1-(d/2R_b)\cos\theta_l\right]$. Bonds above the equator of the nanotube ($\pi/2 < \theta_i < 3\pi/2$) are expanded while those below ($3\pi/2 < \theta_i < \pi/2$) are compressed.

To translate changes in the bond lengths to changes in the overlap integral, we make use of the functional [7]

$$\gamma(a) = \gamma_0 \left(\frac{a_0}{a}\right)^n \exp\left\{n\left[-\left(\frac{a}{a_c}\right)^{n_c} + \left(\frac{a_0}{a_c}\right)^{n_c}\right]\right\} \tag{5.7}$$

where $n = 2, n_c = 6.5, a_0 = 0.153$ nm, and $a_c = 0.218$ nm. We set the value of γ_0 equal to 2 eV in order to recover $\gamma\,(a = 0.142 \text{ nm}) = 2.5$ eV. With this expression and those for the deformed bond lengths, the conductance of the looped nanotube is calculated using the nonequilibrium Green's function formalism [8]. The conductance as a function of energy is obtained from

$$G(E) = \frac{2e^2}{h}\sum_{ij} T_{ij}\left(E\right) \tag{5.8}$$

where $T_{ij}(E)$ is the transmission probability between bands i and j. For scattering Hamiltonians that preserve the nanotube symmetry, T_{ij} is a diagonal matrix. However, for the nanotube loop, the azimuthal symmetry is broken, and

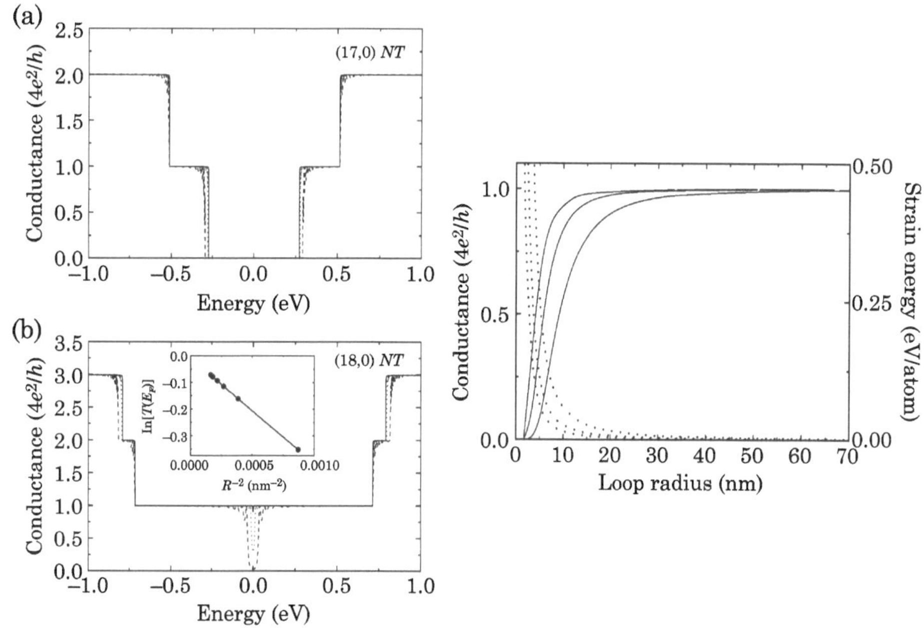

Figure 5.6 Left: calculated conductance as a function of electron energy for loop radii of 67.8 nm (solid line), 16.95 nm (dotted line), and 8.47 nm (dashed line). (a) shows the results for a (17,0) nanotube while (b) is for a (18,0) nanotube. The inset in (b) shows the Fermi level transmission probability as a function of the inverse of the loop radius squared. See text for details. Right: room temperature conductance of metallic nanotubes as a function of loop radius (solid lines), and the associated strain energy (dashed lines). Curves from left to right correspond to (9,0), (12,0), and (18,0) nanotubes. Figure from Ref. [3].

an incoming electron in band i can scatter to any other band, as illustrated in Fig. 5.5. Fig. 5.6 shows the calculated conductance of the nanotube loop as a function of energy for semiconducting and metallic zigzag nanotubes. The semiconducting nanotube shows little change in its electronic structure and the conductance is very close to that of a straight nanotube, even for loop radii as small as 8.47 nm. For the metallic nanotubes, a bandgap opens around the Fermi level, and the conductance at the Fermi level follows the relation

$$T\left(E_F\right) = \exp\left[-\left(\frac{\lambda}{R_b}\right)^2\right] \tag{5.9}$$

with the parameter $\lambda \approx 20$ nm for an (18,0) nanotube. This relationship reflects the fact that while there is a small bend-induced bandgap in the loop, the loop is only of length $2\pi R_b$ and electrons can tunnel through the bandgap, leading to the exponential behavior. The value of λ sets a length scale for the loop radius at which the conductance starts to deviate significantly from that of the straight tube. However, in a finite temperature experimental measurement of conductance, the measured conductance is the total conductance

$$G = \int_{-\infty}^{\infty} G(E) \left(-\frac{\partial f}{\partial E} \right) dE. \tag{5.10}$$

The integrand is peaked at the Fermi level with a width of order kT; at room temperature, this width is comparable to the bandgap that is opened in metallic nanotubes, so the measured conductance will be less sensitive to bending than the length scale λ suggests. In fact, the calculated room-temperature conductance reported in Fig. 5.6 indicates that the loop radius has to be less than 10 nm to see an impact on the conductance. This result is consistent with the experimental data in Fig. 5.4 (60 nm loop radius) which finds little change in the conductance.

The fact that the bend radius where deformations become important is small is encouraging for assembly methods where such topological deformations may be present, because it is unlikely that nanotubes will take such large bending due to the large strain energy cost. To calculate the strain energy we model the C–C bonds as springs with spring constant k. The total strain energy is then given by

$$E_{\text{strain}} = k \sum_{i=1}^{N_B} (a_i - a)^2 \tag{5.11}$$

where N_B is the total number of bonds. We use the value $k = 2000$ eV/(bond nm^2) extracted from Ref. [7]. Fig. 5.6 shows the computed strain energy per atom as a function of loop radius (dotted lines in the right panel). The energy increases rapidly as the loop radius decreases. The rapid increase of the strain energy for $R < 5$ nm sets a lower bound on the loop radii that might appear during device fabrication.

5.2 Uniaxial and Torsional Strain

5.2.1 General Behavior

We begin this section with a qualitative discussion of the effects of uniaxial and torsional strain on the electronic properties of carbon nanotubes. This general discussion is then followed by a more detailed mathematical discussion describing a theory to explain these effects.

Nanotubes show a dramatic and rich bandgap dependence on uniaxial and torsional strain[4,10–15]. At one extreme are armchair nanotubes, which remain metallic under tensile strain, due to preservation of mirror symmetry under uniaxial strain. At the other extreme, the change of bandgap with strain is

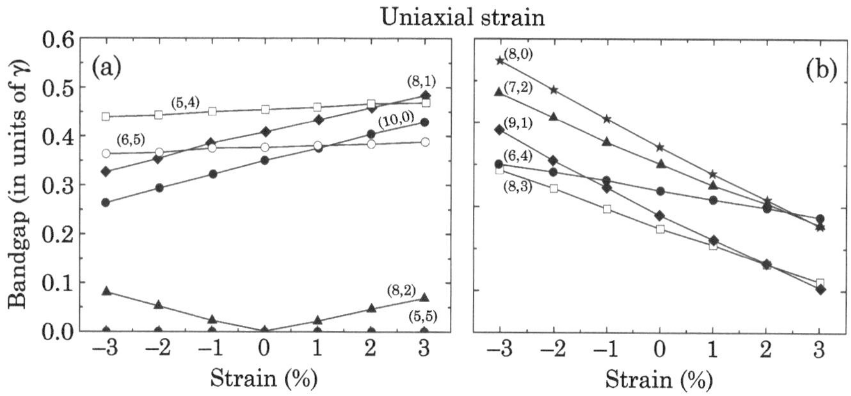

Figure 5.7 Bandgap versus uniaxial strain. (a) Nanotubes satisfying $n - m = 3I$ develop a bandgap except for armchair nanotubes, which remain metallic. Nanotubes satisfying $n - m = 3I + 1$ have a positive slope of $dE_g/d\sigma$. (b) Nanotubes satisfying $n - m = 3I - 1$ have a negative slope. The magnitude of $dE_g/d\sigma$ decreases with chiral angle. All results are within the single π orbital approximation. Figure after Ref. [9].

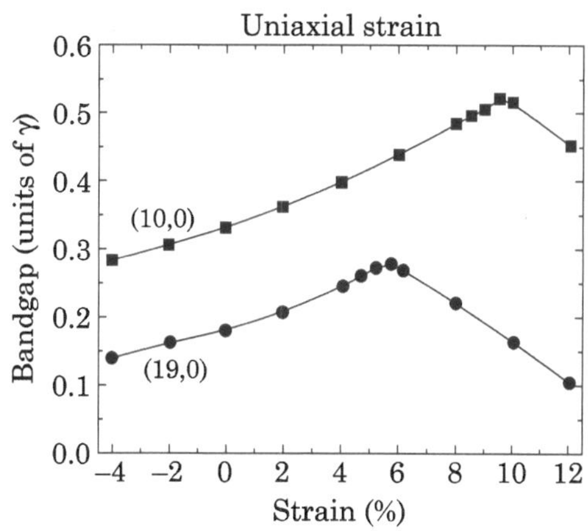

Figure 5.8 Bandap versus uniaxial strain for large values of strain, calculated within the π orbital approximation. Figure after Ref. [9].

largest for zigzag nanotubes. In general, the change of bandgap with uniaxial strain σ for a nanotube is given by (within the uniform π orbital tight binding approximation discussed in Chapter 1)

$$\Delta E_g = \pm 3\gamma \cos(3\chi)\,\sigma \tag{5.12}$$

where χ is the chiral angle. The bandgap versus strain is shown for various chiralities in Fig. 5.7. For zigzag nanotubes, Eq. (5.12) gives a bandgap change of $3\gamma\sigma$; this corresponds to a change of 75 meV at 1% strain. As the chiral angle increases, the absolute value of ΔE_g decreases, and becomes zero for armchair nanotubes (chiral angle of 30 degrees). The sign of ΔE_g has an interesting dependence on $(n-m)$ for a general (n,m) nanotube. The slope is positive for $n-m=3I+1$, and negative for $n-m=3I-1$. That is, if $n-m=3I+1$, the bandgap increases for extension and decreases for compression.

With increasing strain, a reversal of the change in bandgap occurs as shown in Fig. 5.8. For example, the bandgap of both the (10,0) and (19,0) nanotubes increases at small strain but begins to decrease after ~6% and ~9.5% strain respectively. The reason for this is as follows (this will be discussed in more detail further below): consider the two lowest conduction subbands. Because of strain, the lowest subband increases in energy, but the second lowest subband decreases in energy. The reversal in bandgap change corresponds to the crossover point between these two bands [14].

The change in bandgap in the presence of torsional strain is quite the opposite of tensile strain (Fig. 5.9) [14]. The bandgap change is largest for armchair nanotubes, with ΔE_g being approximately $3\gamma\sigma$, independent of diameter (Fig. 5.10 (a)). As the chiral angle decreases, the absolute value of ΔE_g decreases and is smallest for zigzag nanotubes. For positive values of the strain, the change in bandgap with torsional strain is negative for $n-m=3I+1$, and positive for $n-m=3I-1$, in contrast to the uniaxial case discussed above. Finally, small changes similar to the uniaxial strain case result from a four orbital treatment for quasi-metallic nanotubes [14].

5.2.2 Theory

We now discuss in detail the origin of these effects using a tight-binding description of the nanotube. As mentioned in Chapter 1, the electronic structure of carbon nanotubes can be obtained from the graphene tight-binding Hamiltonian

$$E(\vec{k}) = \left|H_{AB}\left(\vec{k}\right)\right| \tag{5.13}$$

with

$$H_{AB}(\vec{k}) = \sum_{j=1,2,3} \gamma e^{i\vec{k}\cdot\vec{r}_j}. \tag{5.14}$$

In the presence of deformations, the distance between carbon atoms is changed, causing a change in the value of the overlap integral γ. We use the same model as in Eq. (5.7) to model the dependence of the overlap integral on the bond length, but somewhat simplified to ease the calculations (this is the original model due to Harrison [16]):

$$\gamma_j = \gamma\left(\frac{a}{r_j}\right)^2 \tag{5.15}$$

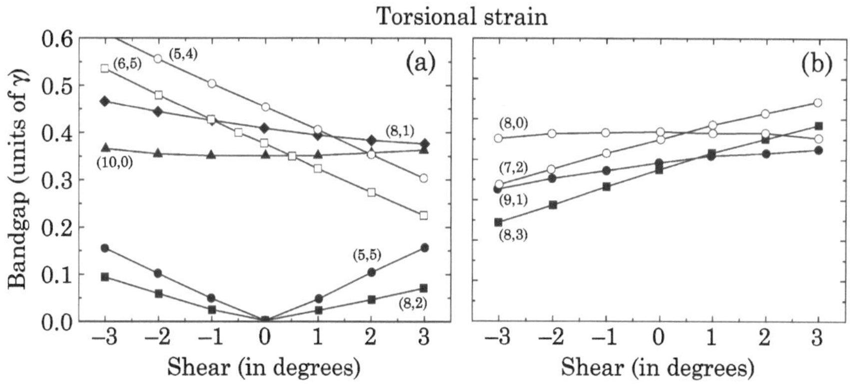

Figure 5.9 Bandgap change with torsional strain. (a) Armchair nanotubes have the largest change in bandgap under torsional strain, and zigzag tubes are the least affected. $dE_g/d\sigma$ is negative for nanotubes satisfying $n - m = 3I + 1$. (b) $dE_g/d\sigma$ is positive for nanotubes satisfying $n - m = 3I - 1$. All calculations are within the single π orbital approximation. Figure after Ref. [9].

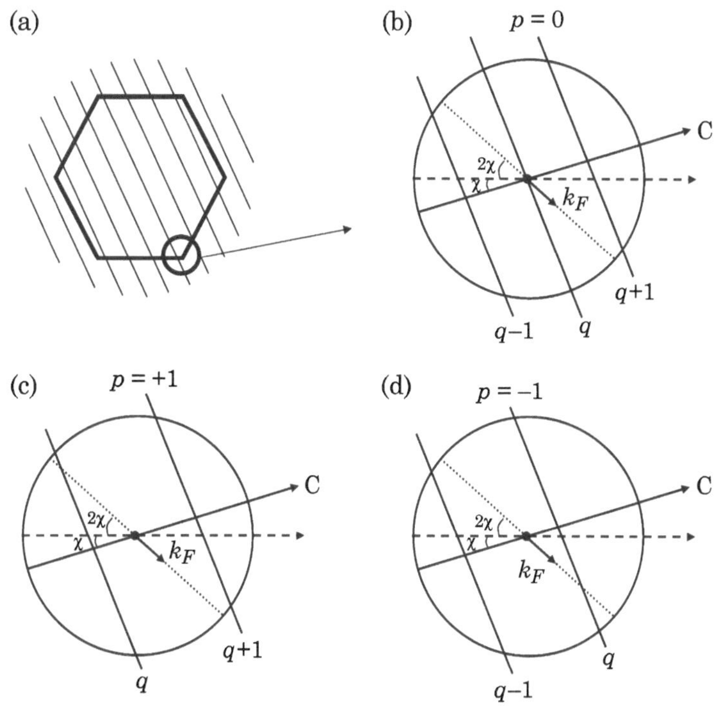

Figure 5.10 (a) Brillouin zone of graphene and the lines of quantized wavevectors for a carbon nanotube. (b), (c), and (d) illustrate the displacement of the Fermi points under uniaxial strain, for three types of nanotubes with $n - m = 3q + p$. Figure after Ref. [15].

where γ is the value of the overlap integral when the bond length is equal to the unperturbed carbon–carbon spacing a. Uniaxial and torsional deformations of the nanotube can be modeled as two-dimensional deformations of the graphene sheet by the strain tensor ε_{ij} such that

$$\vec{r} = \vec{r}_0 + \vec{u}(\vec{r}_0) \tag{5.16}$$

with

$$\varepsilon_{ij} = \frac{\partial u_i}{\partial x_j}. \tag{5.17}$$

Here $\vec{r}_0$ are the unperturbed atomic positions and the subscripts i or j = x or y, the two directions in the graphene plane. For small strains, the new atomic positions can be written as

$$\vec{r} = \left(\overleftrightarrow{I} + \overleftrightarrow{\varepsilon}\right) \vec{r}_0. \tag{5.18}$$

We note that these perturbed atomic displacements in principle distort the first Brillouin zone. However, a transformation of the k-space allows the Brillouin zone to remain unchanged [15]. To see this, consider that the Hamiltonian matrix elements in the old k-space would be given by

$$H_{AB}(\vec{k}_0) = \sum_{j=1,2,3} \gamma_j e^{i\vec{k}_0 \cdot \vec{r}_j} \tag{5.19}$$

where $\vec{k}_0$ now refers to the original k-space (i.e. for the unperturbed nanotube) but $\vec{r}_j$ are the new atomic positions. The transformation

$$\vec{k} = \left(\overleftrightarrow{I} + \overleftrightarrow{\varepsilon}\right)^T \vec{k}_0 \tag{5.20}$$

leads to the matrix elements

$$H_{AB}(\vec{k}) = \sum_{j=1,2,3} \gamma_j e^{i\vec{k} \cdot \vec{r}_{0j}}; \tag{5.21}$$

thus, the analysis can proceed in the new k-space as if the atoms were in their original position, and the impact of the deformation is solely to change the value of the overlap integral. To proceed further, the overlap integral is expanded to first order in the strain. This requires a first order expansion of r_j:

$$r_j \approx a + \frac{r_{0j}^c \left(\varepsilon_{cc} r_{0j}^c + \varepsilon_{ct} r_{0j}^t\right) + r_{0j}^t \left(\varepsilon_{tc} r_{0j}^c + \varepsilon_{tt} r_{0j}^t\right)}{a} \tag{5.22}$$

where the ε_{ij}'s are the strain components in a coordinate system with unit vectors $\hat{c}$ and $\hat{t}$ corresponding to the directions perpendicular and parallel to the nanotube axis [14]. The overlap integral to first order in the strain is then given by the expression

$$\gamma_j = \gamma - 2\gamma \frac{r_{0j}^c \left(\varepsilon_{cc} r_{0j}^c + \varepsilon_{ct} r_{0j}^t \right) + r_{0j}^t \left(\varepsilon_{tc} r_{0j}^c + \varepsilon_{tt} r_{0j}^t \right)}{a^2}. \tag{5.23}$$

These perturbed values of γ, in conjunction with Eq. (5.21) can be used to calculate the perturbed band structure. To understand the impact of the strain on the band structure it is convenient [15] to consider shifts of the Fermi points

$$\vec{k}_F = \vec{k}_F^0 + \Delta \vec{k}_F \tag{5.24}$$

where $\Delta \vec{k}_F$ is of the same order as the strain. By expanding the relation $E\left(\vec{k}\right) = \left| H_{AB}\left(\vec{k}\right)\right|$ to first order in the strain and in $\Delta \vec{k}_F$, one obtains the shift of the Fermi points [15]

$$\begin{aligned} a\Delta k_F^c &= (1+\nu)\,\sigma \cos 3\chi + \tau \sin 3\chi \\ a\Delta k_F^t &= -\,(1+\nu)\,\sigma \sin 3\chi + \tau \cos 3\chi. \end{aligned} \tag{5.25}$$

Here σ is the uniaxial strain and τ is the torsional strain. The shift of the Fermi points is illustrated in Fig. 5.10 for three different types of nanotubes [classified according to $n - m = 3I + p$ with $p = 0$(metallic) or $p = \pm 1$(semiconducting)] and indicate that the new Fermi points can be closer or farther from the allowed quantized lines in the Brillouin zone depending on the nanotube type. This behavior is responsible for the rich dependence of the nanotube bandgap on strain, as we now discuss.

Much like the case of unperturbed carbon nanotubes, the deformed nanotube electronic structure and bandgap can be obtained by quantizing the wavevector along the nanotube circumference and expanding the graphene electronic structure near the Fermi points. As shown in Chapter 1, in such a procedure the bandgap is determined by the distance between the Fermi points and the quantized wavevector line

$$E_g^p = 3a\gamma\Delta k_p \tag{5.26}$$

where as before p is the quantized line index. The distance Δk_p is obtained from the expression

$$\Delta k_p = \frac{1}{R}\left| \frac{\vec{k}_F \cdot \vec{C}}{2\pi} - p \right|. \tag{5.27}$$

For the displaced Fermi points $\vec{k}_F = \vec{k}_F^0 + \Delta\vec{k}_F$ this gives

$$\Delta k_p = \frac{1}{R}\left|\frac{\vec{k}_F^0 \cdot \vec{C}}{2\pi} - p + \frac{\Delta\vec{k}_F \cdot \vec{C}}{2\pi}\right| = \frac{1}{R}\left|3p - (n-m) - R\Delta k_F^c\right|. \quad (5.28)$$

The change in bandgap is thus [15]

$$\Delta E_g = \mathrm{sgn}(2p+1)3\gamma\left[(1+\nu)\sigma\cos 3\chi + \tau\sin 3\chi\right] \quad (5.29)$$

providing a general expression for the effects of uniaxial and torsional strain on the bandgap of any nanotube. A few specific cases are worth mentioning. Zigzag nanotubes which have a chiral angle $\chi = 0$ are insensitive to torsional strain, but show the largest change in bandgap for uniaxial strain. For positive values of the strain, zigzag nanotubes with $p = +1$ or $p = 0$ have an increased bandgap, while those with $p = -1$ show a decrease in bandgap. Note that this implies that metallic zigzag nanotubes open a bandgap under uniaxial strain. Armchair nanotubes($\chi = \pi/6$) behave in a different manner: they are insensitive to uniaxial strain but show the largest bandgap change due to torsional strain. This bandgap change implies that armchair nanotubes, which are metallic in the absence of strain, can open a bandgap under torsional strain.

An interesting effect arises as the strain is further increased because the displaced Fermi points can eventually move closer to new quantized wavenumber lines. Consider the situation of uniaxial strain. A zigzag nanotube with $p = +1$ [e.g. (19,0)] has a bandgap that increases at first with increasing strain since the Fermi points move away from the nearest quantized wavevector line, as illustrated in Fig. 5.11. For larger strain, the Fermi point will eventually come closer to the original second van Hove singularity, leading to a reduction in the bandgap. This happens when the Fermi point is exactly halfway between the two quantized lines or $\Delta k_q^c = \Delta k_{q+1}^c$. This leads to the strain where the maximum bandgap $E_g = 3E_g^0/2$ is attained

$$\sigma_c = \frac{\pi a}{3R(1+\nu)\cos 3\chi}. \quad (5.30)$$

A similar effect occurs for $p = 0$ (e.g. (18,0)) nanotubes except that the critical strain is larger, being equal to $3\sigma_c$. For $p = -1$ [e.g. (17,0)] the bandgap decreases with increasing strain until the Fermi point reaches the original second van Hove singularity, where the bandgap goes to zero ($\sigma = 2\sigma_c$). For any nanotube, increasing the strain leads to periodically increasing and decreasing bandgap as the Fermi points cross different wavevector lines, as illustrated in Fig. 5.11.

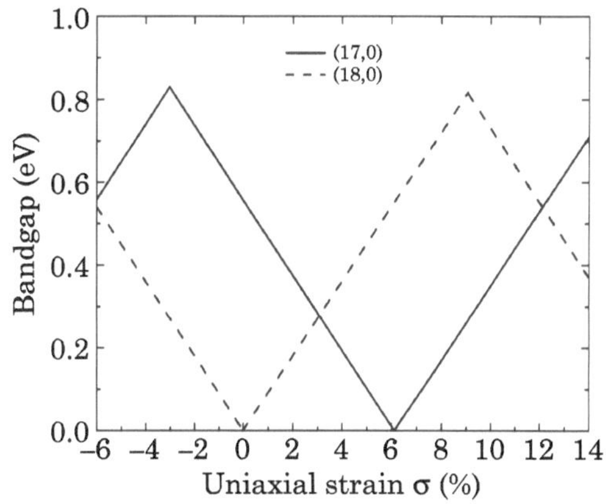

Figure 5.11 Bandgaps of two zigzag carbon nanotubes under uniaxial strain, using the analytical model presented in the text.

5.2.3 Experiments

The first experiment [17] to measure the conductance of a single carbon nanotube in the presence of strain utilized the setup shown in Fig. 5.12. In this experiment, a nanotube is suspended over a trench created in a SiO_2 substrate, and contacted at its two ends by source and drain electrodes. The portion of the nanotube over the trench is deformed with the tip of an atomic force microscope, and the small bias conductance is simultaneously measured. The measurements indicate a drop of the conductance of up to two orders of magnitude when pushing with the tip, while withdrawal of the tip resulted in the conductance returning to its undeformed value. Initially [13,17], it was believed that carbon atoms near the tip underwent a change in bonding from sp^2 to sp^3 character, causing a reduction in conductance. However, ab initio calculations later indicated that the carbon atoms under the tip remain essentially sp^2 coordinated [4]. A key experiment [18] utilized a similar setup to that in Fig. 5.12, but also operated the tip of the atomic force microscope as a local gate. This simultaneous application of strain and gate voltages showed that a bandgap was opening in originally metallic nanotubes. Thus, it is now believed that the modulation of the conductance in these experiments originates from a strain-induced bandgap, as discussed in the previous theory section. We should note that the magnitude of bandgap change is found experimentally to be about 80 meV/(% strain), in good agreement with theoretical predictions.

5.3 Radial Deformation

The electronic properties of nanotubes can also be affected by radial deformation, but again the situation depends sensitively on the type of

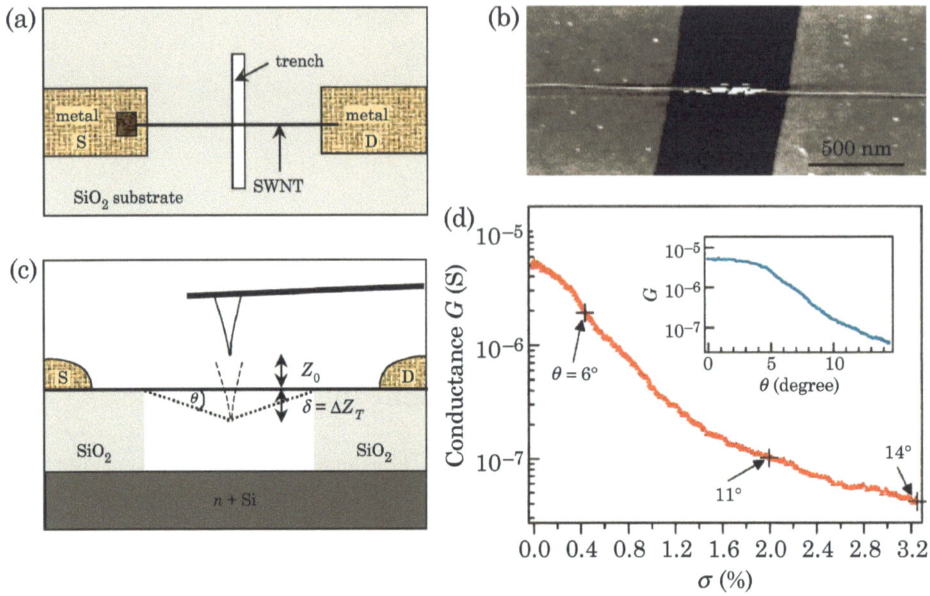

Figure 5.12 A nanotube suspended over a trench for electromechanical measurements. (a) Device viewed from above. The substrate has a trench that is about 500 nm wide and 175 nm deep. The nanotube bridging a pair of metal electrodes is suspended over the trench. (b) An atomic force microscopy (AFM) image of the nanotube suspended over the trench. (c) Side view of (a) which shows the AFM tip pushing the nanotube down into the trench. (d) The experimentally measured conductance versus strain shows a two order of magnitude change in conductance. The diameter of the nanotube is ~3 nm. Figures from Ref. [17].

nanotube and the type of deformation. Some of these effects have already been discussed in Chapter 1 in the context of the change in bandgap due to phonons, where it was shown that the bandgap is modified by a few meVs. However, much larger radial deformations can be induced when a nanotube lying on a hard substrate is deformed by a tip as shown in Fig. 5.1 (c), a situation that has been studied experimentally in Refs. [19,20]. Such a deformation causes a localized, large change in the cross section of the nanotube under the tip. (The situation is different from that of tips deforming *suspended* nanotubes in that the radial deformations discussed here are much more severe.) Modeling has focused on the simpler case of the change in bandgap under uniform radial deformation of infinitely long nanotubes [21–25].

The bandgap change upon radial deformation is very sensitive to the details of the nanotube cross section. The cross section in turn depends on the nanotube–substrate and nanotube–tip interaction. An example of such a deformation is shown in Fig. 5.13 (a), and is characterized by a parameter d_{TB} measuring the distance between top and bottom of the nanotube. If perfect symmetry is maintained upon deformation, then an armchair nanotube does not develop a bandgap upon deformation [21–25] as shown in Fig. 5.13 (a). A change from perfect symmetry causes a bandgap that can be quite large but becomes

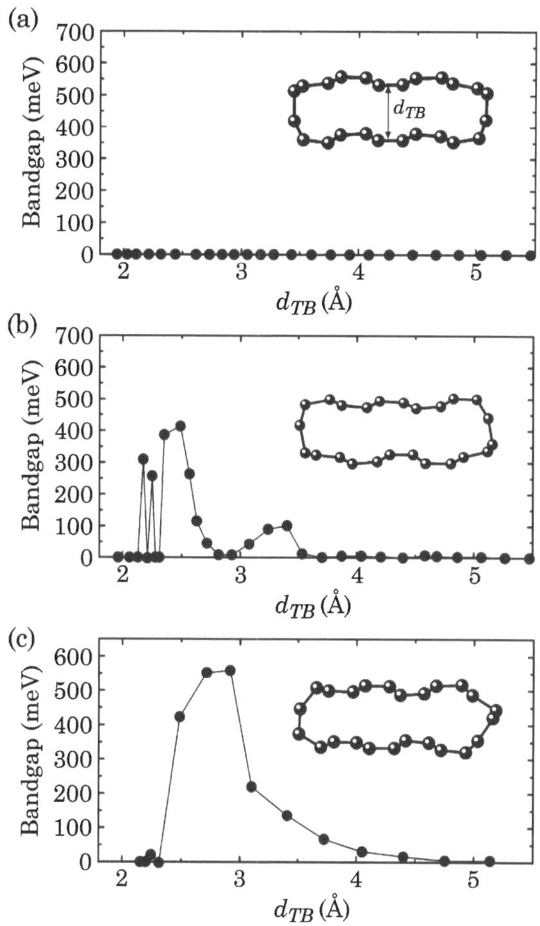

Figure 5.13 Bandgap of a (6,6) armchair nanotube as a function of d_{TB} for different initial conditions. (a) When perfect symmetry is maintained a bandgap does not develop. (b) The bandgap oscillates as a function of d_{TB}. (c) A different initial condition shows that the bandgap increases monotonically as d_{TB} is decreased, and approaches 600 meV. When d_{TB} becomes smaller than approximately 2.2 Å, the bandgap becomes zero for all cases. Figure after Ref. [21].

negligible with increase in deformation, irrespective of symmetry as shown in Fig. 5.13 ((b), (c)). Depending on the symmetry, the bandgap can oscillate with decreasing d_{TB} as shown by the two humps in Figs. 5.13 (b). The opening of a bandgap as d_{TB} is reduced from the value for the unperturbed nanotube arises because of interactions between π orbitals at the top and bottom of the nanotube; as d_{TB} decreases further, the σ–π interactions at nanotube edges become important and are responsible for the bandgap oscillations.

In the case of small diameter zigzag nanotubes satisfying $n-m = 3I$, there is a small curvature-induced bandgap. For example, this bandgap is about 100 meV

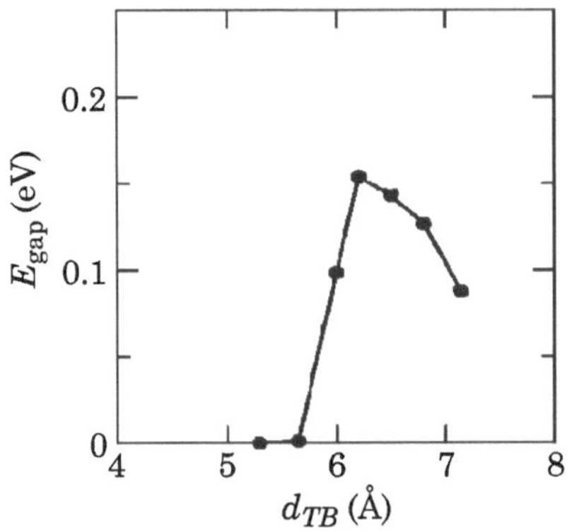

Figure 5.14 Bandgap of a (9,0) zigzag nanotube as a function of d_{TB}. The bandgap of the undeformed nanotube at $d_{TB} \sim 7.2$ Å is induced by curvature. As a result of deformation, the bandgap initially increases before becoming zero at $d_{TB} \sim 5.7$ Å. Figure from Ref. [9].

for a (9,0) nanotube. Upon radial deformation, the curvature-induced bandgap initially increases with decrease in d_{TB}[25], and vanishes for $d_{TB} < 6$ Å, and the nanotube remains metallic with further deformation (Fig. 5.14).

5.4 Devices

5.4.1 Electromechanical Oscillators

An electromechanical oscillator based on a nanotube has been experimentally demonstrated in Ref. [26]. The device consists of a three terminal transistor-type device with a nanotube forming the channel (Fig. 5.15). The nanotube is suspended over a trench between the source and drain, and is capacitively coupled to the gate terminal. Applying an oscillating gate voltage causes the nanotube to mechanically oscillate. When the applied ac frequency matches the resonant frequency of the nanotube, the amplitude of mechanical oscillation becomes large. The coupling between the nanotube displacements and the gate electrostatics arises because a change in the distance between the nanotube and the gate causes a change in the capacitance between the nanotube and gate. For gate-controlled electronic devices, the charge on the carbon nanotube is proportional to the gate voltage: $n = C_g V_g$ where C_g is the gate capacitance. This charge is attracted to the image charge of opposite sign on the gate. The force can be written as

$$F_{\text{el}} = \frac{1}{2} \frac{dC_g}{dz}\bigg|_{z=z_0} V_g^2 \tag{5.31}$$

where z is the distance between the nanotube and the gate and z_0 is the equilibrium nanotube position. The applied gate potential is of the form

$$V_g = V_g^{DC} + \delta V_g \cos(\omega t) \tag{5.32}$$

and consists of a background DC component over which small time-dependent oscillations of frequency ω are superposed. The electrostatic force on the nanotube is then given by

$$F_{\text{el}} = \tfrac{1}{2} C_g' \left(V_g^{DC}\right)^2 + C_g' V_g^{DC} \delta V_g \cos(\omega t) \tag{5.33}$$

where $C_g' = dC_g/dz|_{z=z_0}$. The force on the nanotube thus consists of a static component that can be used to tune the tension and an oscillating component to probe the nanotube resonance. In addition to the electrostatic force, the nanotube experiences an elastic force that tends to bring it back to its equilibrium shape, and is of the form

$$F_{\text{elastic}} = -k\delta z \tag{5.34}$$

where k is the spring constant. The equation describing the classical motion of the nanotube for small oscillations in the presence of the electrostatic and elastic forces is then

$$m\frac{d^2\delta z}{dt^2} + b\frac{d\delta z}{dt} = -k\delta z + C_g' V_g^{DC} \delta V_g \cos(\omega t) \tag{5.35}$$

where m is the mass of the nanotube and a damping term was added. This equation is that for a damped harmonic oscillator with an external forcing term. The time-dependent displacements are given by

$$\delta z(t) = \frac{C_g' V_g^{DC} \delta V_g}{\sqrt{b^2\omega^2 + m^2\left(\omega_0^2 - \omega^2\right)^2}} \cos(\omega t + \phi) \tag{5.36}$$

with $\omega_0 = \sqrt{k/m}$ the undamped resonant frequency and $\phi = \pi/2 + \arctan\left[m\left(\omega^2 - \omega_0^2\right)/b\omega\right]$ the phase shift.

The mechanical oscillations of the carbon nanotube can be detected by monitoring the electronic current that flows through the carbon nanotube. Since the nanotube operates as a field-effect transistor, near the threshold voltage the current is proportional to the charge on the nanotube. The induced charge on the nanotube has two time-dependent components, one due to the change in capacitance between the nanotube and gate $\delta C_g(t)$, and the other due to the change in gate voltage $\delta V_g(t)$,

$$\delta n(t) = C_g \delta V_g \cos(\omega t) + V_g \delta C_g(t). \tag{5.37}$$

The capacitance between the nanotube and the gate is $C_g = 2\pi\varepsilon_0 L/\ln(2z/R)$, and under small perturbations of the nanotube around its equilibrium position

$$\delta C_g(t) = -\frac{C_g^2}{2\pi\varepsilon_0 L z_0}\delta z(t). \tag{5.38}$$

The charge on the nanotube is thus

$$\delta n(t) = C_g \delta V_g \cos(\omega t) + \left(\frac{C_g^2}{2\pi\varepsilon_0 L z_0}\right)^2 \times \frac{V_g^{DC}}{\sqrt{b^2\omega^2 + m^2\left(\omega_0^2 - \omega^2\right)^2}} \delta V_g \cos(\omega t + \phi). \tag{5.39}$$

To detect the change in current due to this change in charge, the nanotube is used as a mixer [26], and the current through a lock-in amplifier is monitored. This current can be written as

$$\delta I = \delta G \delta V_{sd} = \frac{dG}{dV_g}\frac{\delta n}{C_g}\delta V_{sd} \tag{5.40}$$

and with the help of Eq. (5.56) we finally obtain

$$\delta I(t) = \delta V_{sd}\frac{dG}{dV_g}\left[\delta V_g \cos(\omega t) + \left(\frac{C_g^2}{2\pi\varepsilon_0 L z_0}\right)^2 \times \frac{V_g^{DC} C_g^{-1}}{\sqrt{b^2\omega^2 + m^2\left(\omega_0^2 - \omega^2\right)^2}} \delta V_g \cos(\omega t + \phi)\right]. \tag{5.41}$$

The ac current in the nanotube has two components, given by the two terms in the square brackets of Eq. (5.41). The first term is just the gate voltage oscillations, while the second term represents the mechanical oscillations of the nanotube with a resonant frequency $\omega_r = \sqrt{\omega_0^2 - (b/m)^2/2}$. A typical experimental result is shown in Fig. 5.15 (b), where the ac component of source–drain current shows a slowly varying background over which a sharp feature is superposed. This feature is attributed to the nanotube mechanical resonance. A fit of the experimental data with a lorentzian function gives the undamped resonant frequency $\omega_0 = 55$ MHz and a quality factor $Q = f/\Delta f = 80$.

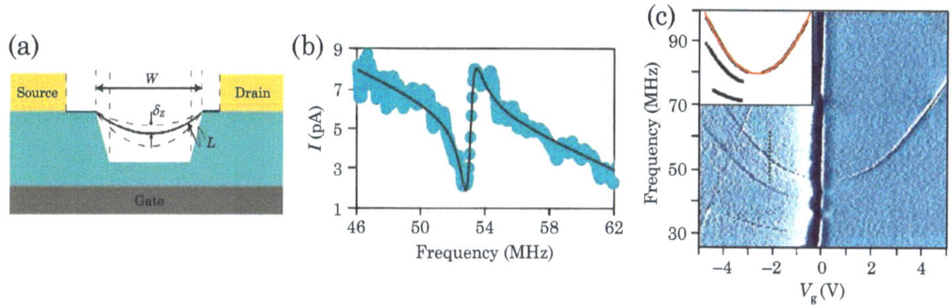

Figure 5.15 (a) Image of a suspended nanotube bridging a trench. The metal electrodes are shown in yellow. A dc gate voltage modulates the tension in the nanotube and an ac gate voltage causes mechanical oscillations shown by δz. When the frequency of the ac voltage matches the natural resonant frequency of the nanotube, δz becomes large. (b) The experimentally measured source–drain current as a function of the frequency of ac voltage applied to the gate, at a dc gate voltage of 2.2 V. The solid black line is a Lorentzian fit giving a resonant frequency of 55 MHz. (c) The experimentally measured resonant frequency as a function of the dc gate voltage. Figure from Ref. [26].

The resonant frequency can be controlled by varying several parameters. For example, because the resonant frequency depends on the total mass of the resonator, the length and diameter of the nanotube can be varied to change the resonant frequency. In addition, the tension in the nanotube also affects the resonant frequency since it enters in the constant k. In the suspended nanotube resonator system, the tension has two components: the bending rigidity of the nanotube and the electrostatic force on the nanotube caused by the dc gate voltage. Fig. 5.15 (c) shows the experimentally observed change in resonant frequency as a function of dc gate voltage. At zero gate voltage the electrostatic force on the nanotube vanishes and the resonant frequency is determined by the bending rigidity of the suspended nanotube. As the dc gate voltage increases several regimes are observed: for small V_g^{DC}, the resonant frequency is proportional to V_g; intermediate values of the dc gate voltage show a linear behavior; larger values of V_g^{DC} give rise to resonant frequencies that increase as $V_g^{2/3}$.

To understand this behavior, we consider the impact of the elastic and electrostatic energies on the tension of a suspended nanotube[27]. For the situation of zero contact capacitance, the energy of the nanotube can be written as

$$E = E_{\text{elas}} + E_{\text{elec}} \tag{5.42}$$

with

$$E_{\text{elas}} = \frac{B\pi R^4}{8}\int_0^L \left(\frac{d^2z}{dx^2}\right)^2 dx + \frac{T}{4}\int_0^L \left(\frac{dz}{dx}\right)^2 dx \tag{5.43}$$

and

$$E_{\text{elec}} = \frac{n^2}{2\left(C_g + \frac{C_g^2}{2\pi\varepsilon_0}\int\limits_0^L z(x)dx\right)} - nV_g. \tag{5.44}$$

In these equations, B is an elastic constant, T is the tension in the nanotube and x is the direction along the nanotube with a profile $z(x)$. The first term in Eq. (5.43) is the energy of a bent rod, while the second term is the contribution due to the tension in the nanotube. The expression in Eq. (5.44) considers a position-dependent gate capacitance. Minimizing the total energy with respect to the functional $z(x)$ we obtain the differential equation

$$\frac{B\pi R^4}{4}\frac{d^3 z}{dz^3} - T\frac{d^2 z}{dz^2} = \frac{n^2}{L^2 R}. \tag{5.45}$$

The solution of this equation with the boundary conditions $z(0) = z(L) = z'(0) = z'(L) = 0$ is

$$z(x) = \frac{n^2}{2TLR\xi}\left[\frac{\sinh \xi L}{\cosh \xi L - 1}(\cosh \xi x - 1) - \sinh \xi x + \xi x - \xi\frac{x^2}{L}\right] \tag{5.46}$$

with $\xi = \sqrt{4T/AR^4}$. From this equation, the tension in the nanotube can be calculated from the expression

$$T = \frac{B\pi R^2}{2L}\int\limits_0^L \left(\frac{dz}{dx}\right)^2 dx. \tag{5.47}$$

(Note that in principle, this expression for the tension should be used directly in Eq. (5.43) before deriving the differential equation. However, by initially assuming a constant tension along the nanotube, the problem is linearized and amenable to analytical solution.) We now consider the limits of small and large tension. In the small tension limit, $\xi \ll 1$, the bending profile is

$$z(x) = \frac{n^2\xi^2}{2TLR}\frac{x^2}{12L}(x - L)^2 \tag{5.48}$$

and the tension is

$$T = \frac{n^4 L^2}{3780 B\pi R^8}. \tag{5.49}$$

The resonant frequency is proportional to $\sqrt{T}$ so in the small tension regime we obtain $\omega_0 \propto V_g^2$ by using the relation $n = C_g V_g$.

For large gate voltages the tension is no longer small and we consider the limit $\xi \gg 1$. In this case the nanotube bending profile is

$$z(x) = \frac{n^2}{2TL^2R} x (L - x) \tag{5.50}$$

and the tension is

$$T = \left(\frac{B\pi}{24L^2} \right)^{1/3} n^{4/3}. \tag{5.51}$$

This functional form for the tension implies that the resonant frequency $\omega_0 \propto V_g^{2/3}$. Both this and the low tension regime are in agreement with the experimental results of Fig. 5.15.

It is interesting to further analyze the oscillating nanotube to obtain values for some of the constants that describe the system. At resonance, it is estimated that the maximum displacement of the nanotube is about 10 nm, and that the electrostatic force due to the gate is 60 fN. From these values, the effective spring constant is calculated to be $k_{\text{eff}} = FQ/z_{\text{max}} = 4 \times 10^{-4}$ N m^{-1}. A similar analysis can be used to estimate the sensitivity of the nanotube oscillator. The smallest detected motion of the nanotube occurs for a driving voltage $\delta V_g = 1$ mV, which corresponds to a maximum nanotube displacement of 0.5 nm at resonance and a force sensitivity of 1 fN Hz$^{-1/2}$. Without much optimization, this nanotube device has a sensitivity that is within a factor of ten of the highest force sensitivities observed at room temperature. The limit on the force sensitivity is determined by the thermal fluctuations of the nanotube according to the formula $\sqrt{4k_BTk/\omega_o Q} = 20$ aN Hz$^{-1/2}$. The measured value is about 50 times larger, possibly due to the limited room temperature transconductance of the nanotube transistor. In any case, low temperature operation, even without improving the device characteristics (Q) could give even lower force sensitivities.

A consequence of the high sensitivity of the nanotube oscillator is that air drag can significantly impact the device performance. Indeed, the device discussed above was operated in a vacuum chamber at pressures less than 10^{-4} Torr. Increasing the pressure leads to a significant degradation of the device performance due to an increase in damping. This effect is illustrated in Fig. 5.16, which shows the quality factor Q as a function of pressure. The quality factor decreases steadily with increase in pressure, and resonance is no longer observed above 10 Torr.

Improvements over the original device discussed above have been reported [28]. The key feature is the coating of the carbon nanotube with a thin layer of metal, dramatically enhancing the resonant frequency, and allowing for room temperature operation in air at atmospheric pressure. With this approach,

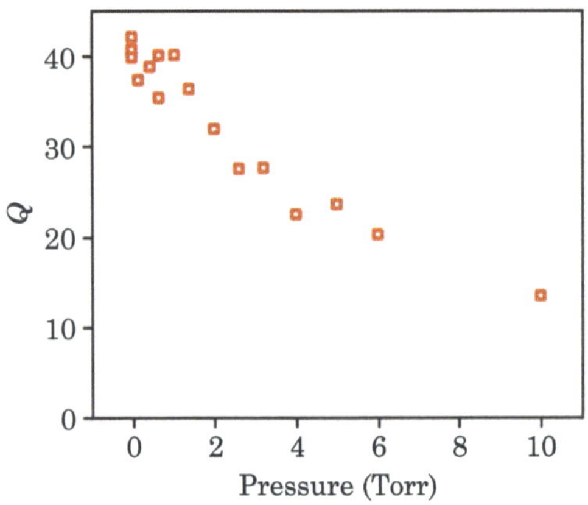

Figure 5.16 Measured quality factor Q for a nanotube oscillator as a function of the pressure in the vacuum chamber. Figure from Ref. [26].

resonant frequencies as high as 1.85 GHz have been measured using a trench width of 300 nm. Fig. 5.17 shows the measured amplitude and phase of a typical device where the nanotube is coated with 2.5 nm of indium; this device has a resonant frequency of 1.33 GHz in vacuum, with a small shift to 1.32 GHz in air. The phase changes by 180 degrees when going through the resonance, as expected from the expression for the phase below Eq. (5.36). Because the mass of the oscillator directly affects the resonant frequency, nanotube resonators can be used as ultrasensitive mass detectors. An example of this application is shown in the right panel of Fig. 5.17. There, the response of a suspended nanotube coated with Fe is presented as a function of frequency, indicating a resonant frequency of 470 MHz. After an additional 2 nm of Fe were deposited on the nanotube, the resonant frequency is decreased to 390 MHz. The estimated added mass is 3.5×10^{-17}g, and from the width of the resonance, one can estimate that the minimum detectable resonance shift is 16 MHz. Assuming that the shift is proportional to the mass of the added material, the predicted mass detection limit is on the order of 10^{-18} g.

An alternative approach that can be used to change the resonant frequency of nanotube resonators is to change the length of the carbon nanotube. While this can be done by making double-clamped nanotube devices with varying channel lengths, an elegant approach to achieve the same effect without the need to make multiple devices is to use the telescoping properties of multiwall carbon nanotubes [29]. Fig. 5.18 shows a conceptual rendition of this idea, where a multiwall nanotube is attached to a fixed contact at one end, and attached to a mobile electrode at the other end. The low friction between the outer and inner shells of the multiwall nanotube allow for telescoping of the inner and outer shells, permitting control of the nanotube length. Instead of exciting the mechanical vibrations of the nanotube with an oscillating gate voltage, the device is operated in a magnetic field. Application of an alternating electrical

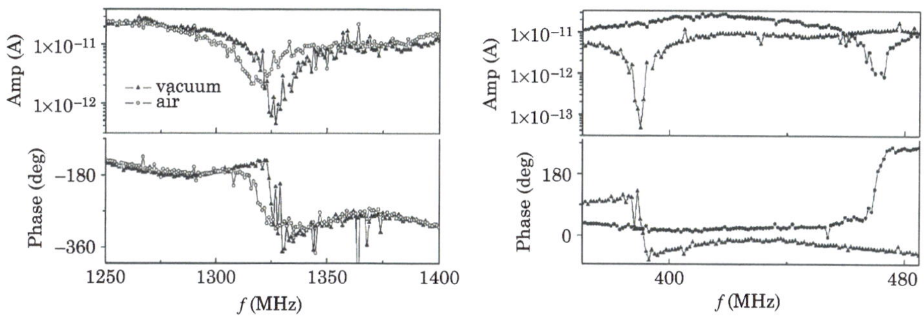

Figure 5.17 Left: amplitude and phase for an indium-coated carbon nanotube resonator in vacuum and in air. The resonant frequency is 1.33 GHz in vacuum and 1.32 GHz in air. Right: electrical response of a Fe-coated nanotube resonator before (circles) and after the addition of an additional 2 nm of Fe (triangles). Figures from Ref. [28].

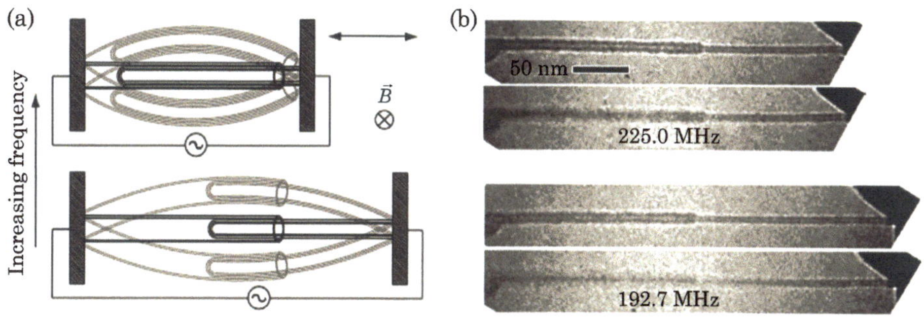

Figure 5.18 (a) Sketch of a telescoping multiwall carbon nanotube, whose length can be changed to tune the frequency of a nanoresonator. The nanotube is attached to a fixed contact at one end and to a movable piezo-controlled electrode at the other end. The device is operated in an external magnetic field, allowing for actuation of the nanotube with the Lorentz force. (b) Transmission electron microscopy images of a telescoping nanotube at two different lengths. In each set of two images, the top image shows the nanotube off resonance (sharp) and the bottom image shows the nanotube at resonance (blurry). Changing the length of the nanotube changes the resonant frequency from 225.0 MHz to 192.7 MHz. Figure from [30].

current through the nanotube leads to a Lorentz force $\vec{F} = e\vec{v} \times \vec{B}$, which is directed vertically for the orientation of the magnetic field shown in Fig. 5.18. Transmission electron micrographs of the oscillating nanotube are shown in the same figure. In these images, the dark region on the right is the movable electrode (the stationary electrode does not appear in the image, being much further to the left). The top two images show a nanotube of length 558 nm before (sharp) and during resonance (blurry) with a resonance frequency of 225.0 MHz. Telescoping of the nanotube to a length of 608 nm (bottom two micrographs) decreases the resonance frequency to 192.7 MHz. The response of the amplitude oscillations fits well to a Lorentzian distribution with values of Q

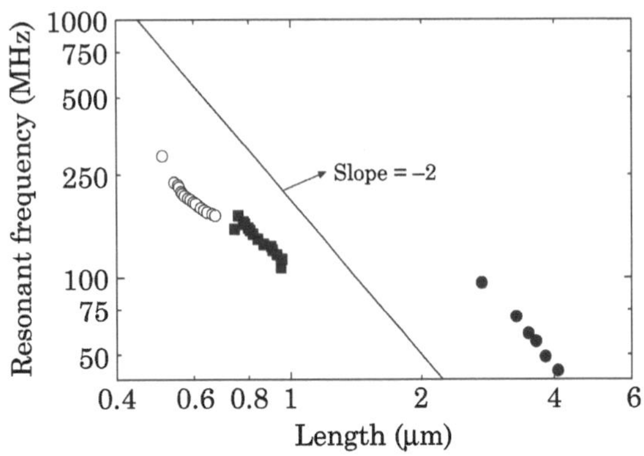

Figure 5.19 Resonant frequency of telescoping nanotube resonator as a function of the nanotube length for three different devices. Figure after Ref. [30].

that can be fairly large. Indeed, for such devices, the quality factor is typically between 100 and 1000, much larger than for doubly clamped devices.

The ability to tune the resonant frequency by changing the tube length is demonstrated in Fig. 5.19. It is quite remarkable that the resonant frequency can be tuned by an order of magnitude, varying between 30 and 300 MHz. The resonant frequency is found to be very sensitive to the length of the nanotube, with the largest change of 1 MHz/nm observed in one device.

The length dependence of the resonance frequency can be predicted by treating the nanotube as a continuum beam [30] with displacements satisfying the differential equation

$$\rho A \frac{\partial^2 z}{\partial t^2} = T \frac{\partial^2 z}{\partial x^2} - EI \frac{\partial^4 z}{\partial x^4} \tag{5.52}$$

where ρ is the mass density, A is the cross-sectional area, T is the tension in the nanotube, E is Young's modulus and I is the areal moment of inertia. By using separation of variables and the initial condition of a straight nanotube, the solution to this equation can be written in the form

$$\begin{aligned} z(x,t) = \sin(\omega t)\,[z_1 \cos(\beta x) + z_2 \sin(\beta x) + z_3 \cosh(\beta x) \\ +\ z_4 \sinh(\beta x)] \end{aligned} \tag{5.53}$$

where the z_i are constants to be determined from the boundary conditions. Substitution of Eq. (5.53) into Eq. (5.52) provides the relationship between β and ω:

$$\beta^2 = \frac{T}{2EI}\left(\sqrt{1 + \frac{4EI\rho A\omega^2}{T^2}} - 1\right). \tag{5.54}$$

To proceed further, we make use of the boundary conditions for a clamped beam $z(0) = z'(0) = z(L) = z'(L) = 0$. This procedure leads to a complicated equation for the variable βL involving trigonometric and exponential functions. This equation provides a numerical value for the product βL, which we write as $\beta L = \alpha$. Thus, Eq. (5.54) can be solved for the frequency to obtain

$$\omega = \frac{\alpha^2}{L^2}\sqrt{\frac{EI + \alpha^{-2}TL^2}{\rho A}}. \quad (5.55)$$

Assuming that the tension in the tube is due to the van der Waals interaction between tube shells [29], one can show that the term proportional to the tension in Eq. (5.55) is negligible compared to the term from the Young's modulus. Therefore, the resonant frequency behaves as $\omega \sim L^{-2}$ as can be seen from the experimental data presented in Fig. 5.19.

5.4.2 Torsional Actuators

The ability to impart rotational motion to small mechanical components is important for many applications such as rotating micro-mirrors and magnetic resonance force microscopy. Rotational motion of small components may also be exploited for detection of fluid motion in microfluidics, for general sensors or for the transmission of electromagnetic radiation. Most approaches to enable torsional actuators use conventional Micro ElectroMechanical Systems (MEMS) made of silicon. Because of their favorable mechanical properties, stability and resistance to wear, carbon nanotubes are attractive materials for torsional actuators. In addition, the impact of torsional strain on the electronic properties of carbon nanotubes can be used to measure the torsional motion. One of the first realizations [31] of torsional actuators with carbon nanotubes used a geometry where a suspended multiwall carbon nanotube is clamped to two electrodes, with a metal plate attached to the nanotube. Rotation of the metal plate is induced by applying voltages to three stator electrodes. Applied dc voltages up to 50 V between the bottom stator gate and the rotor plate generates rotations up to 20 degrees, which can be operated in ac mode with resonance frequencies as large as hundreds of megahertz. Torsional spring constants for such devices range between 10^{-15} to 10^{-12} N m. These fairly large torsional spring constants imply that large angular rotations of the nanotube are difficult to induce. To circumvent this problem, it has been demonstrated that a freely rotating outer nanotube shell can be created by shearing the outer walls of the nanotube by application of large stator voltages. Stator voltages larger than 80 V are found to shear the nanotube past its elastic limit, leading to failure of the outer walls and a dramatic increase in the rotational freedom of the rotor plate. With this approach, the rotor plate can be positioned at any angle between 0 and 360 degrees with an appropriate combination of stator

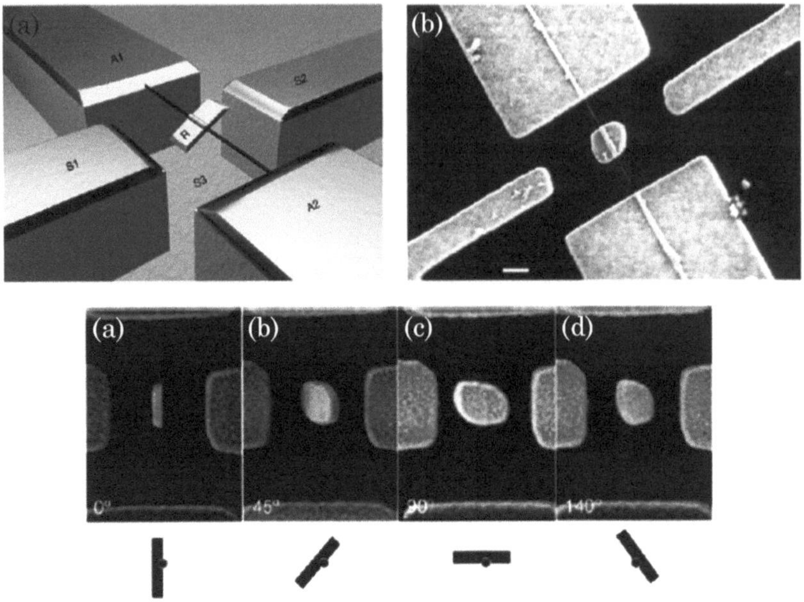

Figure 5.20 Top left: drawing of a rotational actuator using a single multiwall carbon nanotube. A metal plate rotor (R) is attached to a multi-walled carbon nanotube (MWNT) which acts as a support shaft and is the source of rotational freedom. Electrical contact to the rotor plate is made via the MWNT and its anchor pads (A1, A2). Three stator electrodes, two on the SiO2 surface (S1, S2) and one buried beneath the surface (S3), provide additional voltage control elements. Top right: scanning electron microscope (SEM) image of nanoactuator just prior to HF etching. Scale bar, 300 nm. Bottom: series of SEM images showing the actuator rotor plate at different angular displacements. The MWNT, barely visible, runs vertically through the middle of each frame. The schematic diagrams located beneath each SEM image illustrate a cross-sectional view of the position of the nanotube/rotor-plate assembly. Scale bar, 300 nm. Figures and caption from Ref. [31].

voltages. The images at the bottom of Fig. 5.20 show the controlled motion of the rotator plate through a complete cycle using a sequence of quasi-static stator voltages. Such cycles have been repeated thousands of times with no apparent wear or degradation in performance.

While these initial experiments used scanning electron microscopy to detect the motion of the rotor plate, the impact of torsional strain on the nanotube can also be detected by monitoring the current that flows through the nanotube [32] since torsional strain modifies the electronic properties of the nanotube, as discussed in Section 5.2. In these experiments, the tip of an atomic force microscope is used to apply a controlled force to the rotor plate and thus a controlled torsion angle (Fig. 5.21); simultaneous electrical measurements of the low bias conductance give the relative change in resistance as a function of torsion angle plotted in the same figure. These measurements indicate that for this particular nanotube, the resistance increases and decreases with increasing applied strain. A simple model to describe this effect considers that the change

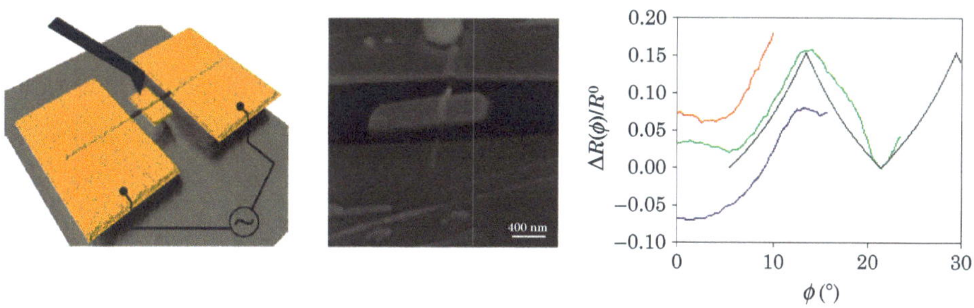

Figure 5.21 The left and middle panels show a schematic representation and a scanning electron micrograph of a nanotube torsional electromechanical device where the tip of an atomic force microscope is used to apply force to a "pedal" attached to the carbon nanotube. The rotation of the pedal causes a torsional strain in the nanotube, leading to a change in the resistance as shown in the right panel, where different curves represent repeated experiments and the smooth line is the theoretical prediction. See text for details. Figures from Ref. [32].

in the resistance is due to the change in bandgap of the nanotube due to the distortions, and a concomitant change in the Schottky barrier for charge injection at the contacts [32]. We consider the case of a metallic nanotube with torsion angles ϕ and $-\phi$ in the two segments between the contacts and the rotor plate. Assuming that the transport in the nanotube is ballistic, the conductance is given by

$$\begin{aligned} G &= \frac{4e^2}{h} \int_{-\infty}^{\infty} T(E) \left(-\frac{\partial f}{\partial E}\right) dE \\ &= \frac{4e^2}{h} \int_{-\infty}^{E_v(\phi)} \left(-\frac{\partial f}{\partial E}\right) dE + \frac{4e^2}{h} \int_{-\infty}^{E_c(\phi)} \left(-\frac{\partial f}{\partial E}\right) dE \end{aligned} \tag{5.56}$$

where we used the relation $E_g(\phi) = E_g(-\phi)$ (appropriate for a metallic nanotube) and assumed perfect transmission. Integration of Eq. (5.56) with the Fermi level in the middle of the bandgap provides the relative change in resistance

$$\frac{\Delta R}{R} = e^{\frac{\Delta E_g(\phi)}{2kT}} - 1. \tag{5.57}$$

As discussed in the context of Fig. 5.11, $\Delta E_g(\phi)$ increases at first for a metallic nanotube, reaches a maximum and then decreases, with this behavior periodically repeated. Thus, from Eq. (5.57) we expect that the relative change in resistance will also have this behavior, which is confirmed by the experimental data presented in Fig. 5.21.

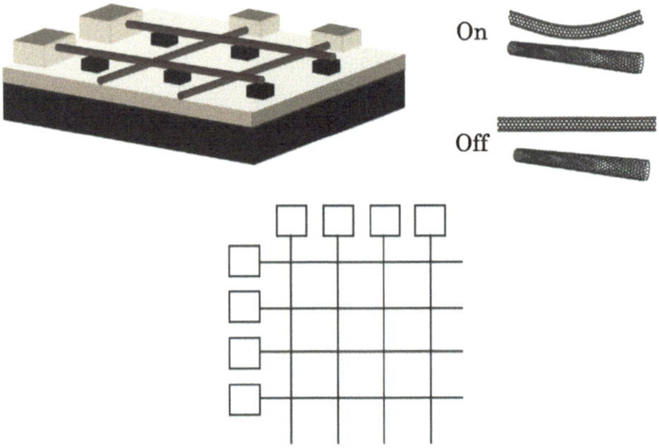

Figure 5.22 Top left: representation of a nanotube memory device using suspended nanotubes. Each nanotube is connected to electrical contacts at its ends. The bottom nanotubes sit on a thin insulating layer and the top nanotubes are suspended by being placed on a series of nonconducting supports. Application of a voltage to the bottom and top nanotubes causes the top nanotube to bend and make contact with the bottom nanotube (right). Bottom: top view of the device indicating how each bit of the memory can be addressed with the crossbar architecture.

5.4.3 Nanotube Memory

Another exciting development in the application of nanotube electromechanical response has been the fabrication of memory devices [33,34]. The general idea is illustrated in Fig. 5.22. There, a crossbar architecture is created using perpendicular arrays of carbon nanotubes. The bottom nanotubes sit on an insulator, while the top nanotubes are suspended on a series of nonconducting supports. Every nanotube of the two arrays is electrically contacted at its two ends, and application of a voltage causes an electrostatic force on the suspended nanotube, causing it to bend downwards and eventually contact the bottom nanotube. The conductance between the separated and contacted nanotubes varies by orders of magnitude; ON and OFF states correspond to high and low conductance states due to the two mechanical states of the suspended nanotube. Each crossing point between nanotubes forms a bit in the nanotube memory device.

The operation of this device requires the presence of two well-defined stable states: one where the top tube is suspended and far from the bottom nanotube, and one where the two nanotubes are in contact. The stability of these two states is determined by the energetic competition between elastic, electrostatic and van der Waals interactions. The total energy is

$$E_{\text{tot}} = E_{\text{elas}} + E_{\text{elec}} + E_{vdW} \tag{5.58}$$

where E_{elec} and E_{vdW} are the electrostatic and van der Waals interaction energies between the two nanotubes, and E_{elas} is the elastic energy due to deformation of the suspended nanotube. From linear elasticity theory (and from the earlier discussion on nanotube oscillators) we expect that the elastic energy will be of the form $E_{\text{elas}} = B(\delta z)^2$ where B is a positive constant and δz is the displacement of the central portion of the tube with respect to the straight suspended tube configuration. The important point is that the elastic energy increases monotonically with increase in bending, and therefore favors the OFF state. The electrostatic energy is attractive or repulsive depending on the relative sign of the potential on each nanotube, with a value that depends monotonically on the separation between the nanotubes.

The van der Waals interaction energy between the nanotubes can be computed using a Lennard-Jones potential:

$$E_{vdW} = \sum_{i,j} \left(\frac{C_{12}}{r_{ij}^{12}} - \frac{C_6}{r_{ij}^6} \right) \tag{5.59}$$

where C_6 and C_{12} are positive constants and r_{ij} is the distance between atoms i and j. To illustrate the behavior of this energy with nanotube separation L, we consider the simplified situation where only two atoms at the crossing point between the nanotubes interact:

$$E_{vdW} = \frac{C_{12}}{L^{12}} - \frac{C_6}{L^6}. \tag{5.60}$$

This energy is strongly repulsive for very short separations, and approaches zero for large separations; it has a minimum at a (small) separation $L = (2C_{12}/C_6)^{1/6}$ where the energy is negative. The van der Waals energy thus favors close proximity of the nanotubes.

The sum of the three energy components computed using more sophisticated models [33] is plotted in Fig. 5.23 as a function of the separation between the nanotubes. For an initial state where the nanotubes are separated and no voltage is applied to them, the energy has a minimum at a separation distance of about 1.4 nm, representing the OFF state of the device. For large enough opposite values of the voltage on each nanotube, a minimum in the total energy develops at small separations, stabilizing the ON state of the device. Note that in this regime, the OFF state is at a higher energy and there is no barrier to prevent the system from moving to the lower energy configuration. A similar effect occurs if one considers switching from the ON state, where a large enough voltage of the same sign applied to the two nanotubes leads to an energy minimum for the separated state.

Because the current between the top and bottom nanotubes occurs through a tunneling process with a short decay length, the ratio of ON to OFF currents can be orders of magnitude. Furthermore for a short height of the nonconducting supports, the strain in the nanotube in the ON state is less than the elastic limit, but larger than can be sustained by traditional materials indicating that the

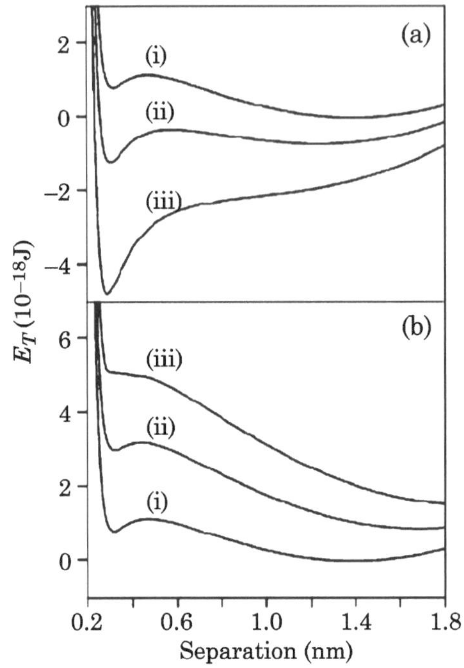

Figure 5.23 Total energy of the top nanotube (van der Waals + electrostatic + elastic) as a function of the distance from the bottom nanotube. (a) shows the curves for switching ON with the voltages applied to the two nanotubes equal to (i) 0 V for both, (ii) +3 V and −3 V, and (iii) +4.5 and −4.5 V. (b) shows curves for switching OFF with voltages of (i) 0 V, (ii) +15 V, and (iii) +20 V for both nanotubes. Figure from Ref. [33].

unique mechanical strength of nanotubes may provide an advantage over other materials [33].

A proof-of-principle single bit device based on the above theoretical proposal has been realized using carbon nanotube ropes. A dark-field optical micrograph of the device is shown in Fig. 5.24. Two crossing 50 nm thick nanotube ropes are attached to Cr/Au electrodes, and lie on an oxidized silicon substrate. Measuring the current between electrodes 1 and 2 or 3 and 4 provides a measure of the electrical characteristics of each nanotube separately; as shown in Fig. 5.24, both of the nanotube ropes show linear I–V curves. When the current through the nanotube junction is measured using electrodes 2 and 3, the I–V curve is linear in the ON state, but nonlinear and much lower in the OFF state; the ON/OFF conductance ratio is as large as five orders of magnitude. Furthermore, reversible switching of a similar device over several days has been demonstrated (Fig. 5.24).

The idea of using the electromechanical deformation of nanotubes for memory applications is now being explored [34] using a variation of the crossing nanotube approach. In this modification, the channel consists of a network of nanotubes lying between source and drain electrodes and the channel is

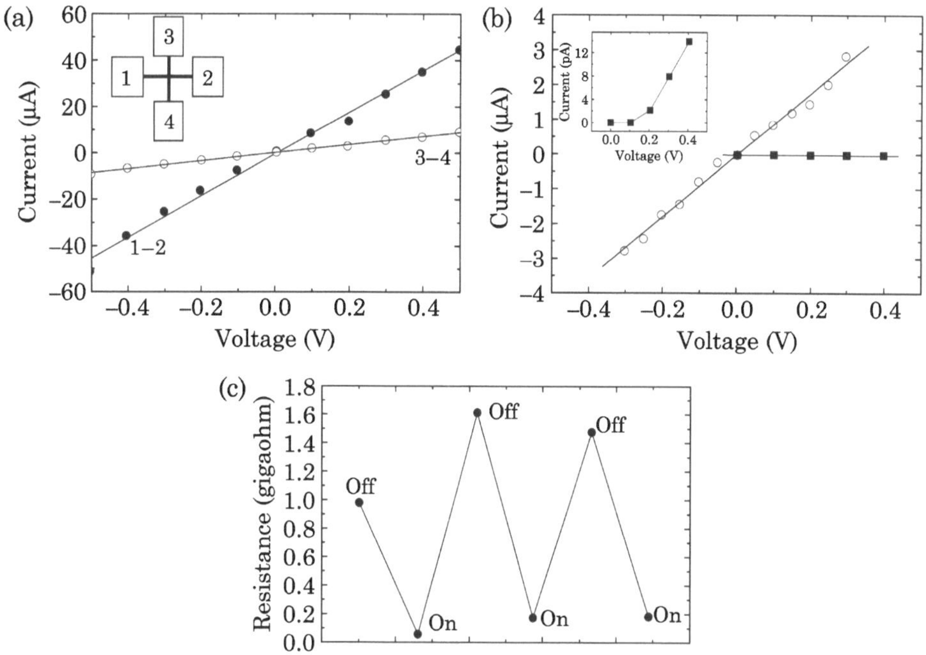

Figure 5.24 Experimental realization of one bit of the suspended nanotube memory device. (a) is the *I–V* curve for the bottom (filled circles, electrodes 1 and 2) and top (open circles, electrodes 3 and 4) nanotubes. The inset shows a sketch of the fabricated device. The nanotubes in this case are thick nanotube ropes. (b) shows the *I–V* curves for the nanotube junction (measured between electrodes 2 and 3) in the ON (circles) and OFF states (squares). The inset shows the OFF state *I–V* curve, indicating tunneling-like behavior. (c) Reversible switching of a nanotube device over repeated cycles. Figure after Ref. [33].

capacitively coupled to a gate electrode, as illustrated in Fig. 5.25. Binary 0 and 1 states are determined by whether or not the nanotube network makes contact to the electrode. Binary 0 is represented by the state where there is a gap between the nanotube network and electrode as shown in Fig. 5.25 (b). Application of a voltage to the electrode pulls the nanotube fabric down such that it contacts the electrode, and this state represents binary 1 (Fig. 5.25 (c)), as there is a large current flowing between the interconnect and the electrode. This current is nonzero only when the nanotubes make contact to the electrode. Advantages of this memory device are that it should be both nonvolatile and fast. Once a 1 is written, the voltage applied to the electrode can be removed; van der Waals interactions between the nanotubes and electrode hold the memory in the 1 state for extended periods of time. The access time of this nonvolatile nanotube RAM is predicted to be comparable to that of conventional DRAM / SRAM [33,34].

The current–voltage characteristics [35] of such a device are shown in Fig. 5.26. Initially, the nanotube network is suspended and the resistance between the electrode and the nanotube is high ($> M\Omega$). As a critical voltage of

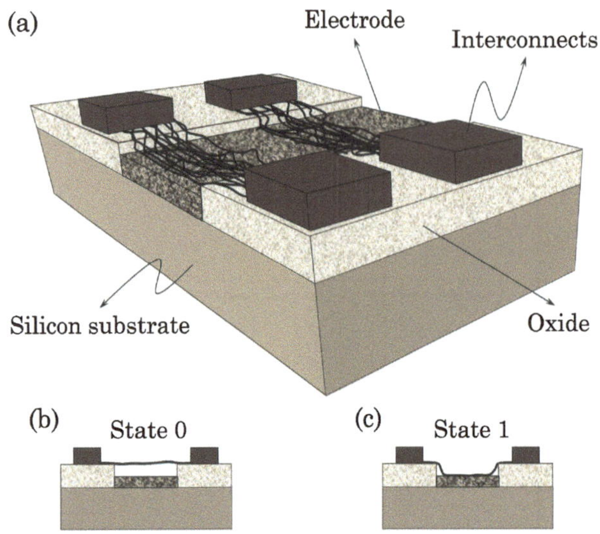

Figure 5.25 Random Access Memory (RAM) made with a nanotube fabric. (a) The nanotube memory array. (b) A binary 0 state corresponds to the nanotube suspended and not making contact to the electrode. (c) A binary 1 state corresponds to the nanotube making contact with the electrode. The length of the suspended nanotubes is 130 nm.

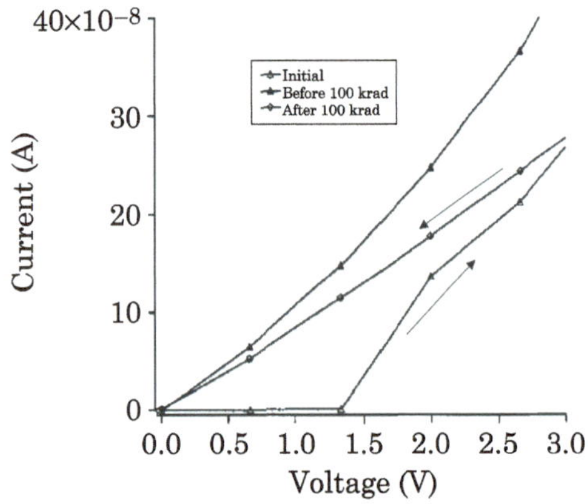

Figure 5.26 Current–voltage characteristic of a nanotube memory device. The open triangles (lower curve) show the initial writing stage: the nanotube network is initially suspended, and the conductance is low. When the applied voltage exceeds 1.5 V, the conductance increases substantially, signaling contact between the nanotube network and the bottom electrode. Once this state has been written, its conductance remains high, even as the applied voltage is decreased to zero (middle curve). The top curve shows that the device is tolerant to radiation. Figure adapted from Ref. [35].

1.5 V is reached in the current–voltage measurements, the electrode/nanotube resistance decreases significantly (lower trace). This ON state of the memory is stable, even when the applied voltage is reduced to values below the turn-on voltage (middle trace). It is also interesting to note that the nanotube device has good radiation tolerance.

References

1. A. Rochefort, P. Avouris, F. Lesage and D. Salahub, "Electrical and mechanical properties of distorted carbon nanotubes", *Phys. Rev.* B, Vol. 60, p. 13824, 1999.
2. M.B. Nardelli and J. Bernnholc, "Mechanical deformations and coherent transport in carbon nanotubes", *Phys. Rev.* B, Vol. 60, p. R16338, 1999.
3. F. Léonard, F.E. Jones, A.A. Talin and P.M. Dentinger, "Robustness of nanotube electronic transport to conformational deformations", *Appl. Phys. Lett.*, Vol. 86, p. 093112, 2005.
4. A. Maiti, A. Svizhenko and M.P. Anantram, "Electronic transport through carbon nanotubes: Effects of structural deformation and tube chirality", *Phys. Rev. Lett.*, Vol. 88, p. 126805, 2002.
5. A.A. Farijian, H. Mizuseki and Y. Kawazoe, "Electronic transport properties of a metal–semiconductor carbon nanotube heterojunction", *Physica* E, Vol. 22, p. 675, 2004.
6. L.F. Chibotaru, S.A. Bovin and A. Ceulemans, "Bend-induced insulating gap in carbon nanotubes", *Phys. Rev.* B, Vol. 66, p. 161401(R), 2002.
7. C.H. Xu, C.Z. Wang, C.T. Chan and K.M. Ho, *J. Phys.: Condens. Matter*, Vol. 4, p. 6047, 1992.
8. F. Léonard, "Quantum transport in nanotube transistors", *Phys. Stat. Sol.* (b), Vol. 239, p. 88, 2003.
9. M.P. Anantram and F. Léonard, "Physics of carbon nanotube devices", *Rep. Prog. Phys.*, Vol. 69, p. 507, 2006.
10. R. Heyd, A. Charlier and E. McRae, "Uniaxial-stress effects on the electronic properties of carbon nanotubes", *Phys. Rev.* B, Vol. 55, p. 6820, 1997.
11. Yu.N. Gartstein, A.A. Zakhidov and R.H. Baughman, "Mechanical and electromechanical coupling in carbon nanotube distortions", *Phys. Rev.* B, Vol. 68, p. 115415, 2003.
12. A. Kleiner and S. Eggert, "Band gaps of primary metallic carbon nanotubes", *Phys. Rev.* B, Vol. 63, p. 73408, 2001.
13. L. Liu, C.S. Jayanthi, M. Tang, S.Y. Wu, T.W. Tombler, C. Zhou, L. Alexseyev, J. Kong and H. Dai, "Controllable reversibility of an sp2 to sp3 transition of a single wall nanotube under the manipulation of an AFM Tip: A nanoscale electromechanical switch?", *Phys. Rev. Lett.*, Vol. 84, p. 4950, 2000.
14. L. Yang, M.P. Anantram, J. Han and J.P. Lu, "Band-gap change of carbon nanotubes: Effect of small uniaxial and torsional strain", *Phys. Rev.* B, Vol. 60, p. 13874, 1999.
15. L. Yang and J. Han, "Electronic structure of deformed carbon nanotubes", *Phys. Rev. Lett.*, Vol. 85, p. 154, 2000.
16. W.A. Harrison, *Electronic structure and the properties of solids: The physics of the chemical bond*, Freeman, San Francisco, 1990.
17. T.W. Tombler, C. Zhou, L. Alexseyev, J. Kong, H. Dai, L. Liu, C.S. Jayanthi, M. Tang and S-Y. Wu, "Reversible electromechanical characteristics of carbon nanotubes under local-probe manipulation", *Nature*, Vol. 405, p. 769, 2000.
18. E.D. Minot, Y. Yaish, V. Sazonova, J-Y. Park, M. Brink and P.L. McEuen, "Tuning carbon nanotube band gaps with strain", *Phys. Rev. Lett.*, Vol. 90, p. 156401, 2003.
19. S. Paulson, M.R. Falvo, N. Snider, A. Helser, T. Hudson, A. Seeger, R.M. Taylor, R. Superfine and S. Washburn, "*In situ* resistance measurements of strained carbon nanotubes", *Appl. Phys. Lett.*, Vol. 75, p. 2936, 1999.
20. C. Gómez-Navarro, P.J. de Pablo and J. Gómez-Herrero, "Radial electromechanical properties of carbon nanotubes", *Adv. Mat.*, Vol. 16, p. 549, 2004.

21. H. Mehrez, A. Svizhenko, M.P. Anantram, M. Elstner and T. Frauenheim, "Analysis of band-gap formation in squashed armchair carbon nanotubes", *Phys. Rev.* B, Vol. 71, p. 155421, 2005.
22. O. Gülseren, T. Yildirim, S. Ciraci and Ç. Kılıç, "Reversible band-gap engineering in carbon nanotubes by radial deformation", *Phys. Rev.* B, Vol. 65, p. 155410, 2002.
23. Ç. Kılıç, S. Ciraci, O. Gülseren and T. Yildirim, "Variable and reversible quantum structures on a single carbon nanotube", *Phys. Rev.* B, Vol. 62, p. 16345, 2000.
24. J.-Q. Lu, J. Wu, W. Duan, F. Liu, B.-F. Zhu and B.-L. Gu, "Metal-to-semiconductor transition in squashed armchair carbon nanotubes", *Phys. Rev. Lett.*, Vol. 90, p. 156601, 2003.
25. C.J. Park, Y.H. Kim and K.J. Chang, "Band-gap modification by radial deformation in carbon nanotubes", *Phys. Rev.* B, Vol. 60, p. 10656, 1999.
26. V. Sazonova, Y. Yaish, H. Ustunel, D. Roundy, T.A. Arias and P.L. McEuen, "A tunable carbon nanotube electromechanical oscillator", *Nature*, Vol. 431, p. 284, 2004.
27. S. Sapmaz, Ya.M. Blanter, L. Gurevich and H.S.J. van der Zant, "Carbon nanotubes as nanoelectromechanical systems", *Phys. Rev.* B, Vol. 67, p. 235414, 2003.
28. H.B. Peng, C.W. Chang, S. Aloni, T.D. Yuzvinsky and A. Zettl, "Ultrahigh frequency nanotube resonators", *Phys. Rev. Lett.*, Vol. 97, p. 087203, 2006.
29. J. Cummings and A. Zettl, "Low-friction nanoscale linear bearing realized from multiwall carbon nanotubes", *Science*, Vol. 289, p. 602, 2000.
30. K. Jensen, C. Girit, W. Mickelson and A. Zettl, "Tunable nanoresonators constructed from telescoping nanotubes", *Phys. Rev. Lett.*, Vol. 96, p. 215503, 2006.
31. A.M. Fennimore, T.D. Yusvinsky, W.-Q. Han, M.S. Fuhrer, J. Cumings and A. Zettl, "Rotational actuators based on carbon nanotubes", *Nature*, Vol. 424, p. 408, 2003.
32. T. Cohen-Karni, L. Segev, O. Srur-Lavi, S.R. Cohen and E. Joselevich, "Torsional electromechanical quantum oscillations in carbon nanotubes", *Nature Nanotech.*, Vol. 1, p. 36, 2006.
33. T. Rueckes, K. Kim, E. Joselevich, G.Y. Tseng, C-L Cheung and C.M. Lieber, "Carbon nanotube-based nonvolatile random access memory for molecular computing", *Science*, Vol. 289, p. 94, 2000.
34. G. Stix, Nanotubes in the clean room, Scientific American, p. 82, February (2005).
35. M.N. Lovellette, A.B. Campbell, H.L. Hughes, R.K. Lawrence, J.W. Ward, M. Meinhold, T.R. Bengston, G.F. Carleton, B.M. Segal and T. Rueckes, "Nanotube memories for space applications", *Proc. 2004 IEEE Aero. Conf.*, 2004, p. 2300.

6 Field Emission

6.1 Introduction

Electron beams play a central role in many applications and basic research tools. For example, electron emission is used in cathode ray tubes, x-ray tubes, scanning electron microscopes, and transmission electron microscopes. In many of these applications, it is desirable to obtain a high density of narrow electron beams, with each beam tightly distributed in energy. So-called electron guns, which operate on thermionic emission of electrons from hot cathodes, are extensively utilized for this purpose. However, achieving electron beams with narrow energy distributions is difficult because of thermal broadening of the emitted electrons. Thus, field emission from cold cathodes is of much interest, but requires large electric fields that cause migration of atoms at the tip surface, making it difficult to achieve stable operation over long periods of time. Carbon nanotubes may offer a solution to these issues. Indeed, carbon nanotubes have many advantages for cold field emission: the inertness and stability of nanotube tips to long periods of operation compared to metal and diamond tips; low threshold voltage for cold field emission; low temperature of operation; fast response times; low power; and small size. As will be discussed later in this chapter, prototype devices using the superior field emission properties of nanotubes have already been demonstrated. These devices include x-ray tubes [1], scanning x-ray sources [2], flat panel displays [3], and lamps [4].

Before going into the details of field emission, we begin by introducing the early experimental work that established the promise of carbon nanotubes for field emission [5]. Fig. 6.1 shows the experimental setup to measure the field emission from carbon nanotube films. There, a film of carbon nanotubes, with the nanotubes oriented perpendicular to the substrate serves as the electron emitter. A copper grid sits 20 microns above the nanotube film, with the separation provided by a mica sheet. Application of a voltage between the copper grid and the nanotube film creates a beam of electrons that passes through the copper grid and is detected at an electrode 1 cm away from the copper grid. (It should be noted that these experiments are performed under high vacuum where the field emitter device sits in a vacuum chamber with a residual pressure of 10^{-6} Torr.) A current versus voltage curve for such a device is shown in Fig. 6.1, indicating a large increase of the current in the forward bias direction (the emission is diode-like: for negative voltages, very little current flows). To verify that the beam indeed consists of electrons, the beam was deflected in a magnetic field, and the deflection corresponds to that of particles with the free electron mass. The inset in this figure shows a plot of $\log\left(I/V^2\right)$ vs V^{-1}, the so-called Fowler–Nordheim plot (more will be said about Fowler–Nordheim emission below). Importantly, it is found that the field enhancement factor is about 100 times larger than conventional field emitter tips. To further establish the promise of carbon nanotubes for field emission, the stability of the field emitters was studied as a function of time (Fig. 6.1). Current fluctuations were observed to be less than 10%, which is remarkable considering that the current depends exponentially on the voltage. Little degradation was observed, even after operation periods as long as 48 h.

The theory of field emission was originally developed by Fowler and Nordheim [6], and has since been refined to include effects such as details of the tunneling potential and material-specific density of states. However, the basic aspects of field emission can be captured from a simple theory of tunneling across a triangular potential barrier using the WKB approximation (Fowler and Nordheim used a more accurate approach where the wavefunctions are matched at appropriate boundaries). We present the simplified theory here as it serves to illustrate the basic physics behind field emission, and in any case recovers the general behavior. Fig. 6.2 illustrates the situation for field emission from a metallic surface. In the metal (cathode), electronic states are filled up to the Fermi level. Upon application of a voltage to the anode, a linear potential drop occurs across the vacuum gap, leading to a triangular potential energy barrier of height Φ equal to the metal work function. Electrons below the Fermi level tunnel through the triangular barrier (at finite temperature, there is a distribution of electrons above the Fermi level; here we consider the zero temperature limit since it illustrates the main points of the theory). The current density is of the form

$$J = e \int dE_x \, d\vec{k}_\perp v_x(E_x) D(E_x, \vec{k}_\perp) T(E_x, \vec{k}_\perp) f\left(E_x, \vec{k}_\perp\right) \tag{6.1}$$

where $D(E_x, \vec{k}_\perp)$ is the density of electronic states in the metal, $v_x(E_x)$ is the electron velocity in the x direction, $T(E_x, \vec{k}_\perp)$ is the transmission probability across the tunnel barrier, and $f\left(E_x, \vec{k}_\perp\right)$ is the Fermi function. Here, E is the total energy of the electron, E_x is the energy in the x direction and $\vec{k}_\perp$ is the momentum vector perpendicular to the tunneling direction. For a free electron gas we have the relationships

$$\begin{aligned} E_x &= E - \frac{\hbar^2 k_\perp^2}{2m} \\ E_x &= \frac{\hbar^2 k_x^2}{2m} \end{aligned} \tag{6.2}$$

and the velocity is

$$v_x = \frac{1}{\hbar}\frac{\partial E}{\partial k_x} = \frac{1}{m}\sqrt{\frac{2mE_x}{\hbar^2}}. \tag{6.3}$$

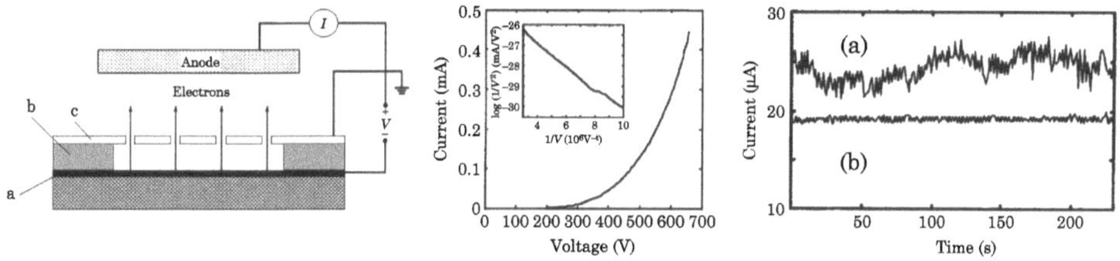

Figure 6.1 The left panel shows a sketch of a field emission electron source. The emitter is a film of aligned carbon nanotubes (a), with the nanotubes aligned perpendicular to a polytetrafluoroethylene substrate. The nanotubes have average diameters of 10 nm and are about 1 micron long. A sheet of mica (b) with a hole of 1 mm diameter is bonded to the nanotube film. The hole is covered with an electron microscopy copper grid (c), which is used to apply the emission voltage. The current is measured at a collector electrode 1 cm away from the grid. The middle panel shows the current versus field emission voltage, with the Fowler–Nordheim plot in the inset. The right panel shows the emitted current as a function of time. The bare fluctuations are on the order of 10% (curve a) but can be reduced to 2% with a feedback system (curve b). Middle and right figures from Ref. [5].

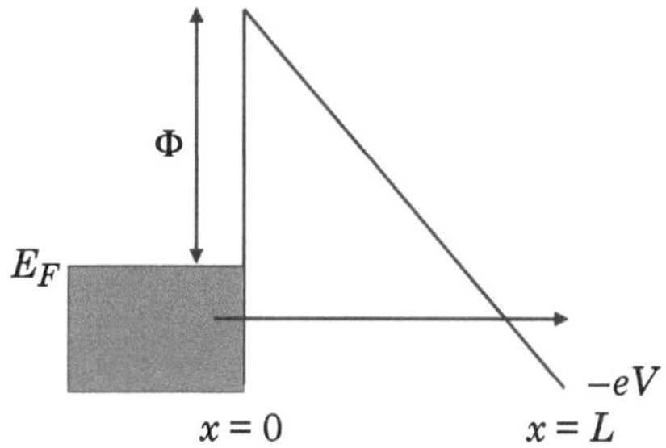

Figure 6.2 Illustration of potential profile for field emission from a metal. Φ is the metal work function, V is the applied potential between the anode and the cathode, and E_F is the metal Fermi level. Electrons in the filled states below E_F (gray shading) tunnel across the triangular potential energy barrier to the anode.

The tunneling probability across the barrier is obtained using the WKB approximation:

$$T(E) = \exp\left[-2\int_{x_1=0}^{x_2(E)} k(x,E)dx\right] \tag{6.4}$$

where x_1 and x_2 are the classical turning points. For a free electron in a potential $V(x) = \Phi - eV\frac{x}{L}$ we have

$$T(E_x) = \exp\left[-2\sqrt{\frac{2m}{\hbar^2}}\int_{x_1=0}^{x_2(E_x)} \sqrt{\Phi - eV\frac{x}{L} - E_x dx}\right]. \tag{6.5}$$

Integration of this equation with $x_2(E_x) = L\,(\Phi - E_x)\,/eV$ gives

$$T(E_x) = \exp\left[-\frac{4}{3}\sqrt{\frac{2m}{\hbar^2}}\frac{L}{eV}(\Phi - E_x)^{3/2}\right]$$
$$\approx \exp\left[-\frac{4}{3}\sqrt{\frac{2m}{\hbar^2}}\frac{L}{eV}\Phi^{3/2} + 2\sqrt{\frac{2m}{\hbar^2}}\frac{L}{eV}\Phi^{1/2}E_x\right]. \tag{6.6}$$

A Taylor expansion around the Fermi level was used to obtain the last approximation, since the tunneling probability decreases rapidly below the Fermi level. The total current is then given by

$$J = e\int dE_x v_x(E_x)T(E_x)\int D(E_x,\vec{k}_\perp)f\left(E_x,\vec{k}_\perp\right)d\vec{k}_\perp. \tag{6.7}$$

For the free electron gas, the last integral at zero temperature is

$$\int D(E_x,\vec{k}_\perp)f\left(E_x,\vec{k}_\perp\right)d\vec{k}_\perp = \frac{2\pi}{\partial E/\partial k_x}\int_0^{\sqrt{\frac{2mE_x}{\hbar^2}}} k_\perp dk_\perp$$
$$= \frac{\pi m}{\hbar}\sqrt{\frac{2mE_x}{\hbar^2}}. \tag{6.8}$$

The current density is then

$$J = e\frac{2\pi m}{\hbar^3}\exp\left(-\frac{4}{3}\sqrt{\frac{2m}{\hbar^2}}\frac{L}{eV}\Phi^{3/2}\right)$$
$$\times\int_{-\infty}^{0} dE_x E_x \exp\left(2\sqrt{\frac{2m}{\hbar^2}}\frac{L}{eV}\Phi^{1/2}E_x\right) \tag{6.9}$$

with the final result

$$J = \frac{e^3V^2}{4L^2\hbar\Phi}e^{-\frac{4}{3}\sqrt{\frac{2m}{\hbar^2}}\frac{L}{eV}\Phi^{3/2}}. \tag{6.10}$$

It is useful to recap the factors that lead to Eq. (6.10):

1. *Tunneling is through a triangular barrier.* The actual shape of the barrier differs from a triangular barrier for many reasons, including the presence of image potentials and the shape of the emitter. As we will see below, this is particularly important for carbon nanotubes because of the small dimensions and high aspect ratio.

2. *The emitter is modeled as a free electron gas.* This obviously neglects the band structure of the emitter, and quantitative comparison with experiments requires the inclusion of these effects. For carbon nanotubes, the electronic structure at the tip is important in determining the field emission properties.
3. *The tunneling barrier height is independent of the applied voltage.* The derivation above assumed that the tunneling barrier height is equal to the metal work function, for any applied voltage. In carbon nanotubes, the barrier height can change by as much as 0.15 eV/(Vμm^{-1}).

Some of these effects are encountered in traditional materials as well, and usually the Fowler–Nordheim model for field emission is generally written as

$$I = aV^2 \exp\left(-\frac{b\Phi^{3/2}}{\beta V}\right) \tag{6.11}$$

where β is the so-called field-enhancement factor according to the relation $F = \beta V/L$ where F is the electric field. This important parameter, which depends on the geometry of the field emission tip, represents the fact that the electric field at the tip can be enhanced from the triangular profile value. In general, large values of β are desirable since it implies that the electric field near the tip is large, and hence electrons can more easily tunnel from the field emission tip. A consequence is that lower threshold voltages are required for field emission; nanotubes, with their high aspect ratio are thus particularly attractive for this reason.

As the discussion of the basic Fowler–Nordheim field emission theory above has indicated, an important question is the validity of the traditional Fowler–Nordheim model to describe field emission from nanotube tips. The unique band structure with van hove singularities in the density of states and localized/quasi-localized states at the tip may necessitate the need for a new model to describe field emission from nanotubes. While both experiments and theory have explored this topic, there is no clear conclusion. However, most experiments show that the *total* field emission current as a function of bias is quite consistent with the Fowler–Nordheim model in single wall nanotubes [7,8], ropes [9,10], films [5,10,11] and patterned films [12]. A field emission tip consisting of a single multiwall nanotube with a radius of 5 nm, attached to a tungsten tip is shown in Fig. 6.3 (a) [8]. Here, one can clearly see the large aspect ratio (length to diameter) achieved experimentally. The current–voltage characteristics are reasonably consistent with the Fowler–Nordheim model over a current window spanning more than two orders of magnitude (Fig. 6.3 (b)). Deviations from the Fowler–Nordheim model for total current have been observed in specific samples because of adsorbates and possibly quantized energy levels in the tip. These topics will be discussed in more detail below.

While the total current can show good agreement with the Fowler–Nordheim model, the total energy distribution of electrons is a more sensitive probe of

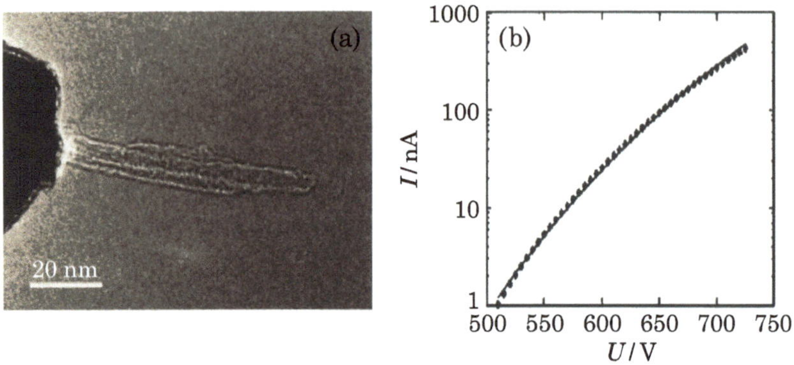

Figure 6.3 (a) Transmission electron microscope image of a nanotube on a tungsten tip. (b) Field emission current as a function of applied bias. Note the good match to the Fowler–Nordheim model Eq. (6.11) (solid line). Figure from Ref. [8].

the electronic structure of the nanotube tip. Experiments and modeling show that the total energy distribution (TED) of the field-emitted current can differ from the Fowler–Nordheim behavior. While experiments on a single nanotube showed reasonable agreement with the Fowler–Nordheim model [8], experiments have shown deviations in the total energy distribution in samples consisting of a bundle of single wall nanotubes [9]. Modeling [13–15] has shown that localized states in the tip cause a significant change in the density of states and the emitted current contains signatures of these states.

The total energy distribution corresponding to the experiment discussed in Fig. 6.3 is shown in Fig. 6.4 (a). The TED reasonably follows the Fowler–Nordheim model over a large energy window. In contrast, field emission from a single wall nanotube bundle shows a total energy distribution that is different from the Fowler–Nordheim model, as compared in panel (b) of the same figure [9]. While the net current agrees with the Fowler–Nordheim model, noticeable deviations are seen in the total energy distribution around the Fermi energy and at energies -0.6 eV and -1.05 eV below the Fermi energy. The deviations around the Fermi energy is attributed to dangling bond states in open nanotubes [14], while the deviations at the other two energies are found to correlate with van Hove singularities in the nanotube density of states.

To gain more insight into the factors that influence field emission from carbon nanotubes, detailed calculations of the properties of field enhancement, potential barriers, and the role of nanotube electronic structure have been performed [16,17]. Fig. 6.5 shows the electric field lines and electric field intensity near the tip of a (5,5) carbon nanotube, obtained by solving Laplace's equation in the presence of the metallic nanotube. The calculations indicate that the electric field is dramatically enhanced near the tip.

What is perhaps more important is that the field enhancement factor β is found to be strongly dependent on the applied field, as shown in Fig. 6.6 (a). One consequence of this effect is that the potential energy barrier through which the

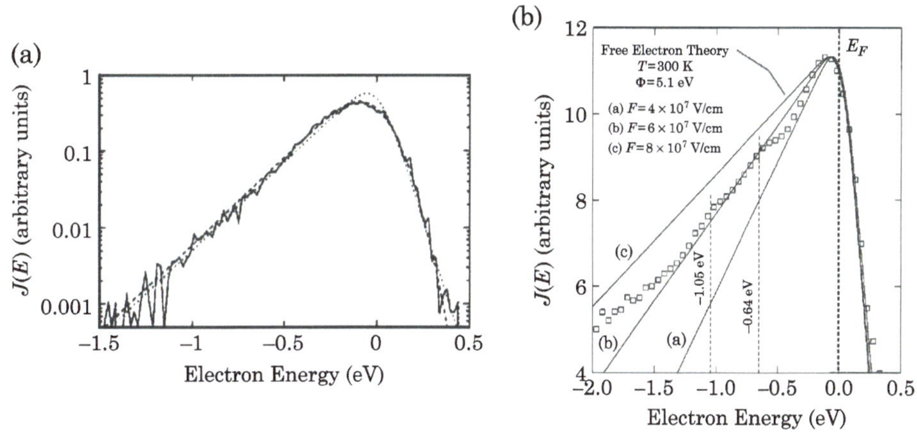

Figure 6.4 (a) Total energy distribution of emitted current corresponding to the single nanotube in Fig. 6.3, at room temperature, applied voltage of 552.8 V, and an emitted current of 11 nA. Figure from Ref. [8]. (b) Log of total energy distribution versus energy for electrons field emitted from a single-wall carbon nanotube rope. The solid lines show the predictions of the Fowler–Nordheim model for three different values of the electric field F. The Fermi energy (E_F) and features representing deviations from the Fowler–Nordheim model at energies of -0.64 and -1.05 eV are marked by dotted lines. Figure from Ref. [9].

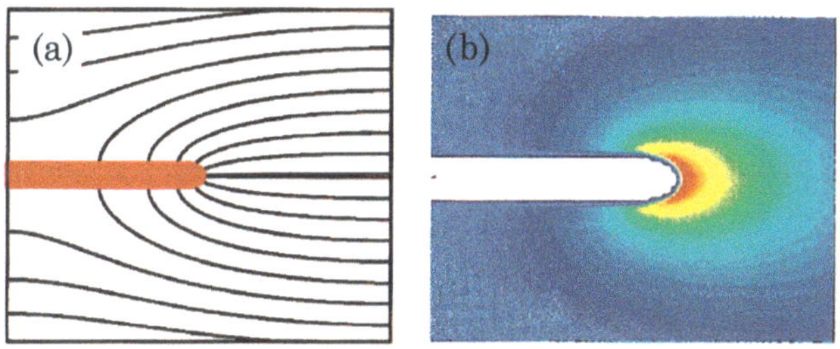

Figure 6.5 Calculated electric field lines and field intensity near the tip of a carbon nanotube. (a) Electric field lines near the tip of a (5,5) nanotube. (b) Field intensity near the tip of a (10,10) nanotube, showing the field enhancement near the tip. Figure from Ref. [16].

electrons must tunnel is dependent on the electric field, and deviates from that of a planar surface (Fig. 6.6 (b)). Because of these effects, the current deviates from the traditional Fowler–Nordheim behavior, and a plot of $\log\left(I/V^2\right)$ vs V^{-1} deviates from a straight line (Fig. 6.6 (c)).

Modeling also provides some detailed information about field emission from individual atomic sites at carbon nanotube tips. In Fig. 6.7, the intensity of electron emission is shown for the atoms at the nanotube tip, as the electric field is increased. Initially, atoms forming a pentagonal ring at the apex start emitting; as the field is increased, more and more atoms contribute to the field emission current. It is interesting to note that there are five other pentagonal rings in the cap of the nanotube, and because these have larger local density

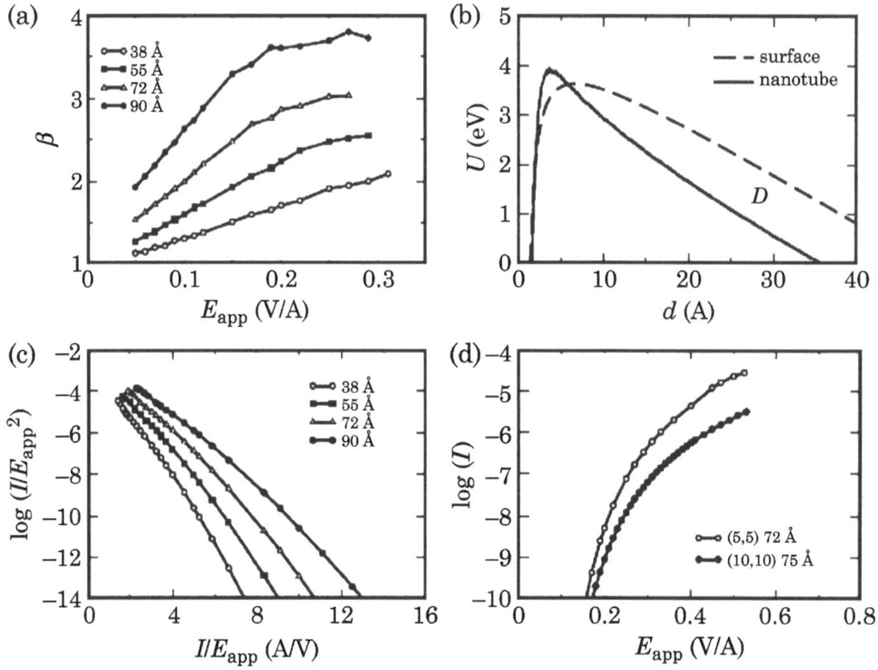

Figure 6.6 Calculated properties of field emission from nanotube tips. (a) shows the field enhancement factor β as a function of the applied field and nanotube length. In (b) the calculated shape of the tunneling barrier for a (5,5) nanotube is compared with that of a planar metallic surface. (c) Fowler–Nordheim plots for (5,5) nanotubes of different lengths. (d) Current versus applied field for (5,5) and (10,10) nanotubes of roughly the same length. Figure from Ref. [16].

of states, it could be expected that they would contribute a significant fraction of the field emission current. While these sites are indeed found to give large field emission at the higher fields (Fig. 6.7 (d)), many other atomic sites also contribute to the emission.

A limitation of the above calculations is that only the cap of the carbon nanotube is considered. However, as the results of Fig. 6.6 indicate, the behavior of the nanotube emission depends on the length of the nanotube. Thus, to make contact with experiments that utilize nanotubes of micron-size length, it is necessary to extend the calculations to much longer nanotubes, while still maintaining an atomistic description of the tip region. Such calculations have been performed [17] by utilizing a hybrid quantum mechanics and molecular mechanics approach, where 8000 atoms near the tip are simulated quantum mechanically, while the rest are treated as electrostatic point charges. This allows for the simulation of a 1 micron-long carbon nanotube in a field emission geometry, as shown in Fig. 6.8. In such a calculation, the induced charge on the nanotube near the tip is obtained by the quantum mechanical calculation, while the charge in the section treated by molecular mechanics is obtained

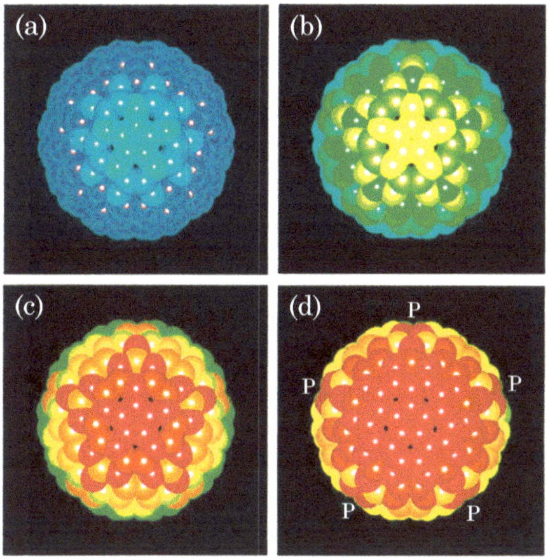

Figure 6.7 Color plots of electron emission from the tip of a (10,10) nanotube (violet = low emission intensity, red = high emission intensity). The electric field increases from panel (a) to (d) with values of 0.2, 0.3, 0.4 and 0.45 V/A. At low fields, only atoms near the apex of the cap emit; as the field is increased, more and more atoms begin to emit. Figure from [16].

as follows: for a metallic carbon nanotube, the density of states is constant around the Fermi level, and a deviation of the electrostatic potential $\delta V(z)$ will induce a charge $\rho(z) \propto \delta V(z)$. Thus, the Poisson equation will be of the form $\nabla^2 \delta V(z) \propto \rho(z)$ or $\nabla^2 \rho(z) \propto \rho(z)$ and the induced charge is of the form

$$\rho(z) \sim \rho(L') \exp\left[-\left(L' - z\right)/\lambda\right] \tag{6.12}$$

where λ is the decay length and $\rho(L')$ is the induced charge at the interface between the quantum mechanical and molecular mechanics regions. The boundary condition on the cathode is satisfied by considering image charges, as shown in Fig. 6.8.

Results of such calculations are shown in Fig. 6.9. First, the induced charge is found to be only a few percent of the total charge at the tip, and is found to oscillate along the length of the nanotube. Most importantly, the induced charge is found to be concentrated near the tip, with a decay length of about 0.5 μm (inset Fig. 6.9 (c)), indicating that simulation of the entire length of the nanotube is necessary to capture all of the electronic charge contributions.

Perhaps the most surprising result from these calculations is that the electric field is found to penetrate the nanotube near the tip region, as shown in Fig. 6.10 (a). Because of this effect, the field enhancement factor is found to be as much as a factor of three larger than what would be predicted without field penetration. Panel (b) in this figure shows a close-up of the electrostatic

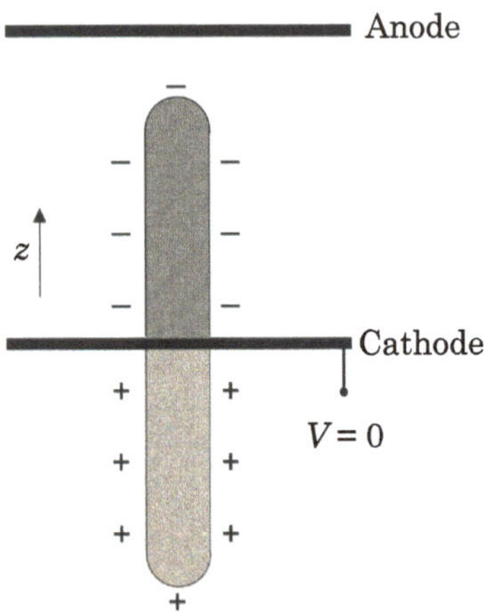

Figure 6.8 Sketch of the system used for calculating the field emission properties of micrometer long carbon nanotubes. After Ref. [17].

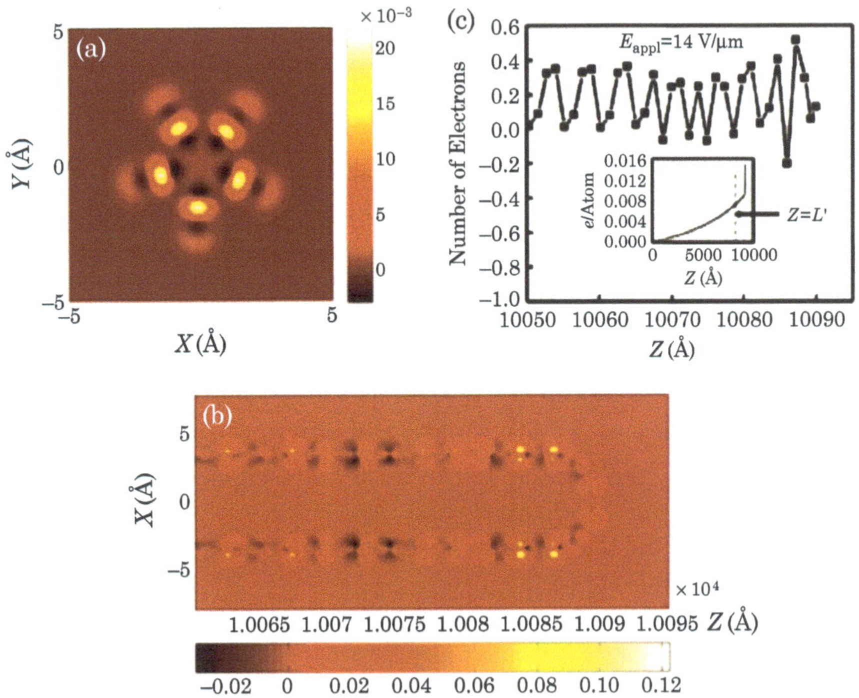

Figure 6.9 (a) Induced charge distribution at the top layer of the nanotube cap. (b) Induced charge distribution in a plane bisecting the nanotube. (c) Number of induced electrons on each layer near the tip region. Excess charge along the entire nanotube is plotted in the inset. Figure from Ref. [17].

potential near the tip, with points marking the Fermi level (−4.5 eV). The Fermi level is below the potential barrier, indicating that the electrons have to

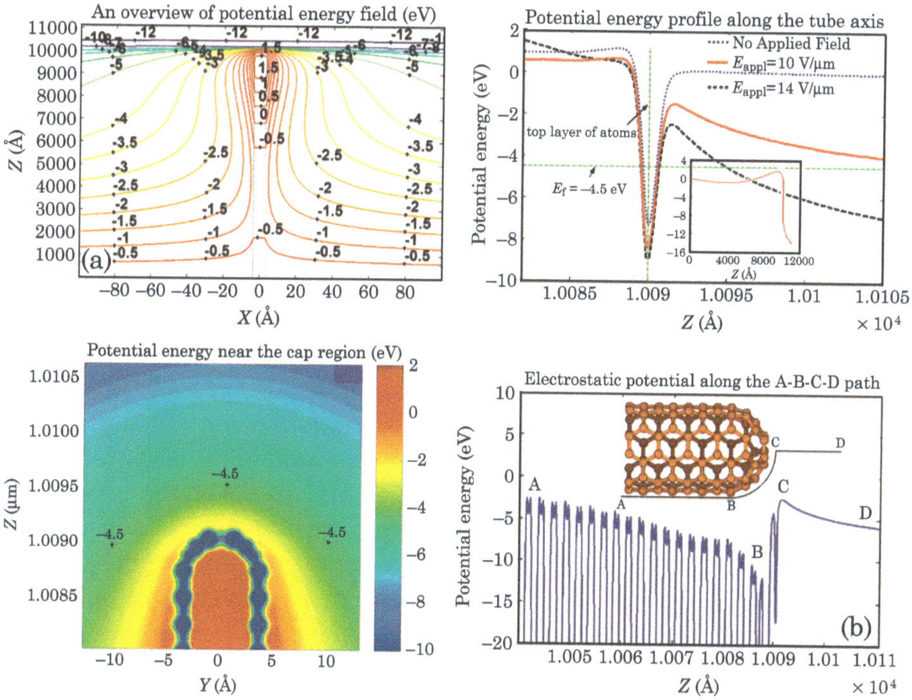

Figure 6.10 Left: potential energy contour plots for a (5,5) nanotube for a field of 14 V μm^{-1}. The axis of the nanotube is at $x = 0$. Right: potential energy profiles near the nanotube cap. Figures from Ref. [17].

tunnel through a barrier. It is also found that the barriers are thicker and higher on the side of the nanotube, indicating that preferential emission from the tip is expected. A closer look at the potential energy profile near the nanotube tip shows that the barrier height depends significantly on the applied field. Indeed, it is found that the barrier height is reduced from 4.5 eV to 2 eV under a field of 14 V μm^{-1}.

The predictions of localized electrical fields at the tip of carbon nanotubes during field emission have been verified using electron holography experiments [18]. In this technique, illustrated in Fig. 6.11, a multiwall carbon nanotube is positioned 6 μm away from a gold electrode, and a high bias voltage is applied between the two to induce field emission. An electron beam perpendicular to the field-emitted electron beam comes from a transmission electron microscope, and is split by a biprism after traversing the nanotube region; the two halves of the beam converge onto an image plane where the hologram is recorded. Interference fringes provide information on the relative phase between the two halves of the beam, which is related to the local electrostatic potentials that the beam traverses. The phase shift acquired by the electrons as they traverse a spatially-dependent potential V is given by the expression

$$\Delta\phi = \alpha \int_{\text{beam path}} V dl \tag{6.13}$$

where the integral is over the path traversed by the electron beam and α is a parameter that depends on details of the transmission electron microscope.

Fig. 6.11 shows an electron hologram image taken from a single carbon nanotube during field emission. The main diffraction pattern comes from the biprism and is not related to the phase change; a more detailed view of the region around the nanotube shows that additional fringes are present with periodicity of about 4 nm, and these carry the information about the electric fields around the nanotube tip. When the phase difference is extracted from these images, one can clearly see the carbon nanotube in the phase image (see top right panels of Fig. 6.11, taken at zero bias voltage), indicating that the electron beam is going through a difference of integrated potential when passing through the nanotube. In the presence of an applied bias between the nanotube and the gold electrode, a spatially varying potential is created, causing a spatially-dependent phase shift, as shown in the right panels of Fig. 6.11 for 70 V and 120 V. In these images, stripes are observed because the phase shift is plotted as modulo 2π; from these, the phase gradient can be calculated and provides smooth images of the electric field in the nanotube region. These phase gradient images clearly indicate that electric fields are concentrated near the tip of the nanotube and not at nanotube defects on the sidewall.

These experiments also provide evidence that current fluctuations during field emission are not caused by large changes in the electric field distribution around the nanotube, but are most probably due to small fluctuations of the local electronic structure, as can arise due to adsorbates. During the electron holography experiments, the field emission current is observed to fluctuate by as much as 80%, but the electron hologram fringe patterns are observed to remain sharp. Over the acquisition time of several seconds, these patterns would appear blurry even for small fluctuations of the electric field. Thus, the fluctuations in the electric field amount to only a few percent, much too small to explain the fluctuations in the field-emitted current.

A measure often used to quantify the quality of emission sources is the reduced brightness B_r defined as

$$B_r = \frac{dI}{d\Omega}\frac{1}{V}\frac{1}{\pi r_v^2} \tag{6.14}$$

where I is the emission current, Ω is the solid angle, V the acceleration voltage and r_v the virtual source size. The virtual source size is the area at the tip from which the emission appears to originate when the trajectories of electrons are traced back [19]; its value can be obtained by measuring the diffraction pattern

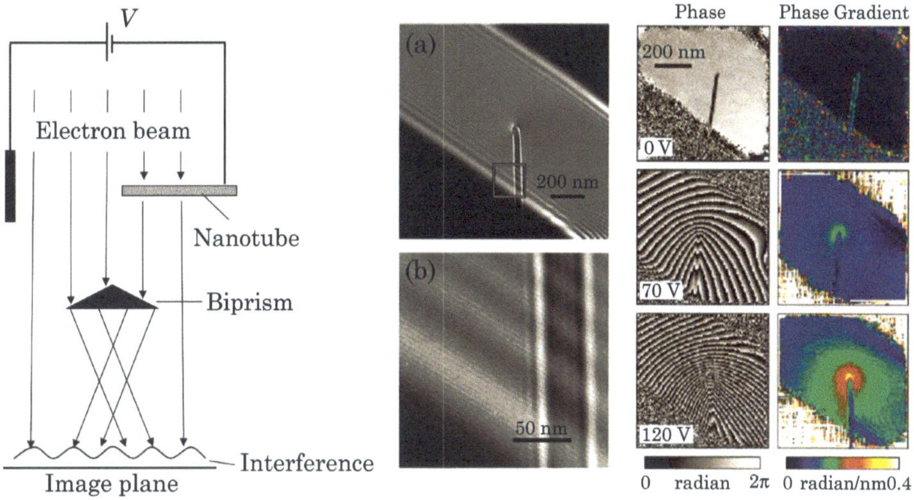

Figure 6.11 Left: schematic of electron holography experiment on a field-emitting carbon nanotube. Middle: electron hologram of a carbon nanotube biased during field emission at 120 V (a) and a more detailed view of the boxed area (b). Right: phase shift and phase gradient from electron hologram experiments at bias voltages of 0, 70 and 120 V. Middle and right figures from Ref. [18].

caused when the electron beam impinges on a sharp edge, as illustrated in Fig. 6.12. In such an experiment, a diffraction pattern consisting of several maxima is observed at a distance z_2 from the sharp edge. Quantitatively, the position of the maxima is given by the expression

$$x(N) - x(0) = z_2\sqrt{\frac{2\lambda}{z_1}}\left(\sqrt{n+\frac{3}{8}} - \sqrt{\frac{3}{8}}\right) \tag{6.15}$$

where

$$x(0) = z_2\sqrt{\frac{2\lambda}{z_1}\frac{3}{8}} \tag{6.16}$$

is the position of the first maximum and λ is the electron wavelength. The virtual source size can be obtained from these expressions [20] as

$$r_v = \frac{\lambda z_2}{\pi x(N)}. \tag{6.17}$$

Fig. 6.12 shows multi-wall carbon nanotube attached to a tungsten tip and the resulting Fresnel oscillations measured due to electron emission from this source near a sharp edge. As shown in the inset of the right panel, up to eight maxima are visible, so that counting from zero we have $x(N) = x(7) = 4.0$ mm;

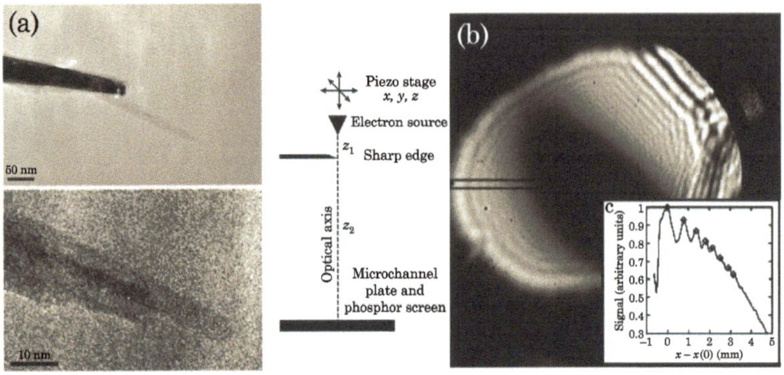

Figure 6.12 (a) Transmission electron microscopy images of a multiwall carbon nanotube at the end of a tungsten tip. Experimental setup to measure the virtual source size of the carbon nanotube emitter. The carbon nanotube is positioned close to a sharp edge, causing a Fresnel diffraction pattern on a screen. (b) The interference pattern from the nanotube on the left, including a line scan (averaged over the rectangle in the main panel) showing the first few maxima. Figure from Ref. [20].

with the value $z_2 = 16$ cm this gives a virtual source size $r_v = 2.1$ nm. This value is somewhat less than the 2.7 nm radius of curvature of this particular carbon nanotube.

To obtain the reduced brightness, the value of $dI/d\Omega = 16\ \mu\text{A sr}^{-1}$ was measured using a Faraday cup for a similar carbon nanotube of $r_v = 2.5$ nm and an acceleration voltage $V = 319$ V; this gives a reduced brightness $B_r \sim 3\times10^9\text{A sr}^{-1}\ \text{m}^{-2}\ \text{V}^{-1}$. This value can be compared [20] with that of traditional Schottky emitters of cold field emission guns, which have reduced brightness values less than $2 \times 10^8\text{Asr}^{-1}\text{m}^{-2}\text{V}^{-1}$. The nanotube emitter therefore has a reduced brightness at least an order of magnitude larger than these conventional sources.

6.2 Adsorbates

Experiments have shown a significant deviation from the Fowler–Nordheim model in the presence of adsorbates. Adsorbates can have many effects, such as changing the work function for emission, the width of the tunneling barrier or the density of electronic states. Furthermore, because large electric fields are applied to the tips during field emission, adsorbates can change their location on the surface or even desorb. These effects can lead to unusual features in the field-emitted current. To probe these effects, controlled experiments have been performed [21], where the field emission properties of clean single wall carbon nanotubes is compared to the field emission in the presence of intentionally introduced adsorbates. These adsorbates are introduced by exposing the nanotube tip to H_20 under a partial pressure of 10^{-7} Torr for five

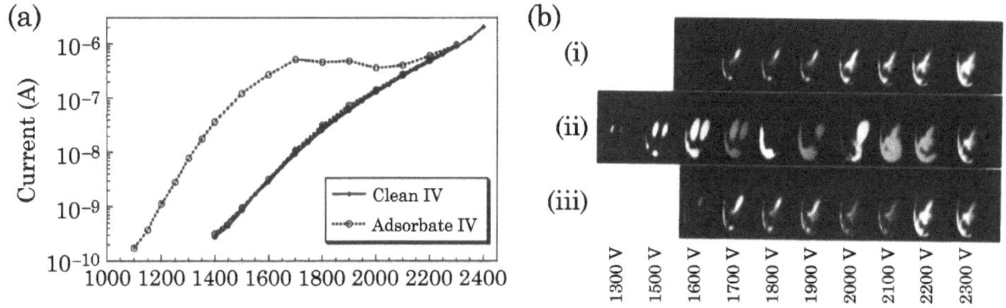

Figure 6.13 (a) Current–voltage characteristic of a single nanotube with and without adsorbates. (b) The field emission pattern of a clean nanotube is stable over the voltage sweep (i) but the pattern of the same nanotube with an adsorbate changes during the current–voltage sweep concurrent with the onset of current saturation (ii). At 2300 V, the effects of the adsorbates disappear, resulting in a clean nanotube field emission pattern. During the downward I–V sweep, the images (iii) match the clean nanotube shown in (i). Figure from Ref. [21].

minutes. As can be seen in Fig. 6.13 there is a qualitative difference between the field emission of the clean and adsorbate-covered nanotube tips. For clean nanotubes, the $I-V$ curve follows the Fowler–Nordheim model; in contrast, the field emission in the presence of adsorbates shows an enhanced current at the lower voltages, followed by a partial saturation and finally a current increase at higher voltages that follows the clean nanotube emission. A downward sweep after such large applied voltages shows current–voltage characteristics of a clean sample. Heating the tip to 900 K under field emission conditions leads to regular field emission behavior without current saturation because of adsorbate desorbtion. These observations are explained by the adsorbate-induced increase in the field emission current at the lower voltages. At intermediate voltages, the adsorbates are displaced into configurations that decrease the tunneling current. Finally, at high enough voltages, the adsorbates desorb from the nanotube tip—the high current behavior and the downward sweep thus follow that of the clean nanotube. This remarkable experiment also showed that adsorbates change the shape of the field emission pattern (Fig. 6.13 (b)). The field emission pattern in a clean sample is shown in panel (i). Introducing water causes a significant deviation in the pattern as shown in panel (ii). The reverse sweep after applying a voltage of 2300 V matches the pattern with a clean tip due to desorbtion of adsorbate as shown in panel (iii).

To further test the hypothesis that adsorbates lead to the behavior reported in Fig. 6.13, the current was measured as a function of time for stepwise increases and decreases in the applied voltage. As shown in Fig. 6.14, when the applied voltage is increased to above the current saturation in Fig. 6.13, the current decreases between the steps, an indication that adsorbates are being re-arranged or removed at the nanotube tip. For the downward sweep, the current is found to increase between steps as adsorbates re-attach and re-occupy configurations with higher tunneling rates.

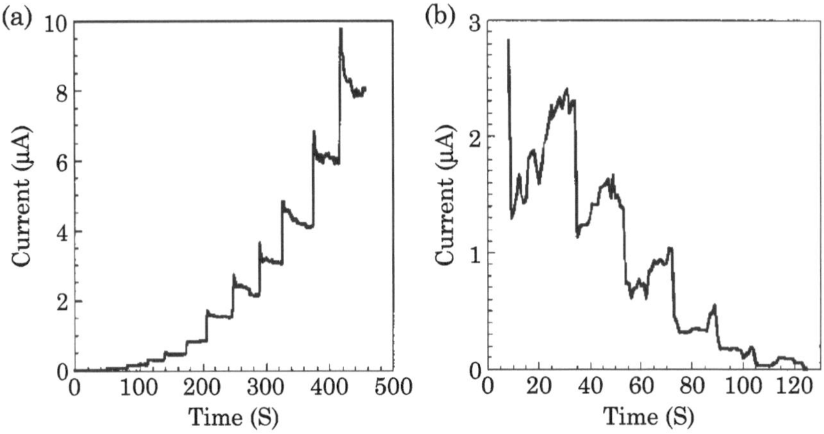

Figure 6.14 Time-dependence of field emission current when the applied voltage is changed in a stepwise fashion. In (a), the applied voltage is increased to above that of the saturation in Fig. 6.13, and the current decreases as adsorbates re-arrange and desorb from the nanotube tip. In (b), the applied voltage is decreased, and the current increases between steps as adsorbates re-attach to the nanotube tip. Figure from Ref. [21].

6.3 Nanotube Arrays

We have so far discussed field emission from a single nanotube or a single bundle. Applications in display devices will however involve carbon nanotube films or arrays because of the need for field emission over large areas. To be competitive with conventional field emitters, the nanotube films should exhibit uniformity with an areal density larger than 10^6 emitters/cm^2 and current densities of 80 μA/cm^2 [22]. However, in exploring high-density nanotube arrays for field emission, it is observed that the emission is not better for high densities, but in fact is maximized at intermediate densities [22]. To probe the origin of this behavior, a scanning field emission technique with a Pt-Ir anode with a tip radius of 2–5 μm was employed to obtain a spatially resolved field emission image from nanotube films of different densities patterned as a grid. Fig. 6.15 shows the printed pattern and higher magnification scanning electron microscope images of patterned nanotube films of low, medium and high areal densities, and the associated scanning field emission scans. At low densities, the emission is inhomogeneous and comes mostly from localized regions in the film, presumably where there are nanotubes with larger length-to-diameter ratios (and thus enhanced β factors). At high densities, some of the patterned lines are visible, but a clear pattern is not obtained. The sample with medium density provides the best image of the emission pattern: the lines, crosslines, and dashes can be seen. The improved emission properties at intermediate densities originate from a combination of two factors. At low densities the pattern has very few efficient emitters while the enhancement factor in the high density pattern is reduced because of electrostatic screening

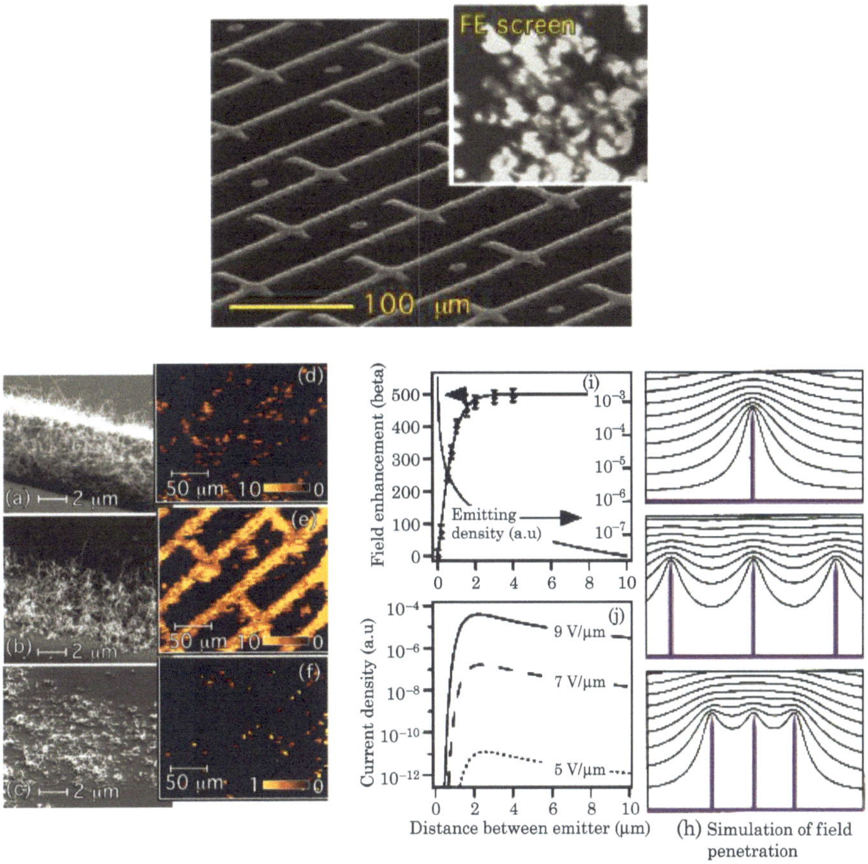

Figure 6.15 The top image shows a low magnification scanning electron microscope image of a carbon nanotube film patterned in the form of lines. Inset is the macroscopic field emission pattern captured on a phosphor screen. The left-most panels show higher magnification scanning electron microscope images of the patterned carbon nanotube films with (a) high, (b) medium, and (c) low density, with the corresponding field emission maps of current density (d)–(f). The scale corresponds to 0–10 μA/pixel for images (d) and (e); and to 0–1 μA/pixel in image (f). (h) Simulation of equipotential lines of the electrostatic field for tubes of 1 μm height and 2 nm radius, for distances between tubes of 4, 1, and 0.5 μm, (i) field enhancement factor and emitting density, and (j) current density, as a function of the distance between nanotube emitters. Figures from Ref. [22].

between neighboring nanotubes. To verify this last hypothesis, simulations of the equipotential lines in an array of carbon nanotubes of different densities have been performed [22]. The calculations indicate that the field penetration diminishes as nanotubes are packed more densely and this affects both β and the total emitted current. Fig. 6.15 shows the field penetration in an array consisting of nanotubes of length 1 μm and radius 2 nm. Clearly the penetration is poor when the spacing between nanotubes is 0.5 μm (bottom) as opposed to 4 μm (top). The corresponding field enhancement factor decreases rapidly as

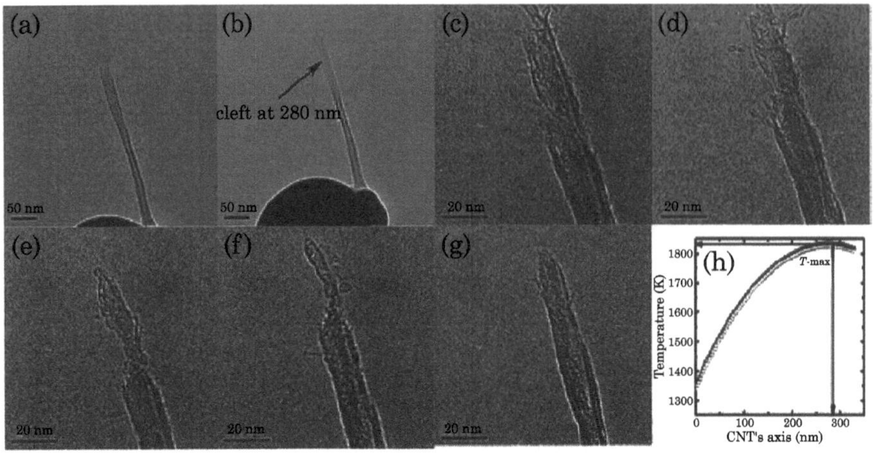

Figure 6.16 Transmission electron microscopy images of the failure process of a carbon nanotube tip during field emission ((a)–(g)) and the predicted temperature profile along the nanotube (h). Figure from Ref. [23].

the nanotube separation is decreased below 2 μm, and the simulated current density peaks at nanotube spacings of 2 μm. This maximum in the current density is a consequence of the interplay between the areal density and field amplification factor β in determining the total emitted current.

6.4 Failure Mechanism

Because of the high electric fields required for field emission, materials for field emission tips must be able to withstand the high temperatures generated by Joule heating and the high stress generated by the electric field. These harsh conditions can lead to degradation of the device performance, and ultimately breakdown. For carbon nanotube tips, it was originally proposed that the maximum temperature reached along the nanotube is at the tip [24], implying that breakdown of the nanotube should occur in a fairly continuous fashion by shortening at the tip. However, experimental work has demonstrated that breakdown of nanotube tips during field emission occurs a few tens of nanometers away from the tip [23]. An example of such an observation is shown in Fig. 6.16. There, a multiwall carbon nanotube attached to a tungsten tip is imaged using a transmission electron microscope, with several images taken in sequence as the field emission voltage is increased. The images indicate that for this nanotube (diameter of 14 nm and length of 330 nm) the initial damage occurs about 50 nm from the tip in the form of a cleave; further increase of the applied voltage leads to further damage in this area eventually causing the nanotube end to be burned off. (It should be noted that a different breakdown mechanism, where the nanotube detaches from the tungsten tip is

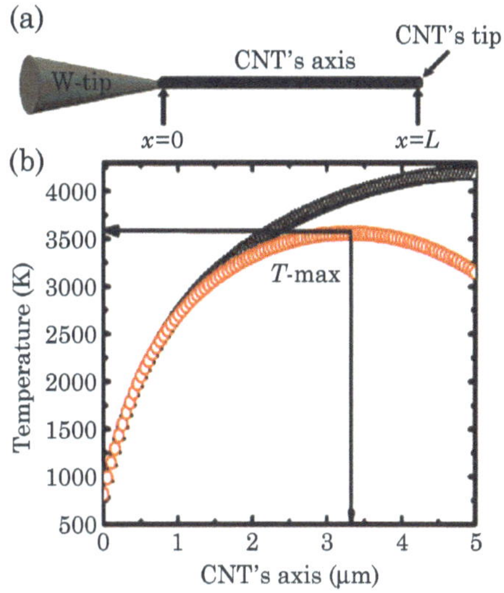

Figure 6.17 (a) Schematic of the field emission tip, consisting of a carbon nanotube of length L attached to a tungsten tip. (b) Calculated temperature distribution along the carbon nanotube in the presence (bottom) and absence (top) of heat loss due to electron emission. Figure from Ref. [23].

also observed for weaker bonds between the nanotube and the tungsten [25].) Similar observations of breakdown away from the tip has been observed in other nanotube samples [23,26]

To understand the origin of the nanotube breakdown away from the tip end, a theoretical model based on Joule heating has been proposed [23]. In this model, a nanotube of length L is attached to a tungsten tip at one end, as illustrated in Fig. 6.17. The temperature profile along the nanotube is determined by the heat conduction equation

$$ms\frac{\partial T}{\partial t} = \kappa v \nabla^2 T + \frac{\dot{d}Q}{dt} \tag{6.18}$$

where T is the spatially-dependent temperature, κ is the thermal conductivity, ν is the sample volume, m is the mass, s is the specific heat and Q is the heat. The heat generated and dissipated in the nanotube comes from two sources: the electrical resistance (Joule heating) and heat radiation. These can be written as

$$\begin{aligned} \frac{dQ_{\text{elec}}}{dt} &= I^2 \rho v \\ \frac{dQ_{\text{rad}}}{dt} &= -\sigma A \left(T^4 - T_0^4\right) \end{aligned} \tag{6.19}$$

where the first relation is simply Joule's law with current I and resistivity ρ, and the second equation is the Stefan–Boltzmann law for radiation, with

σ the Stefan–Boltzmann constant for a body of surface area A emitting into surroundings at temperature T_0.

For one-dimensional heat transport in the steady-state we have

$$\pi R^2 \kappa \frac{\partial^2 T}{\partial x^2} - 2\pi R \sigma \left(T^4 - T_0^4\right) + \frac{I^2 \rho}{\pi R^2} = 0. \tag{6.20}$$

To solve this equation, an expression for the temperature dependence for the resistivity is needed, as well as boundary conditions. An empirical, nonlinear expression for the dependence of the resistivity on temperature has been proposed [27] to reproduce experimental data of resistivity at high temperatures [28,29]:

$$\rho(T) = \rho_0 \left(1 - \alpha T + \beta T^{3/2}\right) \tag{6.21}$$

where ρ_0 is the resistivity at room temperature and α, β are fitting coefficients. The boundary conditions at the interface with the microtip and at the free end are determined as follows. At the nanotube/microtip interface, there is a temperature drop due to the thermal contact resistance equal to

$$\frac{T_0 - T(x=0)}{\lambda} = Q = -\kappa \pi R^2 \frac{\partial T}{\partial x} \tag{6.22}$$

with the parameter λ representing the quality of the thermal contact, and where we have assumed that the microtip is at temperature T_0. This provides an expression for the temperature at the microtip end of the nanotube

$$T(x=0) = \lambda \pi R^2 \kappa \left.\frac{\partial T}{\partial x}\right|_{x=0} + T_0. \tag{6.23}$$

The interesting physics that differentiates carbon nanotubes from conventional field emitters arises at the free end. When an electron is emitted at the free end, it carries energy, and this effectively cools the tip. In conventional metals, cooling due to field emission is typically much smaller than radiative cooling—in carbon nanotubes however, the field emission current density is very large while the surface area for radiative cooling is small, making field emission cooling a dominant factor. To take into account this effect, the heat loss due to the energy carried away by the field-emitted electrons is assumed to occur at the free end of the nanotube, assuming that each electron carries energy $(3/2)\,kT_L$ with T_L the temperature at $x = L$. From Fourier's heat conduction relation one obtains the boundary condition

$$\left.\frac{\partial T}{\partial x}\right|_{x=L} = -\sigma \kappa^{-1} \left(T_L^4 - T_0^4\right) - \frac{3}{2} \kappa^{-1} \frac{k T_L I}{e \pi R^2}. \tag{6.24}$$

With the boundary conditions provided by Eqs. (6.23) and (6.24), and the relationship between the resistivity and temperature provided by Eq. (6.21), the differential equation for the temperature distribution (6.20) can be solved to obtain the spatial dependence of the temperature in the carbon nanotube. Fig. 6.17 shows the calculated temperature distribution for parameters extracted from experiment [23]. In the absence of the tip cooling effect (black triangles) the maximum temperature is reached at the tip of the nanotube. In contrast, in the presence of tip cooling, the maximum temperature is reached significantly away from the tip, with a much reduced maximum temperature. Since it is expected that failure will occur at the point of maximum temperature, this model predicts that carbon nanotube failure during field emission will occur some distance away from the tip end. In fact, application of the above model to the nanotube of Fig. 6.16 predicts a maximum temperature at a point 50 nm from the tip end, in very good agreement with the experiment.

Finally two points should be noted: first, the maximum temperature that is reached along the length of the nanotube decreases with a decrease of the nanotube length, with or without the tip cooling effect. This serves as a mechanism to stabilize the nanotube and prevent additional failure. Second, Joule heating is but one mechanism that can lead to breakdown during field emission (we have already mentioned breakdown at the nanotube/microtip interface above). In addition to Joule heating, there is a large electrostatic force on the carbon atoms during field emission which can lead to large axial and radial stresses. The total breakdown mechanism is probably due to a combination of heating and field effects.

6.5 Devices

Field Emission Displays: displays using the superior field emission properties of carbon nanotubes have been demonstrated by a number of research groups. Compared to the emissive displays based on microfabricated tips [30], carbon nanotube cathodes offer the potential for improved performance (i.e. lower drive voltage, longer lifetime, and a reduced fabrication cost). Fig. 6.18 shows a multicolor display demonstrated in reference [3]. The display consists of carbon nanotube tips on a patterned metal substrate, with the carbon nanotube bundles firmly attached to the metal electrode. The nanotubes are mixed with an organic nitrocellulose to form a paste and then squeezed through a wire mesh to force alignment of nanotubes perpendicular to the metal. The density of nanotubes obtained is about 5×10^6 to 10^7 cm^{-2}, which is about one hundred times larger than the density of microtips in conventional spindt type field emission displays. Since the initial work of reference [3], nanotube-based displays have become an important area of research and development, and Samsung demonstrated a 38-inch color television in 2003.

X-ray tubes & spectrometers: x-ray sources have many medical and industrial applications, in addition to their use as a spectroscopy tool in materials science.

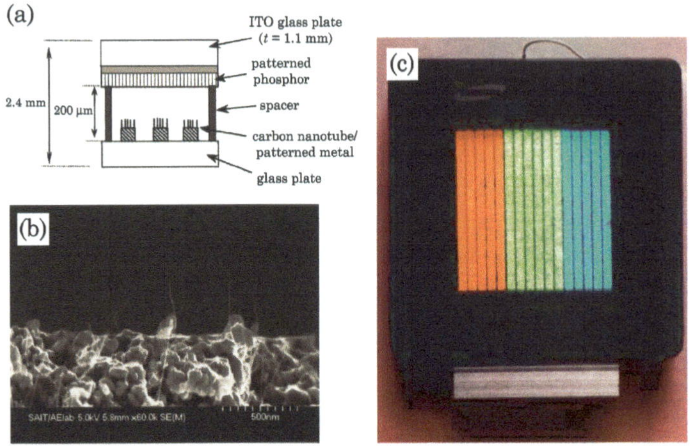

Figure 6.18 (a) Picture representing the setup of a field emission display. Carbon nanotubes are deposited on a patterned metal substrate. Field emitted electrons hit the phosphor screen and cause light emission in a color that depends on the chemical compound on the phosphor screen. (b) Scanning electron microscope image of nanotube bundles projecting from the metal electrode. (c) A sealed carbon nanotube field emission display emitting light in three different colors. The dimension of the display is 4.5 inches. Figure from Ref. [3].

In conventional x-rays tubes, thermionically emitted electrons from a heated metal filament are accelerated and then strike a metal target, which releases x-rays. The high power consumption and small response times of thermionic sources makes x-ray generation using cold field emission attractive. Also, the high temperatures of thermionic sources place limitations on both the size and lifetime of x-ray tubes. Ultra-sharp metal field emission tips obtained by micromachining have suffered from problems related to mechanical and thermal stability due to arcing and cation sputtering [1,31]. Like in field emission displays, the advantage of cold field emission from carbon nanotubes offers a potential to build x-ray tubes that are smaller, portable, use low power, and have long lifetime. Carbon nanotubes overcome many problems associated with metal field emission tips because they are mechanically strong and can withstand higher temperatures.

Many applications of x-ray tubes require currents in the range of 50–500 mA/cm^2. Obtaining such high and stable current densities over macroscopic areas using nanotubes has been challenging. However, recently, current densities as large as 1 A/cm^2 have been obtained [31,32], though their stability over long periods of operation has not been determined. Carbon nanotube x-ray tubes have been demonstrated to be capable of imaging circuit boards [1] and human hands [31]. The basic working principles of these x-ray imaging devices is fairly simple: a high-energy electron beam strikes a metal target, releasing x-rays. A schematic of such a device using a film of carbon nanotubes as the electron emission source is shown in Fig. 6.19. There,

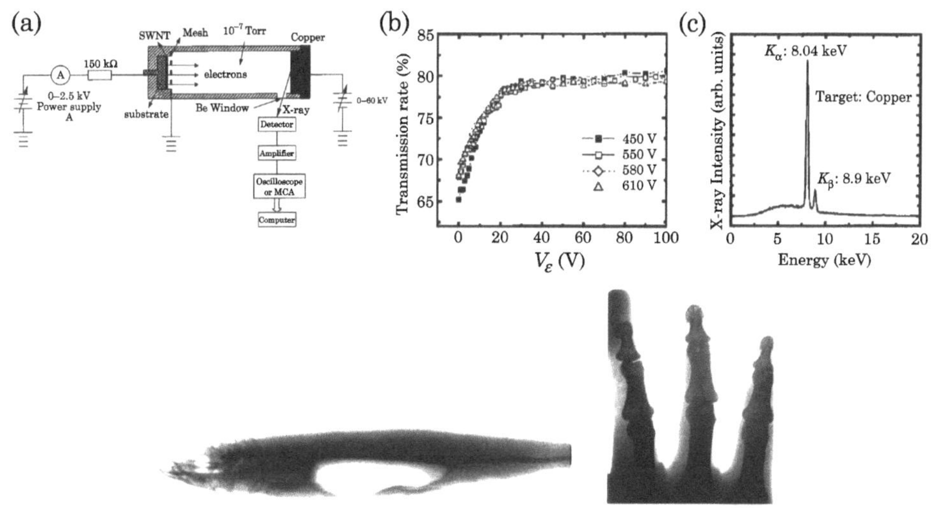

Figure 6.19 (a) Schematic of an x-ray tube using a triode with carbon nanotube field emission. The gate electrode is a metal mesh placed between 50 and 200 μm from the nanotube film. x-rays are produced when the accelerated electron beam from the nanotube film impinges on a copper target. (b) Fraction of field emission current detected at the anode. (c) Energy distribution of the x-ray produced from the copper target, showing the K_α and K_β lines. Bottom: x-ray images of a fish and a human hand. Figures from Ref. [31].

a metal mesh is positioned at a distance of tens to hundreds of microns from the carbon nanotube film, which consists mostly of bundles of single wall carbon nanotubes. A high voltage applied between the metal mesh and the nanotube film causes electron emission; these electrons impinge on a copper target. When the incoming electrons have sufficiently high energy, they can knock out core electrons in the metal atoms, and as higher energy electrons transition to this lower energy level, photons are emitted with energy typically in the keV range. To realize an x-ray source from cold cathode field emission in a triode geometry, it is important that most of the field emission current passes through the metal mesh and impinges on the metal target, so as to prevent excessive heating of the mesh electrode. For the carbon nanotube device in Fig. 6.19, optimization of the mesh density and separation from the nanotube film leads to a high transmission of 80% through the mesh. At an acceleration voltage of 14 kV, these transmitted electrons generate x-ray lines of 8.04 keV and 8.9 keV, corresponding to the K_α and K_β lines of copper.

More recently, a compact multibeam nanotube x-ray source capable of three-dimensional imaging (two dimensional frames at various angles) has been demonstrated [2]. The operation voltage of these x-ray tubes is in the 15–50 kV range, which is comparable to conventional sources. Another recent demonstration has been x-ray tubes for spectroscopy applications using a carbon nanotube field emission source (Fig. 6.20). This x-ray tube uses only 1.5–3 W, and operates on batteries. Using such a small x-ray tube, a field

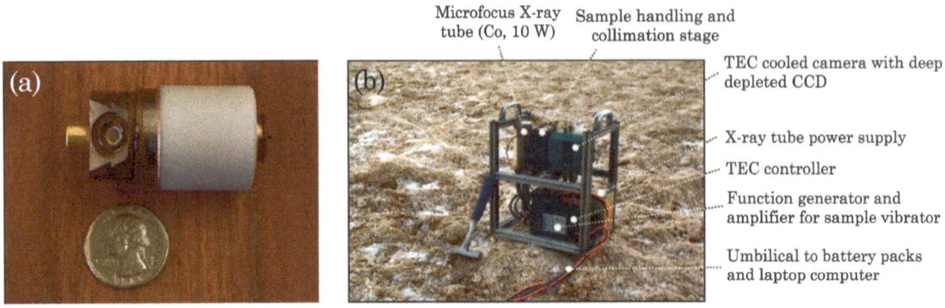

Figure 6.20 (a) A miniature field emission x-ray tube made of a 2 mm diameter multiwall nanotube cathode. Figure from Ref. [32]. (b) CheMin, a portable x-ray spectrometer, with a geologist's hammer for scale. Figure from Ref. [33].

deployable x-ray spectrometer called CheMin that weighs only 15 kg has been built [32]. CheMin has been chosen to be a science instrument to fly in the Mars '09 Science Laboratory mission, and will perform mineralogy experiments on the surface of Mars (Fig. 6.20).

Field emission lamps: Conversion of energy into lighting is a significant portion of the world's energy consumption, with up to 8% of the total energy and 22% of electricity converted into lighting in the U.S. [34]. Currently, 42% of the lighting energy in the U.S. is used by incandescent light bulbs, which are extremely inefficient since most of the energy is wasted as heat. Alternatives to this centuries-old technology are actively sought, and field emission lamps are possible candidates—a sketch of a possible device is shown in Fig. 6.21. There, a cylindrical cathode covered with field emission tips isotropically emits electrons that are captured by a phosphor layer on the cylindrical anode. For this kind of uniform, nondirected illumination, isotropic field emission from a cylindrical surface is needed. Difficulties in depositing or fabricating a high density of field-emitters on a cylindrical surface have prevented the development of such an approach using conventional materials. Carbon nanotubes may provide a solution to this problem because it is possible to cover a metallic cylindrical rod homogeneously with a catalyst and grow nanotubes catalytically. Such an approach has been demonstrated using a Fe–Al–Cr alloy on which a liquid Fe catalyst is applied, and the nanotubes are grown by the catalytic decomposition of acetylene [35]. This results in a metallic cylindrical rod of 1 mm diameter and 7 cm in length homogeneously covered with a tangle of multi-wall carbon nanotubes. The field emission lamp is fabricated by positioning the cylindrical rod in the center of a cylindrical aluminum anode of 21 mm radius and 5 cm length, in a vacuum chamber. Field emission is induced by applying a large voltage between the cylindrical rod and the cylindrical anode, giving a very large current density of 1 mA/cm^2 at a voltage of 1.1 kV. This large current density is due to the large electrical field at the cathode in the cylindrical geometry [35]. Indeed, in the cylindrical geometry, the electric field at the

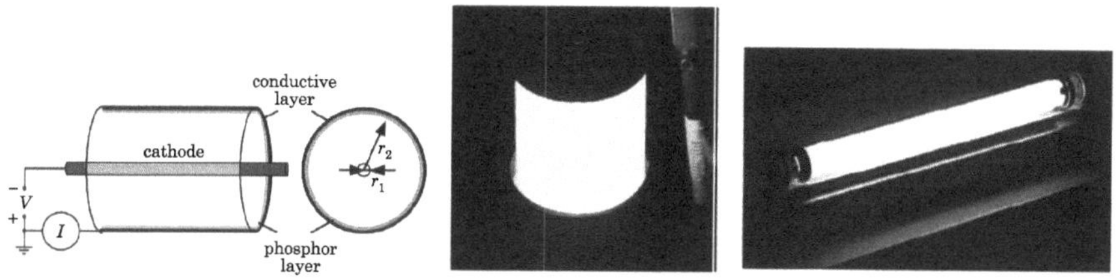

Figure 6.21 Left: sketch of a field emission lamp using a cylindrical cathode covered with carbon nanotubes. Electrons emitted from the cathode strike a phosphor layer on the anode to cause illumination. Middle and right: prototypes of luminescent tubes using cylindrical carbon nanotube cathodes. The device on the right is fully sealed. Figures from Refs [35] and [19].

surface of the cathode is

$$E_{\text{cyl}} = -\frac{V}{R_C \ln\left(R_A/R_C\right)} \tag{6.25}$$

where R_C is the radius of the cathode and R_A is the radius of the anode. This can be compared with the expression $E_{\text{planar}} = -V/d$ for two planar electrodes separated by distance d. The ratio $E_{\text{cyl}}/E_{\text{planar}} \approx \left(R_A/R_C\right)/\ln\left(R_A/R_C\right)$ can be quite large: for the field emission lamp of reference [35] the ratio is about 11, and when combined with the exponential dependence of field emission, can lead to much larger emitted currents for similar physical dimensions.

Fig. 6.21 shows a field emission lamp fabricated using a cylindrical cathode covered with carbon nanotubes. The anode is a glass tube coated with ITO and a phosphor layer on the inside surface. The luminescence of this lamp is 10 000 cd/m^2, comparable to a commercial fluorescent tube. (The power consumption is however much higher, mostly due to the inefficient phosphor utilized in this prototype device.) The carbon nanotube field emission lamp has many advantages over conventional fluorescent lamps: it is better for the environment because it contains no mercury, it starts up instantly, and the light intensity can be easily varied by controlling the field emission voltage. Further technological progress in this area has recently allowed the fabrication of a fully-sealed prototype device (see Fig. 6.21 and [19]).

In addition to cylindrical illumination, field emission lamps also have potential for large area flat illumination. Triode structures seem to be the most promising for this application and attempts have been made at fabricating such structures with carbon nanotube field emitters [36]. The biggest challenge in large area applications is that the suspended metal mesh used for the gate vibrates under the large applied electric fields causing nonuniformity in the electron emission current due to variations in the distance between the gate and the cathode, arcing, and increased leakage current. To address these issues it has been proposed to use a metal mesh patterned with trenches and holes and bonded to the cathode plate [36], in addition to being coated with SiO_2 to reduce

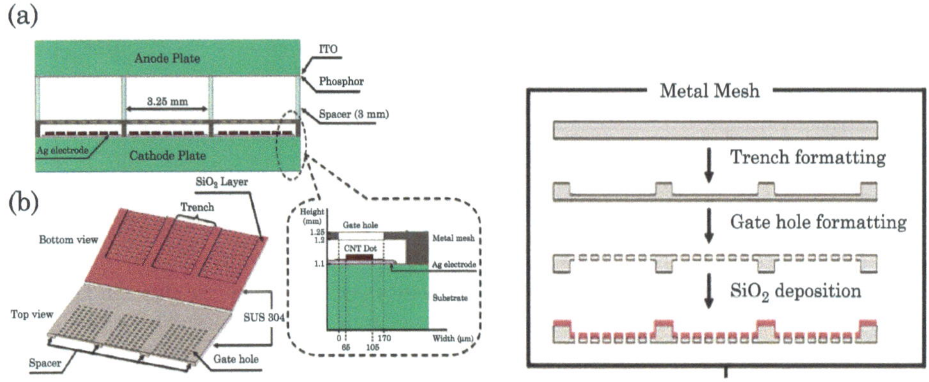

Figure 6.22 (a) Sketch of a triode field emission lamp for large area illumination using carbon nanotube emitters. (b) Sketch of the metal mesh and fabrication process (right). Figures from Ref. [36].

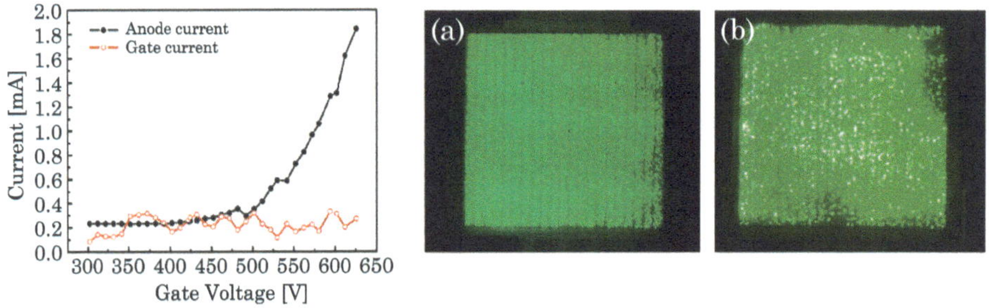

Figure 6.23 Left: Field emission and gate leakage currents as a function of the gate voltage in a triode field emission lamp. (a) and (b): Illumination of the green phosphor at the anode using the field-emitted current. The image labeled (a) has a bonded metal mesh as the gate, while that in image labeled (b) has an unbonded metal mesh. Figures from Ref. [36].

the leakage current. In this design (Fig. 6.22), trenches in a stainless steel metal mesh of 150 micron thickness and 64 cm^2 area are created, and the mesh is perforated with 170 micron diameter holes using a wet etching process. A layer of SiO_2 is deposited on the mesh using plasma-enhanced chemical vapor deposition. The cathode consists of a soda-lime glass on which a thin layer of Ag is deposited; a paste of multiwall carbon nanotubes is printed onto the Ag layer in the form of dots with 40 micron diameter and 6–8 micron thickness. Finally, the metal mesh is bonded to the cathode using glue, and the phosphor-coated ITO anode is combined with the cathode and the metal mesh using 3 mm spacers.

Fig. 6.23 shows the anode current as a function of the gate voltage, for an anode voltage of 4.5 kV. The anode current shows a turn-on behavior with a turn-on gate voltage of 160 V; comparing the anode and gate currents, it is seen that the anode current is about 8 times larger than the gate leakage current.

When impacting the phosphor-coated anode, the large field-emitted current leads to a uniform screen illumination as shown in Fig. 6.23. For comparison, a similar device was fabricated, but without bonding the metal mesh to the cathode and showed a much less uniform emission pattern due to the increased vibrations of the metal mesh. The brightness of the uniformly illuminating device is found to be 6000 cd/m^2. In addition to improving the uniformity of the emission, the bonded metal mesh also reduces the arcing and allows for stable operation of the device even under the very large anode voltages applied.

References

1. H. Sugie, M. Tanemura, V. Filip, K. Iwata, K. Takahashi and F. Okuyama, "Carbon nanotubes as electron source in an x-ray tube", *Appl. Phys. Lett.*, Vol. 78, p. 2578, 2001.
2. J. Zhang, G. Yang, Y. Cheng, B. Gao, Q. Qiu, Y.Z. Lee, J.P. Lu and O. Zhou, "Stationary scanning x-ray source based on carbon nanotube field emitters", *Appl. Phys. Lett.*, Vol. 86, p. 184104, 2005.
3. W.B. Choi, D.S. Chung, J.H. Kang, H.Y. Kim, Y.W. Jin, I.T. Han, Y.H. Lee, J.E. Jung, N.S. Lee, G.S. Park and J.M. Kim, "Fully sealed, high-brightness carbon-nanotube field emission display", *Appl. Phys. Lett.*, Vol. 75, p. 3129, 1999.
4. M. Croci, I. Arfaoui, T. Stöckli, A. Chatelain and J-M. Bonard, "A fully sealed luminescent tube based on carbon nanotube field emission", *Microelectronics Journal*, Vol. 35, p. 329, 2004.
5. W.A. de Heer, A. Chatelain and D. Ugarte, "A carbon nanotube field emission electron source", *Science*, Vol. 270, p. 1179, 1995.
6. R.H. Fowler and L.W. Nordheim, "Electron emission in intense electric fields", *Proc. Roy. Soc. Lond.*, Vol. A119, p. 173, 1928.
7. A.G. Rinzler, J.H. Hafner, P. Nikolaev, L. Lou, S.G. Kim, D. Tomanek, P. Nordlander, D.T. Colbert and R.E. Smalley, "Unraveling nanotubes: Field emission from an atomic wire", *Science*, Vol. 269, p. 1550, 1995.
8. N. de Jonge, M. Allioux, M. Doytcheva, M. Kaiser, K.B.K. Teo, R.G. Lacerda and W.I. Milne, "Characterization of the field emission properties of individual thin carbon nanotubes", *Appl. Phys. Lett.*, Vol. 85, p. 1607, 2004.
9. D. Lovall, M. Buss, E. Graugnard, R.P. Andres and R. Reifenberger, "Electron emission and structural characterization of a rope of single-walled carbon nanotubes", *Phys. Rev.* B, Vol. 61, p. 5683, 2000.
10. J.-M. Bonard, K.A. Dean, B.F. Coll and C. Klinke, "Field emission of individual carbon nanotubes in the scanning electron microscope", *Phys. Rev. Lett.*, Vol. 89, p. 197602, 2002.
11. L.A. Chernozatonskii, Yu.V. Gulyaev, Z.Ja. Kosakovskaja, N.I. Sinitsyn, G.V. Torgashov, Yu.F. Zakharchenko, E.A. Fedorov and V.P. Val'chuk, "Electron field emission from nanofilament carbon films", *Chem. Phys. Lett.*, Vol. 233, p. 63, 1995.
12. Q.H. Wang, A.A. Setlur, J.M. Lauerhaas, J.Y. Dai, E.W. Seelig and R.P.H. Chang, "A nanotube-based field emission flat panel display", *Appl. Phys. Lett.*, Vol. 72, p. 2912, 1998.
13. D.L. Carroll, P. Redlich, P.M. Ajayan, J.C. Charlier, X. Blase, A. De Vita and R. Car, "Electronic structure and localized states at carbon nanotube tips", *Phys. Rev. Lett.*, Vol. 78, p. 2811, 1997.
14. A. De Vita, J.C. Charlier, X. Blase and R. Car, "Electronic structure at carbon nanotube tips", *Appl. Phys.* A, Vol. 68, p. 283, 1999.
15. Ch. Adessi and M. Devel, "Theoretical study of field emission by single-wall carbon nanotubes", *Phys. Rev.* B, Vol. 62, p. R13314, 2000.

16. A. Buldum and J.P. Lu, "Electron field emission properties of closed carbon nanotubes", *Phys. Rev. Lett.*, Vol. 91, p. 236801, 2003.
17. X. Zheng, G. Chen, Z. Li, S. Deng and N. Xu, "Quantum-mechanical investigation of field emission mechanism of a micrometer-long single-walled carbon nanotube", *Phys. Rev. Lett.*, Vol. 92, p. 106803, 2004.
18. J. Cummings, A. Zettl, M.R. McCartney and J.C.H. Spence, "Electron holography of field-emitting carbon nanotubes", *Phys. Rev. Lett.*, Vol. 88, p. 065804, 2002.
19. N. de Jonge and J.-M. Bonard, "Carbon nanotube electron sources and applications", *Phil. Trans. R. Soc. Lond.* A, Vol. 362, p. 2239, 2004.
20. N. de Jonge, Y. Lamy, K. Schoots and T.H. Oosterkamp, "High brightness electron beam from a multi-wall carbon nanotube", *Nature*, Vol. 420, p. 393, 2002.
21. K.A. Dean and B.R. Chalamala, "Current saturation mechanisms in carbon nanotube field emitters", *Appl. Phys. Lett.*, Vol. 76, p. 375, 2000.
22. L. Nilsson, O. Groening, C. Emmenegger, O. Kuettel, E. Schaller, L. Schlapbach, H. Kind, J-M. Bonard and K. Kern, "Scanning field emission from patterned carbon nanotube films", *Appl. Phys. Lett.*, Vol. 76, p. 2071, 2000.
23. W. Wei, Y. Liu, Y. Wei, K. Jiang, L.-M. Peng and S. Fan, "Tip cooling effect and failure mechanism of field-emitting carbon nanotubes", *Nano Lett.*, Vol. 7, p. 64, 2007.
24. P. Vincent, S.T. Purcell, C. Journet and V.T. Binh, "Modelization of resistive heating of carbon nanotubes during field emission", *Phys. Rev.* B, Vol. 66, p. 075406, 2002.
25. J.-M. Bonard, C. Klinke, K.A. Dean and B.F. Coll, "Degradation and failure of carbon nanotube field emitters", *Phys. Rev.* B, Vol. 67, p. 115406, 2003.
26. M. Doytcheva, M. Kaiser and N. de Jonge, "*In-situ* transmission electron microscopy investigation of the structural changes in carbon nanotubes during electron emission at high currents", *Nanotechnology*, Vol. 17, p. 3226, 2006.
27. N.Y. Hua, J.C. She, J. Chen, S.Z. Deng, N.S. Xu, H. Bishop, S.E. Huq, L. Wang, D.Y. Zhong, E.G. Wang and D.M. Chem, "Mechanism responsible for initiating carbon nanotube vacuum breakdown", *Phys. Rev. Lett.*, Vol. 93, p. 075501, 2004.
28. S.T. Purcell, P. Vincent, C. Journet and V.T. Binh, "Hot nanotubes: Stable heating of individual multiwall carbon nanotubes to 2000 K induced by the field emission current", *Phys. Rev. Lett.*, Vol. 88, p. 105502, 2002.
29. P.G. Collins, M. Hersam, M. Arnold, R. Martel and Ph. Avouris, "Current saturation and electrical breakdown in multiwalled carbon nanotubes", *Phys. Rev. Lett.*, Vol. 86, p. 3128, 2001.
30. A.A. Talin and D. Jaskie, "Field emission displays: a critical review", *Solid. State Elec.*, Vol. 45, p. 963, 2001.
31. G.Z. Yue, Q. Qiu, Bo Gao, Y. Cheng, J. Zhang, H. Shimoda, S. Chang, J.P. Lu and O. Zhou, "Generation of continuous and pulsed diagnostic imaging x-ray radiation using a carbon-nanotube-based field emission cathode", *Appl. Phys. Lett.*, Vol. 81, p. 355, 2002.
32. P. Sarrazin, D. Blake, L. Delzeit, M. Meyyappan, B. Boyer, S. Snyder and B. Espinosa, "Carbon-nanotube field emission x-ray tube for space exploration XRD/XRF instrument", *Adv. X-Ray Anal.*, Vol. 46, p. 232, 2004.
33. P. Sarrazin, D. Blake, S. Feldman, S. Chipera, D. Vaniman and D. Bish, "Field deployment of a portable x-ray diffraction/x-ray fluorescence instrument on Mars analog terrain", *Adv. X-Ray Anal.*, Vol. 48, p. 194, 2005.
34. Basic research needs for solid state lighting, Report of the Basic Energy Sciences Workshop on Solid State Lighting, Office of Science, United States Department of Energy (2006).
35. J.-M. Bonard, T. Stöckli, O. Noury and A. Châtelain, "Field emission from cylindrical carbon nanotube cathodes: Possibilities for luminescent tubes", *Appl. Phys. Lett.*, Vol. 78, p. 2775, 2001.
36. W.-S. Cho, H.-J. Lee, Y.-D. Lee, J.-H. Park, J.-K. Kim, Y.-H. Lee and B.-K. Ju, "Carbon nanotube-based triode field emission lamps using metal meshes with spacers", *IEEE Elec. Dev. Lett.*, Vol. 28, p. 386, 2007.

7 Optoelectronic Devices

7.1 Introduction

From the discussion in the previous chapters, it is clear that there has been much work to establish the basic physics that governs electronic devices made with carbon nanotubes. As these electronic devices become better understood, they can be utilized as the fundamental building blocks to achieve additional functionality; an example of an emerging area of research in this direction is that of opto-electronics with carbon nanotubes. The interest in carbon nanotube opto-electronics arises because nanotubes have several properties that make them excellent opto-electronic materials. For example, an important characteristic of opto-electronic materials is the presence of a direct bandgap, which allows electronic transitions between the valence and conduction bands to proceed without the intervention of phonons. As the band structures of Fig. 1.6 indicate, nanotubes are unique materials in this aspect because all of the bands have a direct bandgap. Thus, for a single nanotube, there are multiple bands that can participate in direct opto-electronic events, spanning a wide range of energies. By combining multiple nanotubes of different bandgaps, it should be possible to obtain a nearly continuous response over a broad spectral range. In addition, a problem that affects traditional bulk materials is the presence of defects, which lead to nonradiative processes and significantly reduced device efficiency. Nanotubes, with their low defect density, should be less sensitive to this problem. A further advantage of nanotubes is related to the temperature dependence of the carrier concentration. For a three-dimensional semiconductor, the carrier density peaks slightly above the band edge, because the density of states vanishes at the band edge. Varying the temperature leads to a change of peak position and thus affects the device properties. For nanotubes however, the density of states has a singularity at the band edge and the carrier density always peaks at the band edge.

At the time of writing of this book, experimental and theoretical works on the opto-electronic properties of nanotubes are just starting to emerge. In this chapter, we first discuss the optical properties of carbon nanotubes, and then discuss three areas of opto-electronics: photoconductivity, electro-luminescence and optical switching.

7.2 Optical Properties

7.2.1 Selection Rules

In the simplest, single-particle picture, the optical properties of carbon nanotubes can be understood by considering the Hamiltonian in the presence of optical radiation

$$H = \frac{1}{2m}\left(\vec{p} + e\vec{A}\right)^2 + U \tag{7.1}$$

where $\vec{p}$ is the electron momentum, m its mass, $\vec{A}$ the magnetic vector potential of the impinging optical radiation and U includes all other interactions in the system. As long as the photon flux is not too large, the above Hamiltonian can be expanded to first order in the magnetic vector potential to give

$$H = H_0 + \frac{e}{m}\vec{p}\cdot\vec{A} \tag{7.2}$$

where H_0 is the Hamiltonian in the absence of light and where we have used the gauge $\nabla\cdot\vec{A} = 0$. In this formalism the electron–photon interaction is considered a perturbation and is given by

$$H_{\text{el–ph}} = \frac{e}{m}\vec{p}\cdot\vec{A}. \tag{7.3}$$

The magnetic vector potential for a monochromatic plane wave has the time and spatial dependence

$$\vec{A}(\vec{r},t) = \hat{e}\left(\frac{2\,\hbar^2\,\sqrt{\mu\varepsilon}}{N\varepsilon c\hbar\omega}I\right)^{1/2} e^{i\left(\vec{k}\cdot\vec{r}-\omega t\right)} \tag{7.4}$$

where $\hat{e}$ is the direction of the light polarization, I is the photon flux and ω the optical frequency. $\vec{k}$ is the optical wavevector, oriented in the propagation direction, i.e. perpendicular to the electric and magnetic fields. To proceed further we consider the time-averaged transition probability between initial and final states within Fermi's golden rule

$$\Gamma_{i\to j} = \frac{2\pi}{\hbar}\delta\left(E_f - E_i - \hbar\omega\right)\left|\langle f|\,H_{\text{el–ph}}\,|i\rangle\right|^2. \tag{7.5}$$

The delta function in this equation indicates energy conservation; for band-to-band transitions due to optical absorption, this implies that an electron from the valence band is excited to the conduction band across the bandgap. To calculate the matrix elements we focus on two special cases corresponding to light polarized parallel ($\hat{e} = \hat{z}$) and perpendicular ($\hat{e} = \hat{\phi}$) to the nanotube axis, as illustrated in Fig. 7.1. The matrix elements due to the electron–photon interaction in a carbon nanotube can be expressed as

$$\begin{aligned}\langle q,J,c|\,H_{\text{el–ph}}\,|q',J',v\rangle &= \left(\frac{2\,\hbar^2\,\sqrt{\mu\varepsilon}}{N\varepsilon c\hbar\omega}I\right)^{1/2}\\ &\quad\times\frac{e}{m}\langle q,J,c|\,(\vec{p}\cdot\hat{e})\,e^{i\vec{k}\cdot\vec{r}}\,|q',J',v\rangle\end{aligned} \tag{7.6}$$

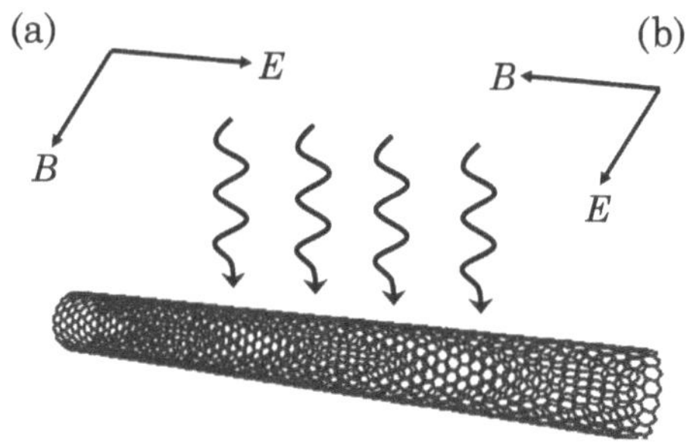

Figure 7.1 Illustration of a carbon nanotube illuminated by light polarized (a) along the nanotube axis, and (b) perpendicular to the nanotube axis.

where $|q, J, c\rangle$ represents a wavefunction for an electron with axial wavevector q in conduction subband J, and $|q, J, v\rangle$ is for the valence band, such that $E_c(q) = E_v(q') + \hbar\omega$. Because the nanotube radius is much smaller than optical wavelengths, the exponential factor in Eq. (7.6) is assumed to be equal to 1. For the two special cases of interest we therefore have

$$\langle f| H_{\text{el-ph}} |i\rangle = \left(\frac{2\,\hbar^2 e^2 \sqrt{\mu\varepsilon}}{m^2 N \varepsilon c \hbar\omega} I \right)^{1/2} \times \begin{cases} \langle q, J, c| p_z |q', J', v\rangle & \text{parallel polarization} \\ \langle q, J, c| p_\phi |q', J', v\rangle & \text{perpendicular polarization.} \end{cases} \tag{7.7}$$

Recall from Chapter 1 that the wavefunctions for carbon nanotubes are given by

$$|q, J, h\rangle = \psi^h_{Jq}(\vec{r}) = \frac{1}{\sqrt{2}} \left[\phi_{AJq}(\vec{r}) + \lambda^h_{Jq} \phi_{BJq}(\vec{r}) \right]. \tag{7.8}$$

The momentum matrix elements for *parallel polarization* are thus

$$\langle q, J, c| p_z |q', J', v\rangle = \delta_{J,J'} \delta_{q,q'} \text{Re} \left(\lambda^v_{Jq} \sum_{i=1,2,3} \alpha_i e^{-iq\delta z_i} e^{-iJ\delta\phi_i} \right). \tag{7.9}$$

To obtain this last expression we assumed that the momentum matrix elements between localized orbitals vanishes unless the orbitals are located on nearest-neighbor atoms. The quantities δz_i and $\delta\phi_i$ represent the axial and azimuthal coordinate differences between the positions of nearest-neighbor

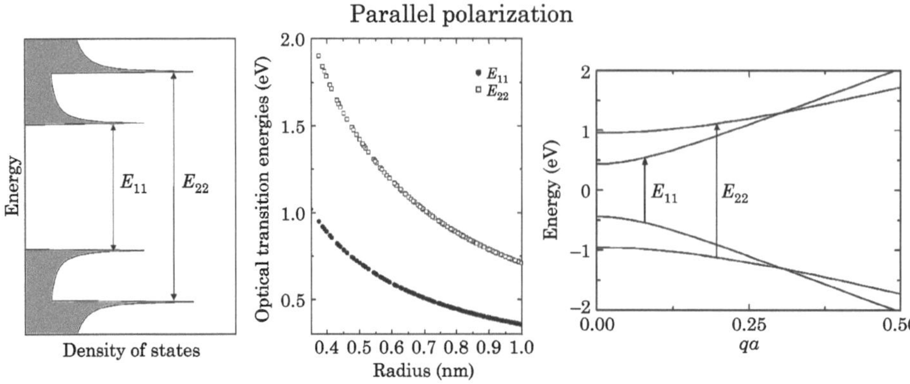

Figure 7.2 Illustration of the first two band-to-band transitions for light polarized along the nanotube axis, and the calculated values of E_{11} and E_{22} as a function of nanotube radius using a one parameter tight-binding model with $\gamma = 2.5$ eV. The right panel uses the band structure of the (17,0) nanotube to illustrate the transistions.

atoms. The coefficients α_i are given by

$$\alpha_i = \int_0^{2\pi} d\phi \int_0^{\infty} r dr \int_{-\infty}^{\infty} dz \chi^*_{\vec{r}_A}(\vec{r})\, i\hbar \frac{d}{dz} \chi_{\vec{r}_A+\vec{\delta}_i}(\vec{r}) \tag{7.10}$$

and can be evaluated with specified forms for the spatial dependence of the orbitals. But what is more important here is the appearance of the conditions $J = J'$ and $q = q'$ in Eq. (7.9), implying that for *parallel polarization* optical transitions will only occur between subbands with the same symmetry and axial wavevector. Thus, as is usually the case in optical transitions, the momentum and energy are conserved. Nanotubes are unique in that all of the subbands have a direct bandgap, and thus conservation of momentum can be satisfied in all subbands without the intervention of phonons. Fig. 7.2 illustrates the first two allowed band-to-band transitions for *parallel polarization*; they are usually referred to as the E_{11} and E_{22} transitions.

For the situation of *perpendicular polarization*, the evaluation of the matrix elements is particularly simple because the angular momentum operator simply changes the value of the quantum number J by one unit:

$$\langle q, J, c | p_\phi | q', J', v \rangle \propto \langle q, J, c | | q', J' \pm 1, v \rangle = \delta_{q,q'} \delta_{J,J'\pm 1} \tag{7.11}$$

where the $\pm$ sign depends on whether the polarization is right- or left-handed. Fig. 7.3 shows the first allowed optical transitions for perpendicular polarization and their calculated values as a function of the nanotube radius.

Before concluding this section, it should be noted that factors other than the selection rules impact the nanotube optical properties. For example,

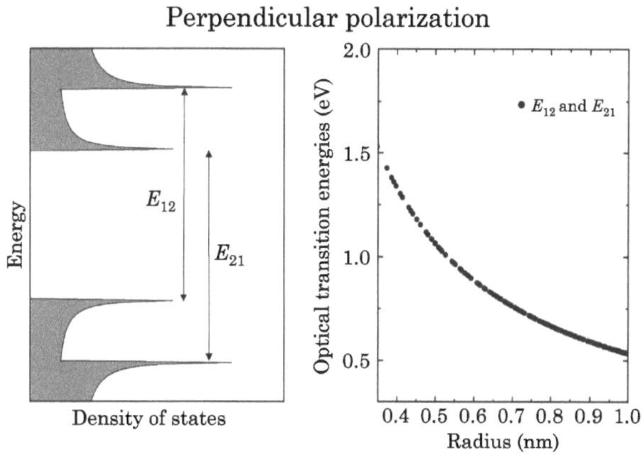

Figure 7.3 Illustration of the first two band-to-band transitions for light polarized perpendicular to the nanotube axis, and the calculated values as a function of nanotube radius using a one parameter tight-binding model with $\gamma = 2.5$ eV.

measurements of the optical absorption cross-section have indicated that induced charges in the nanotube lead to a depolarization effect that can substantially reduce the absorption for light polarized perpendicular to the nanotube axis [1]. Defining the absorption coefficient through the transmittance $T = e^{-\alpha n d}$ where α is the absorption coefficient and n is the density of nanotubes in a film of thickness d, values of $\alpha \sim 10^6$ cm^2/(mole C) have been measured for parallel polarization, but are a factor of five smaller for perpendicular polarization.

7.2.2 Excitons

In most traditional materials, the optical properties are well described by the single particle picture and the band-to-band transitions discussed in the previous section. However, for materials at reduced dimensionality such as polymer chains and nanowires, many-body effects dominate the optical properties; carbon nanotubes are no exception. The most important signature of many-body effects in optical properties are excitons, which are electron–hole pairs bound by the Coulomb interaction. Fig. 7.4 provides a simple picture of excitons. As shown in this figure, the excitation of an electron across the bandgap by the absorption of a photon leaves behind a positively charged hole. The attractive Coulomb interaction between the electron and the hole can lead to a hydrogen-like bound state where the electron and hole are separated by the exciton radius. Much like the hydrogen atom, the attractive potential can lead to quantized energy levels, and the difference between the free electron energy and the energy levels is called the exciton binding energy. (In reality, the electron–hole interaction also contains a repulsive interaction as will be discussed further below.)

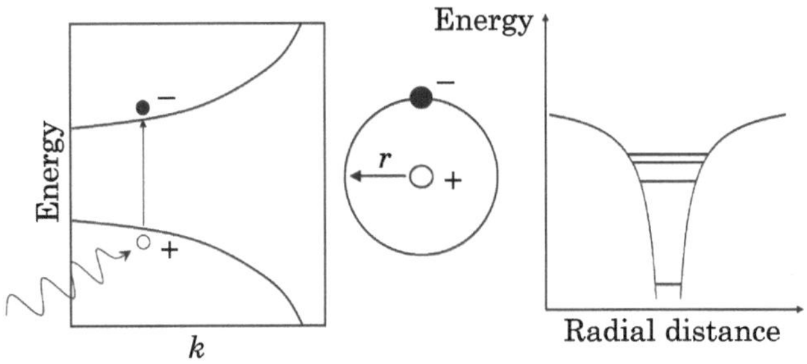

Figure 7.4 Illustration of exciton formation. The left panel shows a simplified band structure and the excitation of an electron from the valence to the conduction band upon photon absorption. The middle panel indicates that the electron can interact with the hole left behind through the Coulomb interaction, which can lead to a hydrogen-like bound state between the two particles, and quantized energy levels in the Coulomb potential (right panel).

A simple illustration of the binding energy of excitons and their importance in different materials can be obtained by considering the expression for the lowest energy level of a hydrogen atom

$$E_b^H = -\frac{m_0 e^4}{8h^2 r^2 \varepsilon_0^2} = -13.6 \text{ eV} \tag{7.12}$$

where m_0 is the free electron mass, ε_0 is the permittivity of free space and r is the Bohr radius. The binding energy is severely reduced in solid materials for at least two reasons: first, the effective mass is typically smaller than the free electron mass, and second, screening of the Coulomb interaction leads to a much higher dielectric constant. Thus, in a simple model assuming the same exciton radius we have

$$E_b = -13.6 \text{ eV}\frac{m^*}{m_0}\left(\frac{\varepsilon_0}{\varepsilon}\right)^2. \tag{7.13}$$

As an example, for GaAs, the electron effective mass $m^* = 0.067m_0$ and $\varepsilon = 12.85\varepsilon_0$ and the exciton binding energy can be estimated to be about 5 meV; thus in traditional semiconductors, excitons tend to be relevant only at temperatures much below room temperature. Note that screening of the Coulomb interaction between the electron and the hole (as expressed by the dielectric constant) plays a crucial role in significantly decreasing the binding energy. However, in quasi-one-dimensional materials (π-conjugated polymer chains are a prime example [2]) electrostatic screening is weak and the binding energy is significantly increased. The weak electrostatic screening has already been discussed in Chapter 4 in the context of the properties of carbon nanotube p–n junctions; additionally, theoretical work has proposed that the dielectric constant of semiconducting carbon nanotubes should be equal to 1 [3].

Evidence for the importance of excitons in carbon nanotubes has come from both experiment and theory. On the theoretical front, *ab initio* calculations of the optical spectra of carbon nanotubes including electron–hole interactions within the Bethe–Salpeter approach has indicated a large exciton binding energy in semiconducting carbon nanotubes and even excitonic effects in metallic nanotubes [4,5]. Fig. 7.5 shows the calculated optical properties of a (8,0) carbon nanotube with (solid line) and without electron–hole interactions (dashed line). The peaks labeled A B and C, corresponding to photon energies of 2.54, 2.66 and 3.7 eV, are the band-to-band absorption peaks expected from the single-particle picture. Clearly, in the presence of electron–hole interactions the optical absorption spectrum is completely altered: each band-to-band transition gives a series of sharp excitonic lines, labeled $A_1{}', A_2{}', A_3{}', B_1{}', B_2{}', C_1{}', C_2{}'$. Most importantly, one can see that the lowest-energy excitons coming from the different band-to-band peaks corresponds to exciton binding energies of about 1 eV, a value that is much larger than traditional semiconductors. Thus, in semiconducting carbon nanotubes, excitons dominate the optical spectra. The distance through which the electron–hole pair extends along the carbon nanotube is shown in Fig. 7.5 ((c)–(d)), where the position of the hole is fixed at the dot ($z = 0$) and the amplitude of the two-particle wavefunction is plotted. The exciton radius is found to be about 2.5 nm. In addition to the bound excitons (with energies below the single-particle gap) carbon nanotubes also exhibit resonant excitons (with energies above the single-particle gap), for example the C' excitons. These resonant excitons are found to have a spatial extent comparable to that of the bound excitons.

It is also interesting to discuss the role of excitons in metallic carbon nanotubes. In general excitons are not expected for metals because of the strong electrostatic screening. The situation is entirely different in metallic carbon nanotubes where excitons can be found with fairly large binding energies. Fig. 7.6 shows the calculated optical spectra for (3,3) and (5,0) carbon nanotubes using the same *ab initio* approach as for the (8,0) nanotube. Here the (5,0) nanotube is metallic because of strong curvature effects at this small diameter. The results for the (3,3) nanotube show an exciton peak below the band-to-band peak, with an exciton binding energy of 0.1 eV. The presence of a bound exciton, especially with such a large binding energy is quite striking. It has been proposed that the presence of a single bound exciton in this system originates from the metallic screening of the Coulomb interaction, which leads to an attractive delta-function potential along the nanotube axis, and in one dimension the attractive delta-function potential has a single bound state [4]. The calculations for the (5,0) nanotube indicate a different behavior: there, excitons are found to have limited influence on the optical spectrum. As we mentioned earlier in this section, the electron–hole interaction also contains a repulsive contribution. In the (5,0) nanotube, the attractive part of the interaction is suppressed because of symmetry, and the repulsive term dominates, precluding the appearance of bound exciton states [4].

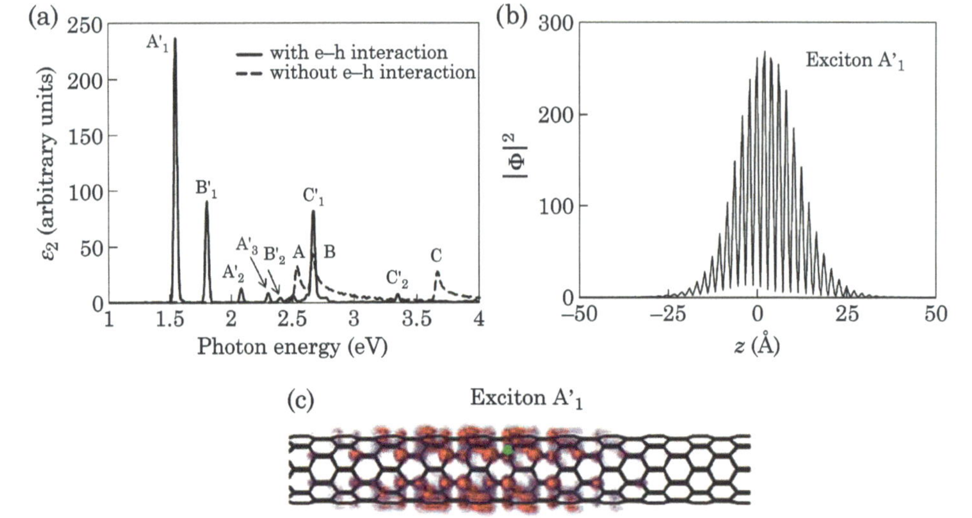

Figure 7.5 (a) Calculated optical absorption spectra with (solid line) and without (dashed line) electron–hole interactions for a (8,0) nanotube. (b) and (c) Representation of the $A_1{}'$ exciton wavefunction on the nanotube and as a function of distance along the nanotube axis. The hole position is fixed at the dot in (c) and is at $z = 0$ in (b). Figure from Ref. [6].

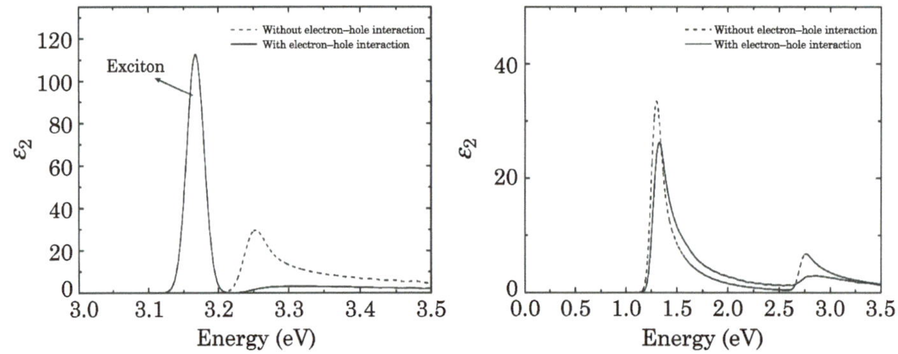

Figure 7.6 Calculated optical absorption spectra for metallic carbon nanotubes. The left panel, for a (3,3) carbon nanotube shows a single bound exciton state, while the right panel, for a (5,0) nanotube shows no bound exciton states. Figures from Ref. [4].

While ab initio calculations can provide detailed results for specific nanotubes, their range of applicability is currently limited to the smaller nanotube radii because of the large computational demands; alternatively, semi-empirical, tight-binding models [7,8] are useful to gain physical insight on the role of excitons and to study trends such as the scaling of the exciton binding energy with nanotube radius. One popular such model is called the Pariser–Parr–Pople (PPP) approach, which has been used extensively to describe excitonic effects in polymer chains. In second quantization, the

Hamiltonian for this model takes the form [7]

$$H = -\gamma \sum_{\langle i,j\rangle,\sigma} c^{\dagger}_{i,\sigma} c_{j,\sigma} + U \sum_{i} n_{i,\uparrow} n_{i,\downarrow} + \frac{1}{2} \sum_{i,j} V_{ij} \left(n_i - 1\right)\left(n_j - 1\right). \quad (7.14)$$

The first term in this equation is the standard tight-binding model that we have so far employed to describe the electronic structure of carbon nanotubes, with $c^{\dagger}_{i,\sigma}$ the operator that creates a π electron of spin σ on a carbon atom at position i. The second and third terms are the new terms that describe electron–electron and electron–hole interactions. $n_i = \sum_{\sigma} c^{\dagger}_{i,\sigma} c_{i,\sigma}$ is the number of electrons on site i; the U term therefore describes the on-site Coulomb interaction and the last term describes the long-range Coulomb interaction. This contribution is parametrized as

$$V_{ij} = \frac{U}{\kappa\sqrt{1 + \varsigma r^2_{ij}}} \quad (7.15)$$

where, in analogy with polymer chains, the value $\varsigma = 0.6117$ is used [7].

Several important conclusions have been reached from this semi-empirical model. The first is that the band degeneracy from the single-particle picture has importance consequences for the exciton structure. As shown in Fig. 7.7, semiconducting carbon nanotubes have doubly-degenerate conduction and valence subbands, denoted as a, b and a', b'. In the single-particle picture, one has the four degenerate optical transitions $a \to a', a \to b', b \to a', b \to b'$; but in the presence of the electron–electron and electron–hole interactions, the degeneracy of these four levels is lifted, and four excitonic states are created. In turns out [7] that only the highest energy excitonic state is optically allowed, and this state is referred to as the "bright" exciton. The other three excitonic levels are not optically allowed, and are called "dark" excitons. This excitonic structure has important consequences for the quantum efficiency of photoluminescence: as an exciton is created in the highest energy state, it can quickly relax into the optically-forbidden dark exciton states, preventing the re-emission of the photon. This is particularly true since the separation between the dark and bright excitonic states are found to be several times kT [7] which prevents thermal population of the bright exciton.

The tight binding approach allows for calculation of excitonic properties over a broad range of nanotube diameters and environments [8]. Fig. 7.8 shows the calculated exciton binding energy of carbon nanotubes as a function of the dielectric constant of the environment in which the nanotubes are embedded. As expected, the environment screens the electrostatic interaction and decreases the binding energy; the trend indicates a power law relationship, with an exponent that is independent of the nanotube radius. In fact, it is found (Fig. 7.8) that a scaling relationship exists between the exciton binding energy

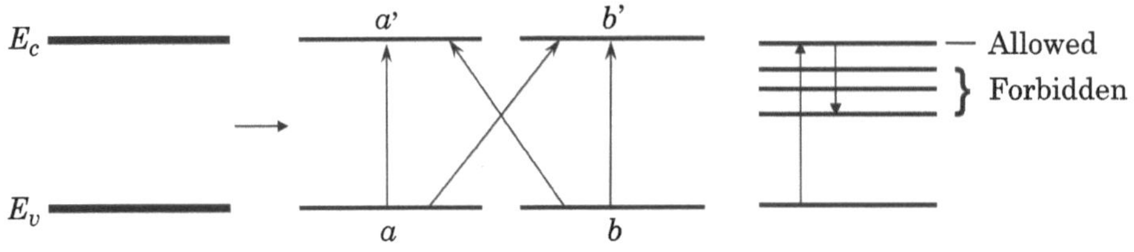

Figure 7.7 Left: illustration of the doubly degenerate conduction and valence bands in a single-wall carbon nanotube. Middle: the four possible optical transitions from a single-particle picture. Right: in the presence of the electron–hole interaction, the four degenerate single-particle transitions are split. The highest energy state is optically allowed, while the lowest three are not. Excitation of an electron to the highest energy state is followed by relaxation into the lower energy states, which prevents photoluminescence. Figure from Ref. [7].

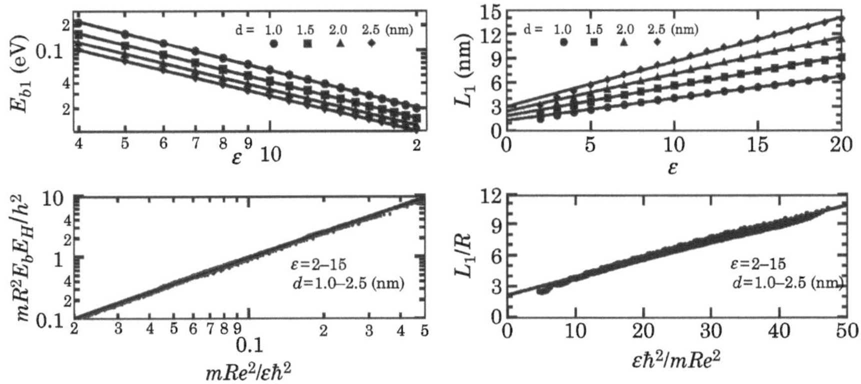

Figure 7.8 Scaling relationships for excitons in semiconducting carbon nanotubes. The left panels show the dependence of the exciton binding energy E_{b1} on the dielectric constant of the surrounding medium ε, and the scaling with nanotube radius R and effective mass m. Figures from V. Perebeinos.

and the nanotube radius R, nanotube effective mass m, and dielectric constant of the environment ε. Indeed in the range of nanotube diameters between 1 and 2.5 nm, and for dielectric constants between 2 and 15, the scaling

$$mR^2E_b = A_b\left(\frac{mR}{\varepsilon}\right)^{1.4} \tag{7.16}$$

is found to provide an excellent description of the calculated exciton binding energies. This scaling relationship is consistent with the decrease of the binding energy with increase of the dielectric constant of the surrounding medium, with a dependence $E_b \sim \varepsilon^{-1.4}$. This dependence is weaker than that in bulk materials, where the dependence is proportional to ε^{-2}.

A heuristic argument [8] to understand the scaling relationship is based on writing an expression for the exciton binding energy as a function of the exciton size L

$$E_L = \frac{\hbar^2}{2mL^2} - \frac{e^2}{\varepsilon R} f\left(\frac{L}{R}\right) = \frac{\hbar^2}{2mR^2} g\left(\frac{L}{R}, \frac{me^2 R}{\hbar^2\,\varepsilon}\right). \tag{7.17}$$

The first term in this equation is the exciton kinetic energy while the second term is the potential energy. The exciton size L^* is obtained by minimizing the energy, i.e. is the solution of $\partial E_L/\partial L = 0$. This gives an exciton binding energy

$$E_b = E_L\left(L^*\right) = \frac{\hbar^2}{mR^2} h\left(\frac{me^2 R}{\hbar^2\,\varepsilon}\right) \tag{7.18}$$

where h is the new scaling function. For consistency with Eq. (7.16), the function h satisfies a power law $h \sim x^\alpha$ and the dependence of the binding energy on the parameters is given by

$$E_b \approx A_b R^{\alpha-2} m^{\alpha-1} \varepsilon^{-\alpha} \tag{7.19}$$

where $\alpha = 1.4$ for carbon nanotubes. This exponent can be compared with bulk materials, where the potential energy is known to be proportional to L^{-1} and the exciton size is proportional to ε/m, giving a binding energy proportional to m/ε^2. The scaling exponent in this case is equal to $\alpha = 2$.

To obtain an explicit dependence of the binding energy on the nanotube radius, the effective mass needs to be expressed in terms of the nanotube radius. In the Introduction chapter, we showed that the effective mass for the lowest energy subband is given by $m^* = 2\,\hbar^2/(9a\gamma R)$; therefore the explicit radius dependence of the exciton binding energy is

$$E_b \propto R^{-1}\varepsilon^{-1.4}. \tag{7.20}$$

This relation has been verified by calculations over a range of nanotube radii [7].

The right panels in Fig. 7.8 show the calculated values of the exciton size L for several nanotubes as a function of the dielectric constant of the environment. A linear dependence of L on ε is found, and a scaling relationship

$$\frac{L}{R} = A_L + B_L \frac{\varepsilon}{Rm} \tag{7.21}$$

is obeyed, consistent with the expectations from the general scaling in Eq. (7.17). Quantitatively, it is found that the exciton size is in the range 1–15 nm, a value that impacts many factors in excitonic physics, such as the radiative lifetime and exciton–exciton interactions.

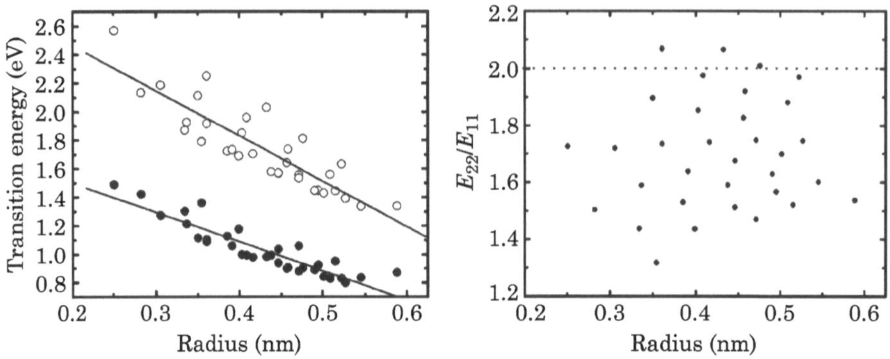

Figure 7.9 Left: measured optical transition energies in semiconducting carbon nanotubes from photoluminescence experiments. Data from Ref. [9]. Right: extracted values of the ratio of the second to the first transition energies from the left panel. In a simple tight-binding model, the ratio is predicted to be equal to 2, as indicated by the dotted line.

Many experiments have now shown evidence for excitonic effects in carbon nanotubes [9,10]. One of the first experiments to provide data on optical transition energies in individual carbon nanotubes combined photoluminescence and Raman scattering data to assign optical transition energies to specific carbon nanotubes [9]. In that experiment, individual carbon nanotubes are dispersed in aqueous micelle-like suspensions, and their optical luminescence is measured as a function of a broad range of excitation wavelengths, providing optical transition energies. Raman spectroscopy on the same samples allow determination of the (n,m) indices, and thus the diameter of the nanotubes. Fig. 7.9 plots the measured optical transition energies as a function of the nanotube diameter.

Two distinct lines are observed, corresponding to the first two subbands in the optical spectrum. While each of these two lines shows a linear dependence with nanotube diameter, the slope of these two lines are not equal, and as shown in the right panel of Fig. 7.9, the ratio of the two transition energies for a given nanotube is not equal to 2, as would be expected from a simple tight-binding model of the nanotube electronic structure. While the large variations of the values between different nanotubes can be explained in part by considering more refined tight-binding models [11], in the limit of large nanotube diameters these deviations vanish, and the experimental data in Fig. 7.9 seem to converge to a ratio value clearly distinct from 2. This effect, often referred to as the "ratio problem" in carbon nanotubes, can be explained on the basis of the excitonic theory, since the first two optical transition energies are determined by the position of the two lowest energy excitons. Further experimental evidence for the excitonic picture comes from two-photon experiments [10]. As illustrated in Fig. 7.10, in a two-photon experiment, optical selection rules are different from single-photon excitation, and the optical excitations can be to the so-called 2p exciton or to the continuum; the excited state then decays to the lowest

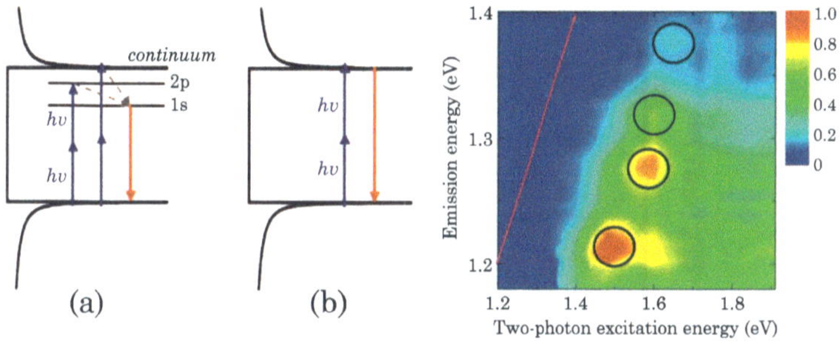

Figure 7.10 Illustration of two-photon excitation (blue lines) and subsequent luminescence (red lines) in the exciton picture (a) and the band-to-band picture (b). The right panel shows a color map of the measured emission intensity as a function of the two-photon excitation energy. The solid red line is the prediction from the band-to-band picture. Figures from Ref. [10].

energy exciton (1s), from which luminescence can be observed. The important point is that, in contrast to a two-photon excitation in the band-to-band picture (Fig. 7.10 (b)), the emission and excitation energies are different in the excitonic picture. The right panel in Fig. 7.10 shows a contour plot of the measured emission intensity as a function of two-photon excitation energy and emission energy, for a sample of carbon nanotubes embedded in a polymer matrix. In the band-to-band picture, the excitation and emission energies would be equal and peaks originating from different carbon nanotubes are expected to lie along a line of slope equal to 1 (solid red line in the figure). Clearly, the observed peaks deviate from that line, with an emission energy much lower than the excitation energy. This result provides a strong support for the excitonic picture.

7.2.3 Excitons in Electric Fields

In optoelectronic devices such as photodetectors, electric fields in the form of applied voltages or band-bending due to doping are usually utilized to separate the charge carriers. In the exciton picture, these electric fields oppose the Coulomb attractive interaction between the electrons and the holes, and may cause dissociation of the exciton. Thus, understanding the impact of electric fields on excitons is important for device design, and to evaluate when the single particle, band-to-band picture might be appropriate. To this end, theoretical calculations [12] have considered this problem by solving the Bethe–Salpeter equation in the presence of a static uniform electric field along the nanotube axis.

Fig. 7.11 shows the shift in the exciton binding energy for several semiconducting carbon nanotubes as a function of the applied electric field. It is found that the shift is always positive, indicating an increase of the exciton binding energy; the actual values are quite small however, on the order of meVs.

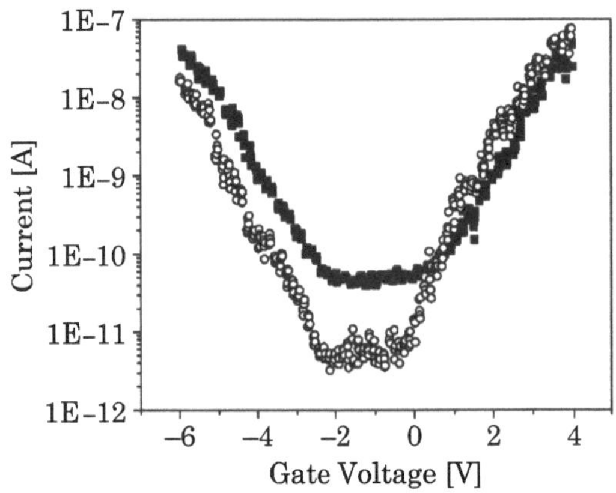

Figure 7.22 Current versus gate voltage for a carbon nanotube transistor without (open circles) and with infrared illumination from a Ti:sapphire laser (filled squares). Figure from Ref. [15].

responsitivity can be compared to those found in other materials: commercial GaAs and InGaAs *p–i–n* photodiodes have peak responsitivities of 0.5 A/W and 0.9 A/W, respectively, while quantum dot infrared photodetectors based on InAs/GaAs stacked layers have achieved peak responsitivities of 0.21 A/W [16].

Experimental evidence for photo-induced currents in nanotube devices has been recently reported, using lasers to illuminate individual nanotubes. Fig. 7.22 shows the measured current versus gate voltage characteristics of a single carbon nanotube transistor in the dark and when illuminated with a Ti:sapphire laser. The device is a Schottky-barrier transistor with ambipolar characteristics; under illumination the OFF current is increased by an order of magnitude, while a shift of the threshold voltage to more positive values of the gate voltage is also observed. This second effect is believed to arise from photovoltages generated in the SiO_2 layer [17]. The photocurrent shows a peak with photon energy centered at around 1.45 eV (Fig. 7.23). This optical transition energy is believed to arise from the second subband of the carbon nanotube (or due to excitons near that subband). A particularly interesting aspect of the photoconductivity is the sensitivity of the photoresponse to the polarization of the incoming light, as the symmetry considerations of the previous sections have described. Thus as the polarization angle of the incoming light is varied, the photocurrent is found to oscillate with a period of 180 degrees, with the maximum value reached when the light is polarized along the nanotube axis (the nanotube is tilted by about 30 degrees from the reference angle in this figure). The presence of a nonzero photocurrent for light polarized perpendicular to the nanotube axis originates from a depolarization effect due to the metallic electrodes, which have a spacing comparable to the optical wavelength. The induced photocurrent in this device is found to depend linearly on the intensity of the optical radiation, see Fig. 7.23 (c).

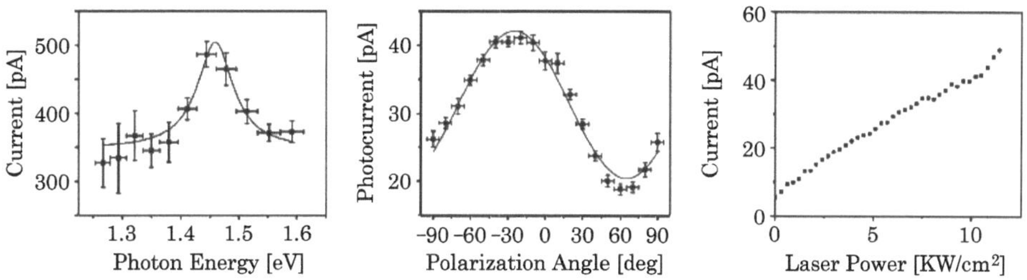

Figure 7.23 The left panel shows the measured photocurrent across a single nanotube when illuminated with a Ti:sapphire laser, as a function of photon energy. The middle panel indicates photocurrent as a function of the polarization angle, while the right panel shows the linear dependence of the photocurrent on laser power. Figure from Ref. [15].

The experimental development and fundamental study of photoconductivity in carbon nanotubes has recently progressed to more complex devices, most notably p–n junctions [18–20]. Bulk photoconduction devices rely extensively on p–n junctions, as the electric field at the junction is efficient in separating electron–hole pairs. These p–n junctions may be even more important in carbon nanotube photoconductive devices because of the need to dissociate excitons. The electronic properties of carbon nanotube p–n junctions were discussed extensively in Section 4.2. In particular, it was shown that suspending the carbon nanotube can lead to ideal diode behavior (see Fig. 4.4). These ideal diodes serve as excellent testbeds for studying carbon nanotube photovoltaic devices. To test their properties, a continuous wave laser diode with variable output power up to 10 mW is used to illuminate single carbon nanotube p–n junctions with infrared light of 0.8 eV energy (1.5 μm wavelength). The power delivered to the nanotube is estimated to be on the order of W/cm^2. The current–voltage characteristics of the device under illumination is shown in Fig. 7.24. One can clearly see the photocurrent generation that leads to a downward shift of the current–voltage curve. This shift is found to be proportional to the light intensity, as shown in the inset of this figure. Much like was discussed in the theoretical section above, the diode is found to operate as a photovoltaic device in the fourth quadrant of the current–voltage plot. The optical absorption in such devices may be dominated by excitons [19,20]. However, the large photocurrent generated here, with optical intensities about 10^4 times less than laser studies indicates that if excitons are present, their dissociation is effective. This may arise because of the relatively large electric field of at least 1 V/μm in this device.

Important parameters for photovoltaic devices are the short-circuit current (I_{sc}) and open-circuit voltage (V_{oc}). Fig. 7.25 shows a plot of the current–voltage characteristics in the fourth quadrant for a particular device and light intensity, from which the values of $I_{sc} \approx 1$ pA and $V_{oc} \approx 0.18$ V can be read-off. The maximum power generated in this device occurs at a voltage of 0.1 V, and is a fraction of a pW. From the ideal diode equation and the generated photocurrent,

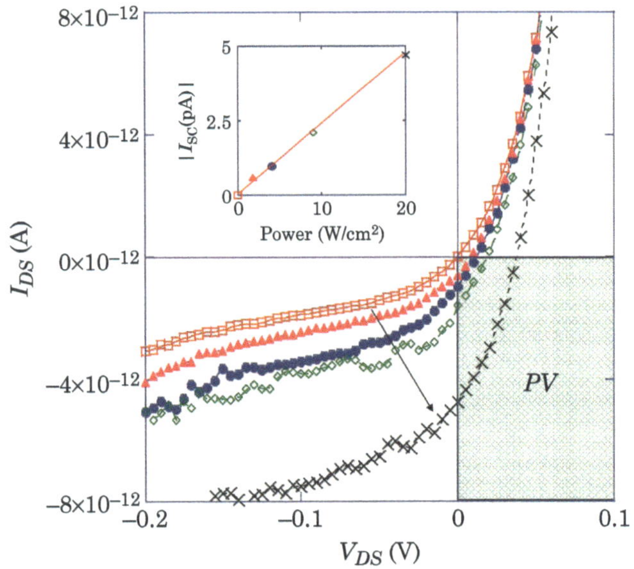

Figure 7.24 Photovoltaic properties of a carbon nanotube p–n junction. The main panel shows the progression of the current–voltage characteristics under increasing light intensities. The inset shows the dependence of the short-circuit current on the light intensity. Figure from Ref. [18].

it is expected that the open-circuit voltage will follow the expression

$$V_{\rm oc} = \frac{kT}{e} \ln\left(\frac{I_{\rm sc}}{I_s} + 1\right). \tag{7.35}$$

Fig. 7.25 indicates that this behavior is satisfied in the ideal nanotube diode.

Typical responsitivities in these devices is found to be around 30 mA/W, which is about a factor of 10 lower than that predicted by the theoretical calculations presented in the previous section. The power conversion efficiency η is estimated to be 0.2%, a value that is believed to be low because of the small fraction of the incident power absorbed [18]. It is important to point out however that much larger values of η should be achievable by appropriately selecting the carbon nanotube for its electronic and optical properties, and with careful device engineering. Further improvement is expected with arrays of carbon nanotubes.

7.3.1 Bolometers

In addition to the conversion of optical radiation to electron–hole pairs to generate photocurrents, optical detection can also be achieved by inducing a change in the resistance of a material due to heating by the incoming light. Vanadium oxide is the prime example of a material that is extensively used

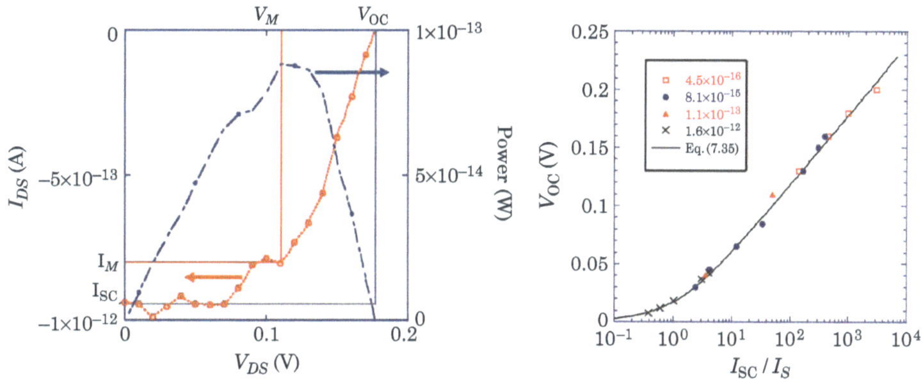

Figure 7.25 Left: current–voltage characteristics and power generation for a carbon p–n junction. The short-circuit current (I_{sc}), open-circuit voltage (V_{oc}), and the current and voltage at the maximum power (I_M, V_M) are indicated in the figure. Right: open-circuit voltage as a function of the light intensity. Figures from Ref. [18].

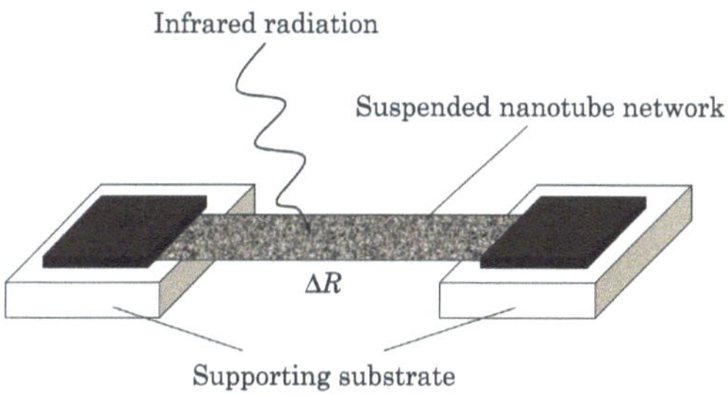

Figure 7.26 Sketch of a bolometer consisting of a network of carbon nanotubes suspended between two electrical contacts.

in bolometer technology. Most experiments on optical detection using carbon nanotubes utilize nanotubes that are in contact with a substrate. In this case, the heat transfer to the substrate precludes the observation of a bolometric response. To circumvent this problem, a device consisting of a suspended network of carbon nanotubes has been fabricated, and has shown a strong bolometric response [21]. As illustrated in Fig. 7.26, the device consists of a 0.5 mm wide ribbon of the single-wall carbon nanotube network suspended between two electrodes, and placed across the 3.5 mm opening of a sapphire ring. A cryostat allows the measurement of the properties of the device at different temperatures.

Upon exposure to infrared illumination of less than microwatt power, the suspended network shows a large change in resistance, with a signal-to-noise ratio of about 100 (Fig. 7.27). Similar experiments performed on a nanotube network in direct contact with the substrate showed little photoresponse; in addition, the photoresponse of the suspended network is well correlated with the absorbance (Fig. 7.27), which is expected for a bolometric response since more power is delivered to the nanotubes at the absorbance peaks.

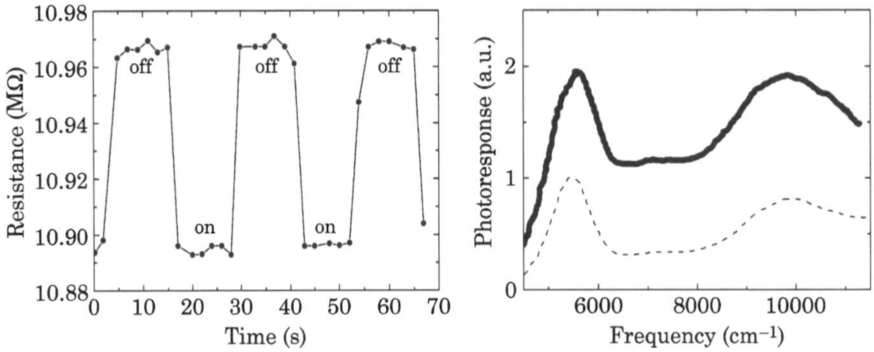

Figure 7.27 Left: resistance of suspended nanotube network at 50 K as infrared radiation is turned on and off. Right: comparison of the photoresponse (solid symbols) and the absorbance (dashed line) of the nanotube network as a function of the optical frequency. Figures after Ref. [21].

An important characteristic of a bolometric device is that the temperature coefficient of resistance (TCR) be as high as possible. The TCR is defined as the change in resistance with temperature,

$$\mathrm{TCR} = \frac{T}{R}\frac{dR}{dT} \tag{7.36}$$

and is typically a few percent in existing bolometer technology. Fig. 7.28 shows the measured resistance of the nanotube network as a function of temperature. The behavior is nonmonotonic: the resistance decreases at low temperature, reaches a minimum and increases at higher temperatures. The lower temperature behavior is typical of semiconductors, while that at high temperatures is characteristic of metals. In most applications of bolometers, a negative TCR value is desirable. The nanotube device achieves this at temperatures less that 230 K, as shown in the bottom left panel in Fig. 7.28. The right panels in this figure compare the temperature dependence of the resistance with the temperature dependence of the responsivity for three devices where the nanotube films have different thicknesses and processing conditions. The strongest responsivity is obtained for the thinnest films (40 nm in this case) and is correlated with the disappearance of the "metallic" increase of the resistance with temperature. From this data, the TCR can be estimated to be between 1% and 2.5%.

7.4 Electroluminescence

Electroluminescence is the process by which current injection causes electron–hole recombination and photon emission. In conventional devices such

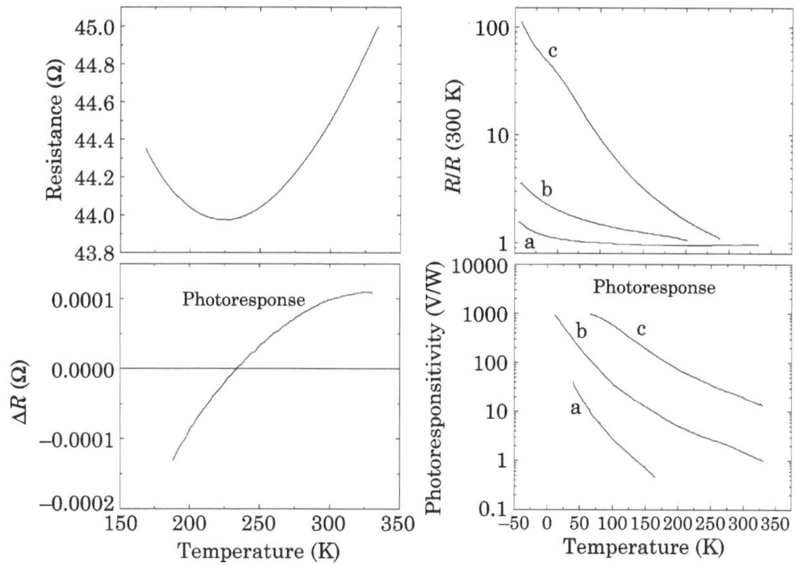

Figure 7.28 Top left: temperature dependence of the resistance of a 1 micron thick nanotube film. Bottom left: the change in resistance with temperature when exposed to infrared light. Top right: temperature dependence of resistance for three nanotube films. (a) 1 micron thickness, purified. (b) 100 nm thickness, purified and annealed in vacuum at 670 K. (c) 40 nm thickness, as-prepared. Bottom right: responsitivity of the nanotube films to infrared radiation. Figures after Ref. [21].

as light emitting diodes, the simultaneous presence of electrons and holes in the same spatial region is achieved by operating a *p*–*n* junction in forward bias, thereby injecting electrons in the *p* region where they recombine with holes. In nanotube field-effect transistors, one can take advantage of the gate modulation of the contacts and the band-bending, to simultaneously inject electrons and holes in the nanotube. Thus, in nanotube devices doping is not required to observe electroluminescence. The situation is illustrated in Fig. 7.29 which shows a sketch of an ambipolar carbon nanotube transistor. The device consists of a single semiconducting nanotube contacted by Ti at the source and drain, creating Schottky barrier contacts. A heavily doped Si backgate, separated from the nanotube by a 150 nm thick oxide controls the device conductance. When operated as a Schottky barrier transistor, the device shows ambipolar behavior, with nearly equal ON currents at positive and negative values of the gate voltage (Fig. 7.29 (b)). However, by choosing the gate–source and drain–source voltages appropriately, the band-bending can be such that the electric field at the source and drain contacts have the same sign. This arises when the gate–source voltage is less than the drain–source voltage. In this regime, the device essentially behaves as a forward biased *p*–*n* junction, with an exponential turn-on of the current with drain–source voltage (Fig. 7.29 (c)). A simulation of the band-bending along the carbon nanotube is shown in Fig. 7.29 (d), with the

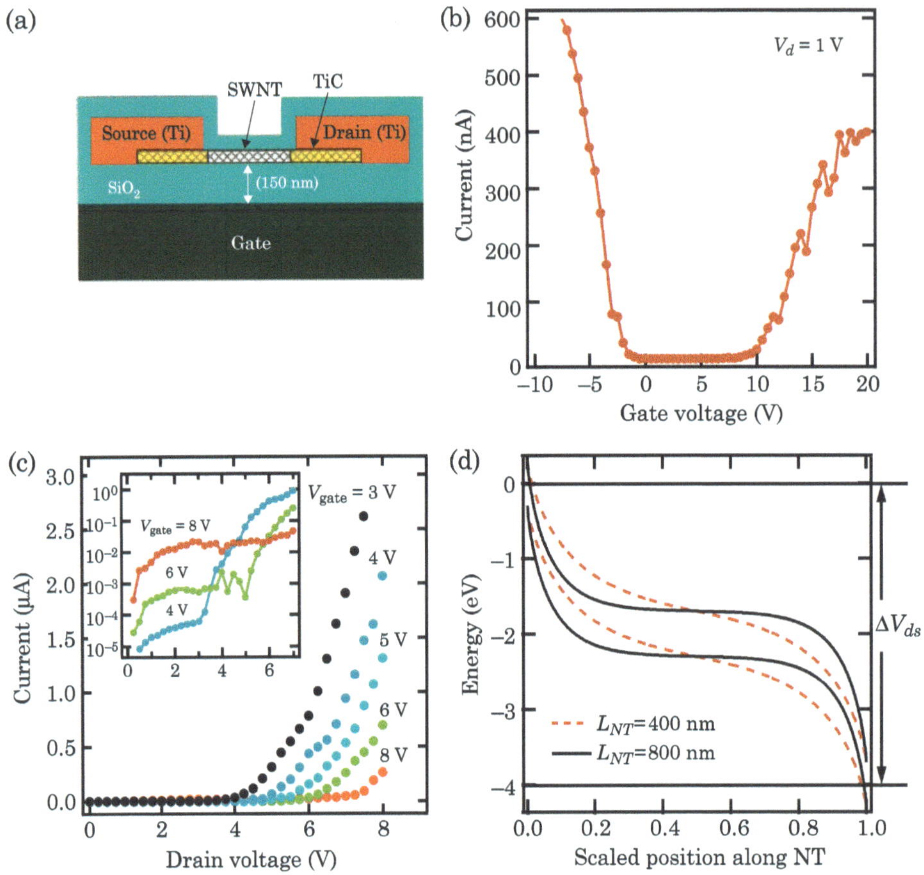

Figure 7.29 (a) Sketch of carbon nanotube device for electroluminescence. (b) Measured transfer characteristics. (c) Output characteristics of the nanotube device. (d) Calculated band-bending for the nanotube device with a source–drain voltage of 4 V and a gate voltage of 2 V. Figure from Ref. [22].

value of the gate–source voltage chosen to be halfway between the source and drain voltages. Because of the relatively sharp band-bending near the contacts, electrons (holes) from the drain (source) can tunnel into the nanotube, and these injected electrons and holes can radiatively recombine in the region of the nanotube where the bands are flat. The resultant light emission is shown in Fig. 7.30.

This image is obtained by combining an image of the optical emission from the carbon nanotube with an infrared image of a portion of the chip where the nanotube device is present. (The optical emission is measured with an infrared camera mounted on the probe station used to measure the electronic properties. Light is detected in the wavelength range 800–1500 nm, and the image is obtained by integrating the emission over 190 seconds.) In the top plane in this image, the blue areas correspond to the silicon dioxide, while the yellow

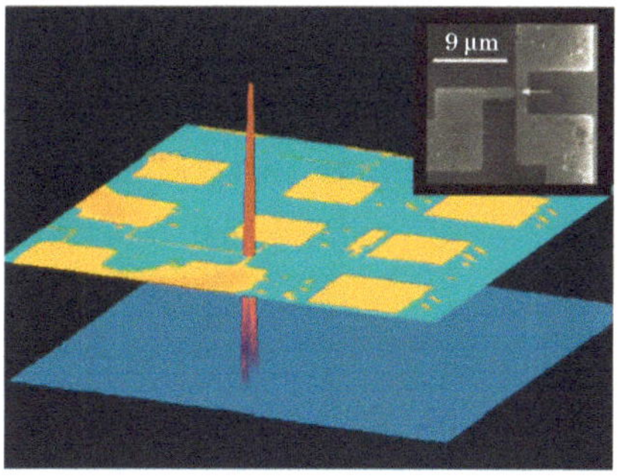

Figure 7.30 Optical emission from a single carbon nanotube, captured by an infrared camera. The image is combined with an infrared image of the chip, showing the contact pads, and electrodes making contact to the nanotube. The inset shows a scanning electron micrograph of the single nanotube device, from which the optical emission peak is observed. Figure from Ref. [22].

areas show the metal pads and lines. The large pads, of area approximately 70 square microns are used to make contact to the probe station, while the thinner lines are the source and drain electrodes for the nanotube device. The important point is that the emission peak is clearly localized at the position of the carbon nanotube. It is worth noting that similar measurements when unipolar electron or hole currents flowed through the device did not show any detectable optical emission, indicating that thermal light emission is not the source of the observed infrared emission.

Further evidence that the optical emission originates from radiative electron–hole recombination in the channel is provided by studying the gate voltage dependence of the optical emission. Fig. 7.31 shows the measured emission intensity when the gate voltage is varied at fixed drain–source current. For the case of a drain–source voltage of 8 V (open circles in the figure) the maximum in the emission intensity is found at a gate–source voltage of 4 V. For a drain–source voltage of 4 V, the maximum is achieved at a gate voltage of 2 V. In both cases, the maximum emission intensity is observed when the gate voltage is halfway between the source and drain voltages, as would be expected from a model where the electron and hole Schottky barriers are equal, and the optical emission is determined by radiative electron–hole recombination in the channel. In fact, a model where the emission intensity is proportional to the lowest of the electron or hole current reproduces well the observed behavior (solid lines in Fig. 7.31).

In a similar manner to the photoconductivity of single carbon nanotubes, the optical emission is also observed to be strongly polarized. For these measurements an infrared polarizer was inserted between the sample and the infrared camera. As shown in Fig. 7.32 (a), the emitted light is significantly more intense along a particular direction, and strongly suppressed perpendicular to that direction. When compared with a scanning electron micrograph of the

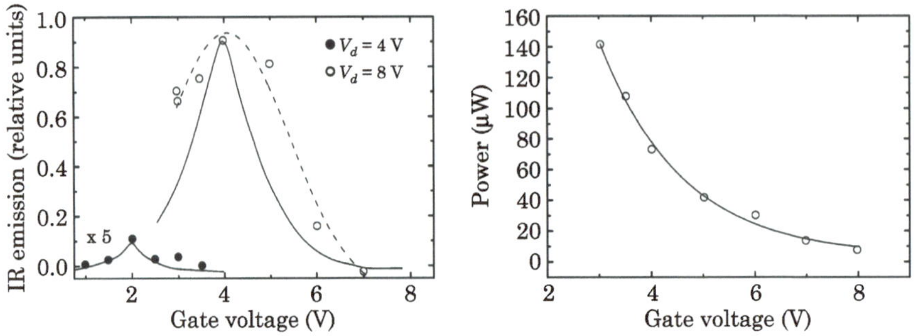

Figure 7.31 Left: emission intensity of infrared radiation from a single carbon nanotube Schottky barrier transistor as a function of the gate voltage. The open circles are for a drain–source voltage of 8 V, and the filled circles are for a drain–source voltage of 4 V. The dashed line is a Gaussian fit as a guide for the eye. The solid lines are the predicted optical intensities assuming that the optical emission is due to the lesser of the hole or electron current. Right: total electrical power in the device as a function of gate voltage. The solid line is an exponential decay fit as a guide for the eye. Figure after Ref. [22].

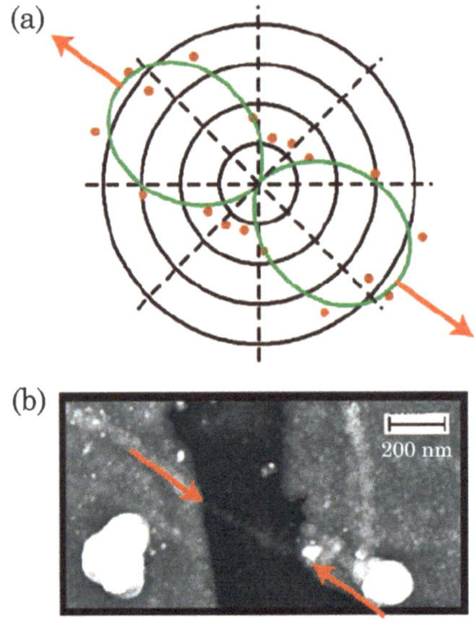

Figure 7.32 Polarization dependence of the optical emission intensity from an ambipolar carbon nanotube field-effect transistor. (a) The data (red dots) are shown in the form of a polar plot of the intensity versus analyzer angle θ (accuracy is about 10°). The green curves are a fit of the data to a $\cos^2\theta$ dependence. (b) A scanning electron microscope image of the emitting device showing the region near the nanotube channel. The whole device is covered by 10 nm of SiO2. The red arrows in (a) and in (b) mark the polarization direction and the orientation of the nanotube, respectively. Unique polarization analysis data were only attained for a 180° sweep of the analysis angle; however, the data are plotted in a standard 360° plot for ease of comparison. Figure and caption from Ref. [22].

emitting nanotube, the optical emission is found to be maximal when the polarizer is oriented parallel to the nanotube axis.

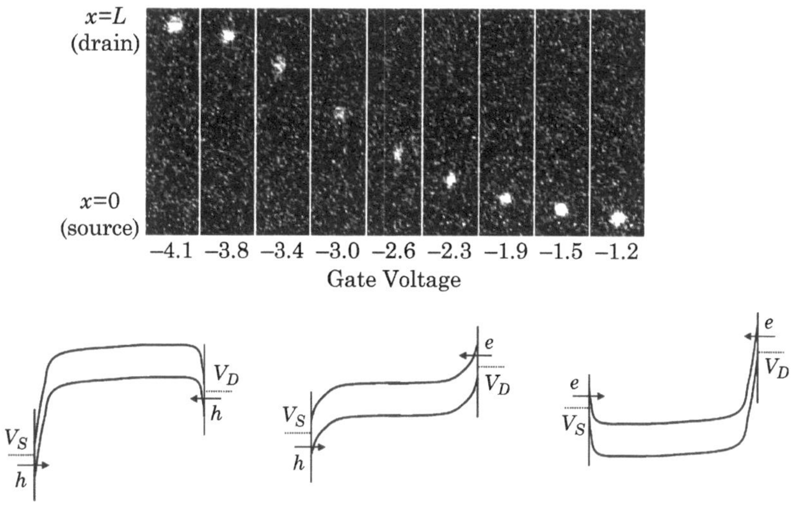

Figure 7.33 Light emission from a carbon nanotube Schottky barrier transistor during a gate voltage sweep. The gate voltage controls the band-bending in the nanotube, and thus the location of the electron–hole recombination region. Figure from Ref. [24]. The bottom panel shows sketches of the band-bending in the nanotube, see text for details.

A particularly striking aspect of this device operation is that the spatial region where recombination occurs can be moved by changing the gate voltage [23] as shown in Fig. 7.33. In this series of images, the device is operated in constant-current mode, and the gate voltage is swept between −30 V and 0 V, and in the reverse direction. At the beginning of the gate voltage sweep, no discernible light emission is observed from the carbon nanotube; as the gate voltage becomes less negative, a clear emission spot appears near the drain contact, and moves along the length of the nanotube as the gate voltage increases further toward 0 V. Once the emission spot reaches the source electrode, it disappears, and further increase in the gate voltage does not lead to optical emission. A reverse gate voltage sweep leads to the emission spot moving from the source to the drain electrode. The key to this observation is that the nanotube is extremely long, 50 microns in this case. The localized and movable light emission can be observed because the spatial extent of the recombination is much smaller than the channel length.

Qualitatively, the operation of this device can be understood by considering the band-bending sketched in Fig. 7.33. In the first image in the sequence in the top panel of Fig. 7.33, the drain–source voltage is −10 V and the gate–source voltage is −40 V. In this situation (bottom left sketch), the electric fields at the source and drain have opposite signs, and holes are injected at both contacts, and no light emission is observed. As the gate voltage is made less negative, the electric fields at the contacts eventually have opposite signs (middle sketch) and electrons and holes are simultaneously injected in the nanotube channel, leading to light emission through radiative recombination. As the gate voltage increases

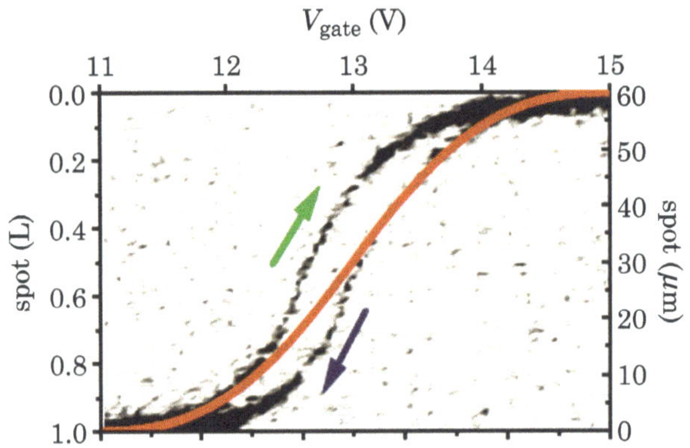

Figure 7.34 Comparison of the calculated and measured position of the light emission spot in a carbon nanotube infrared emitter. The raw data points represent the measured position, with the arrows indicating the direction of the gate voltage sweep. The solid line is that computed from Eq. (7.45) in the text. Figure from Ref. [25].

light spot for the ambipolar device discussed above. In addition to the fixed position of the light emission there are two other differences compared with the nonsuspended device with Ti contacts: the intensity of the light emission is found to be a factor of 1000 larger, and the light emission occurs under conditions of unipolar transport. These differences are believed to arise from impact excitation at the SiO_2/trench interface because of the large electrical field caused by the discontinuity in dielectric constants.

Fig. 7.35 (b) displays the current and optical emission intensity as a function of the gate voltage. The emission intensity is found to depend exponentially on the current (main panel and inset) and to be largest in the unipolar transport regimes of electron or hole conduction, i.e. when $|V_g - V_{th}|$ is large. The measured emission efficiency is about $10^{-3} - 10^{-2}$ photons/electron, which is two to three orders of magnitude larger than the nonsuspended nanotube with Schottky contacts.

Impact excitation occurs when an electron gains sufficient energy to emit an exciton, which then radiatively recombines (see Figs. 7.35 (c) and 7.36). In the presence of high electric fields, the electron can be accelerated to the necessary energies, provided that it does not relax due to scattering with phonons. Since the optical phonon scattering mean free path is on the order of 10 nm in carbon nanotubes, the band-bending must occur over a length scale smaller than this value for the impact excitation process to be efficient. An estimate [26] of the threshold electric field needed can be obtained from the relation $F_{th} = 1.5E_g/l_{op}$; for a carbon nanotube of bandgap 0.5 eV and $l_{op} = 10$ nm, the threshold electric field is 75 V/ μm. This value is about one order of magnitude smaller than bulk materials with the same bandgap, due to the longer optical phonon mean free path in carbon nanotubes.

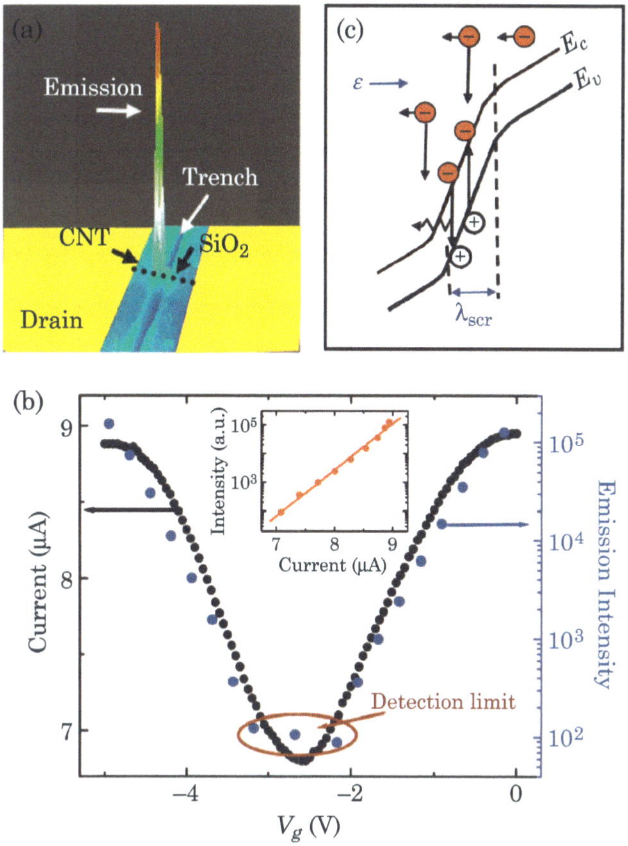

Figure 7.35 (a) Infrared emission image overlaid with an optical image of a carbon nanotube suspended over a trench. The channel length is 26 microns and the trench width is 5 microns. (b) Source–drain current and infrared emission intensity as a function of the gate voltage. Inset shows the exponential dependence of the emission intensity on current. (c) Sketch of the impact excitation process. Band-bending over a short length scale accelerates an incoming electron to energies larger than the bandgap, where it can generate an exciton that decays radiatively. The process can continue as the electron can be re-accelerated by the electric field. Figure from Ref. [26].

The impact excitation rate is given by the exponential relation [26]

$$\Gamma \propto \exp\left(-F_{\text{th}}/F\right) \tag{7.46}$$

where F is the electric field. In the case of the suspended carbon nanotube, the relevant electric field is that at the SiO_2/trench interface, and consists of two components. The first component is the electric field created by the applied drain–source voltage, which can be written as $F_{sd} = \gamma V_{sd}$ where the proportionality factor γ has units of inverse distance, and represents a screening length at the interface. The second component to the local electric field is due

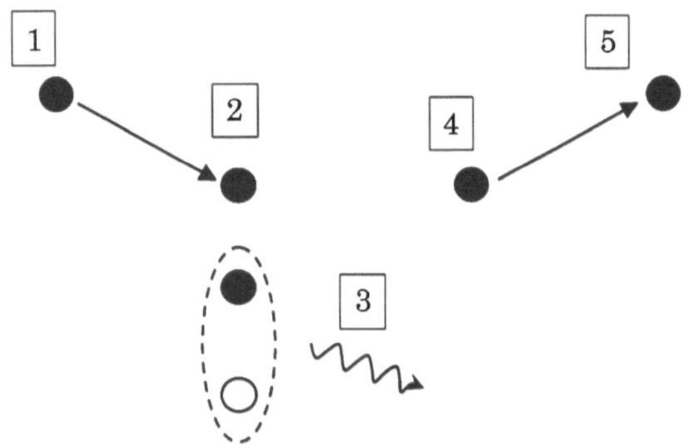

Figure 7.36 Schematic representation of the impact excitation process. An electron with high energy E_e (1) emits an exciton of energy E_{ex}, reducing its energy to $E_e - E_{\text{ex}}$ (2). The exciton radiatively recombines and emits a photon (3). The electron of energy $E_e - E_{\text{ex}}$ can be accelerated by the electric field to a higher energy, and the process can repeat.

to screening of the gate voltage by the discontinuous dielectric constant of the oxide and air at the oxide/trench interface. This contribution is also assumed to be linear, and given by the relation $F_{\text{gate}} = \alpha\,(V_g - V_{\text{th}})$. The impact excitation rate takes the form

$$\Gamma \propto \exp\left(-F_{\text{th}}/\left[\gamma V_{sd} + \alpha\,(V_g - V_{\text{th}})\right]\right). \tag{7.47}$$

Thus, based on this expression, it is expected that increasing the drain–source voltage or the difference between the gate voltage and the threshold voltage leads to exponentially increased light emission. This prediction is in accord with the qualitative results of Fig. 7.35 (b), but can be substantiated by a more quantitative analysis of the experimental data. The data points in Fig. 7.37 (a) show the measured current and light emission intensity as a function of the gate voltage. The general behavior is for the intensity of optical emission to increase exponentially as the current increases roughly linearly away from the threshold voltage. Quantitatively, the emission intensity can be fit accurately using Eq. (7.47) by using the experimentally measured current data points, as indicated by the solid lines in the figure. This procedure also produces excellent agreement for the dependence of the intensity of optical emission on the drain–source voltage (Fig. 7.37 (b)), where a threshold drain–source voltage is necessary to overcome the contribution to the electric field from the gate. It is interesting to note that the onset of light emission is correlated with the presence of negative difference resistance in the current versus drain–source voltage curve.

The exponential dependence of the impact excitation rate on the electric field has also been confirmed by solving the Boltzmann equation in the presence of phonon scattering and impact excitation [26]. The results, plotted in Fig. 7.38 indicate that the simple exponential behavior of Eq. (7.46) is satisfied in carbon nanotubes. The threshold electric field for impact excitation decreases with increase in the nanotube diameter, and the values extracted from

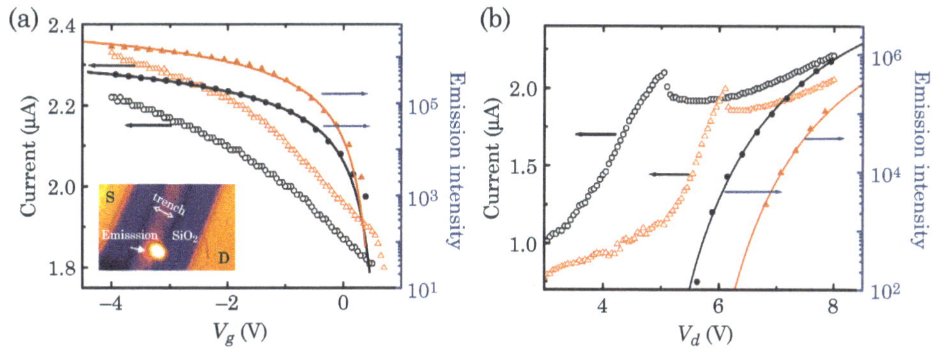

Figure 7.37 Drain–source current and infrared emission intensity for a carbon nanotube suspended over a trench as a function of gate voltage (left) and drain voltage (right). The solid lines are fits to the experimental data using Eq. (7.47). The inset in the left graph is an overlay of the optical emission image and an optical image of the device. Figure from Ref. [26].

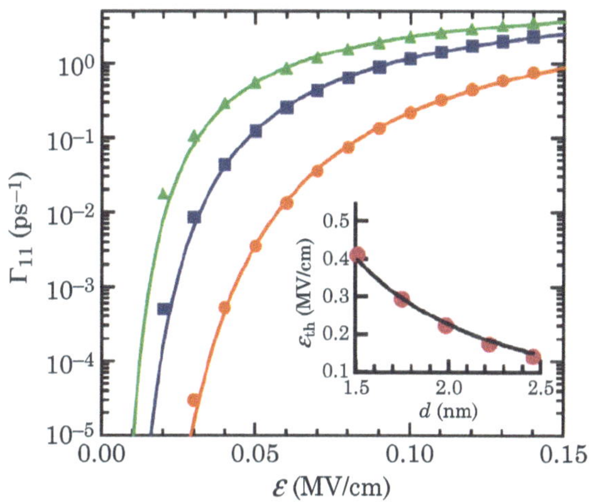

Figure 7.38 Calculated exciton generation rate as a function of the electric field for nanotubes of diameters 2.5, 2.0 and 1.5 nm from left to right. The inset shows the diameter dependence of the threshold electric field for impact excitation as a function of nanotube diameter. Figure from Ref. [26].

these simulations follow the relationship $F_{\text{th}} \sim d^{-2}$ (inset of Fig. 7.38). This relationship arises because $F_{\text{th}} \sim E_g/l_{\text{op}}$ with the bandgap depending inversely on diameter and the optical phonon scattering length being proportional to the nanotube diameter.

7.4.1 Thermal light emission

An interesting discovery in carbon nanotube opto-electronic devices is the appearance of wavelength-specific optical emission from electrically heated quasi-metallic carbon nanotubes [28]. (The term quasi-metallic is used to describe metallic carbon nanotubes that appear to have small bandgaps when measured electrically.) This optical emission differs from the usual broad blackbody radiation in hot materials in that its spectrum is strongly peaked at energies that correspond to the first two band-to-band optical transitions, and is polarized along the nanotube axis. This phenomena is much more pronounced in suspended carbon nanotubes, since heat transfer to the substrate is usually too fast to allow a large enough temperature to build in the carbon nanotubes. Furthermore, because the thermal light emission occurs at large nanotube temperatures, experiments are conducted in an argon environment to prevent oxidation and burning of the nanotube.

Fig. 7.39 (a) shows the measured current versus gate voltage for a carbon nanotube device which shows little gate modulation over a large voltage range. The small current modulation is attributed to the presence of a small gap or to defects in the nanotube. A scanning electron micrograph of the device is shown in Fig. 7.39 (c); in this three-electrode configuration, the nanotube sits on an oxide in one channel and is suspended over a trench in the other channel. When the current is measured as a function of the source–drain voltage, it is found that the suspended portion of the nanotube shows negative differential resistance (Fig. 7.39 (b)), a feature that is not observed in the nanotube sitting on the oxide substrate. This behavior is attributed to self-heating of the carbon nanotube which leads to increased phonon scattering at higher source–drain bias [29]. Analysis of the negative differential conductance using a hot phonon model [29] gives a nanotube temperature of about 1200 K, which is mostly due to electron heating. The observed optical emission from the three-electrode nanotube device is shown in panels (c) and (d) of Fig. 7.39. A few points are worth mentioning: first, the optical emission is more pronounced in the suspended portion of the nanotube; second, the appearance of a peak around 1.6 eV can be clearly seen in the suspended nanotube but is not as discernible when the nanotube is on the oxide substrate; third, the light emission spot is found to be located in the center of the channel.

These observations are consistent with a model where Joule heating of the carbon nanotube leads to a high electronic temperature. In this model, as illustrated in Fig. 7.40, the large electronic temperature leads to a broad Fermi distribution that causes significant occupation of the semiconducting subbands of the metallic nanotube, allowing band-to-band optical transitions to proceed and lead to light emission with specific wavelengths. In a simplified model, the light emission can be calculated from [29]

$$S(E) = \frac{1}{\tau(E)} D_J^B(E) f[E_c(k)]\{1 - f[E_v(k)]\} \tag{7.48}$$

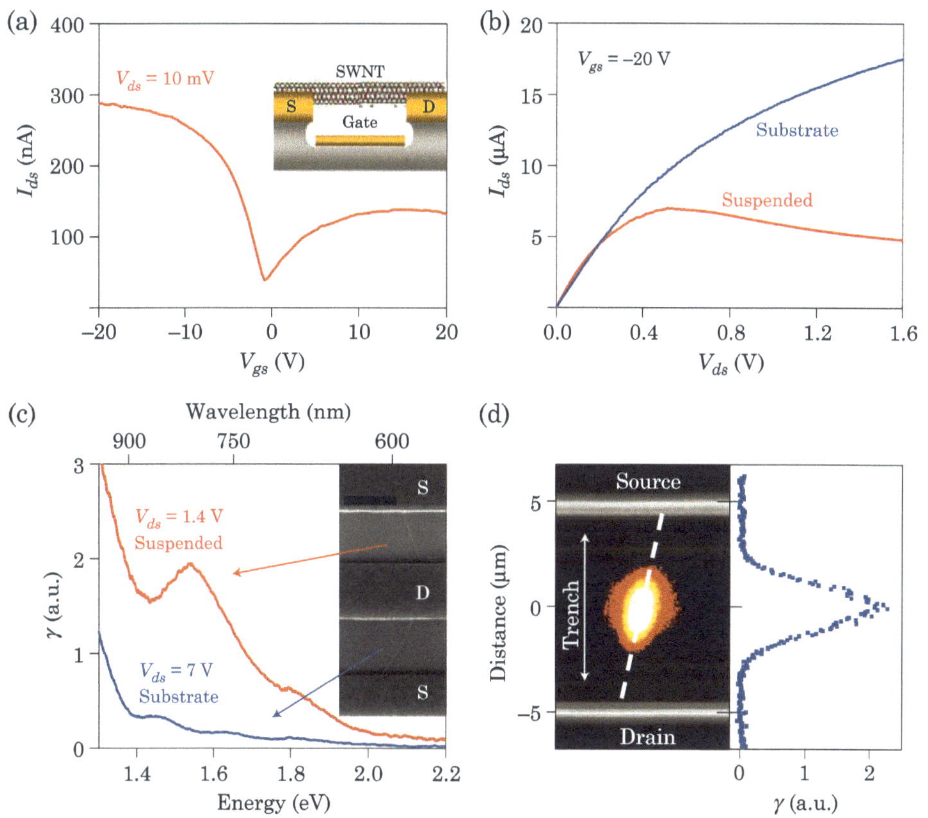

Figure 7.39 Electrical and optical emission characteristics of quasi-metallic carbon nanotubes. (a) shows the current versus gate voltage characteristics of a single carbon nanotube. (b) shows the dependence of the current on the drain–source voltage for the portion of the nanotube on the oxide substrate and the suspended portion. (c) shows a scanning electron micrograph of the nanotube device and the light emission intensity from the suspended and nonsuspended portions of the nanotube. (d) shows an overlay of the measured optical emission and an optical image of the device, as well as the measured optical emission intensity as a function of the position in the channel. Figure from Ref. [28].

where τ is the scattering time, f is the Fermi distribution, and D_J^B is the broadened joint density of states given by

$$D_J^B(E) = \int_{-\infty}^{\infty} D_J(E') \, B(E - E', \sigma) \tag{7.49}$$

where $B(x, \sigma)$ is a broadening function of width σ. The parameter σ is introduced as a fitting parameter and is used to represent the increased broadening of the emission intensity with increase in the source–drain bias. Fig. 7.41 shows a comparison of the measured and computed emission intensity as a function of energy. The behavior is well reproduced using Eq. (7.48) by

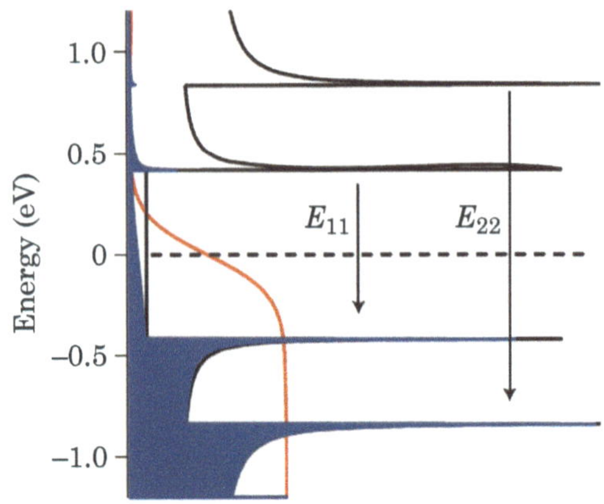

Figure 7.40 Density of states for a metallic carbon nanotube and Fermi distribution at high temperatures. The band-to-band optical transitions that lead to optical emission are indicated in the figure. Figure from Ref. [28].

adjusting the broadening width. Further support for this model is obtained by extracting the temperature as a function of the applied source–drain bias, as indicated in Fig. 7.41 (b). There, the negative differential resistance in the current versus source–drain voltage (inset) is fit to a model of hot phonon scattering [29], from which the nanotube electronic temperature can be calculated. When used in conjunction with Eq. (7.48), the calculated electronic temperature allows a prediction of the emission intensity that agrees very well with experiment (Fig. 7.41 (b)).

The energy dependence of the emission spectrum tends to be fairly broad in these hot nanotubes, on the order of hundreds of meV, and increases with increase in the source–drain voltage. Indeed, the measured energy distribution for a particular suspended nanotube is shown in Fig. 7.41, and shows a broad peak centered at around 0.8 eV with amplitude and width that depend on the source–drain voltage. Calculation of the emission spectra using the temperature from panel (b) of this figure and Eq. (7.48), assuming a nanotube with $E_{11} = 0.8$ eV, gives a good agreement with the measured data. Electron–phonon scattering lifetimes are estimated to be on the order of 10 fs.

Impact excitation, which was discussed earlier in this section, is not believed to be the origin of the light emission in these nanotube devices for several reasons. First, the energy of the emitted light is larger than the energy of the applied source–drain voltage eV_{ds} (for impact excitation, the maximum energy that an electron can gain is eV_{ds}, and this is the maximum energy that can be transferred to a photon). Second, the position of the light emission spot is at the center of the channel, and not at the oxide/trench interface.

Thermal light emission from carbon nanotubes has been demonstrated to be useful for applications in incandescent lighting. Despite much progress in solid-state and fluorescent lighting, illumination with incandescent light bulbs still

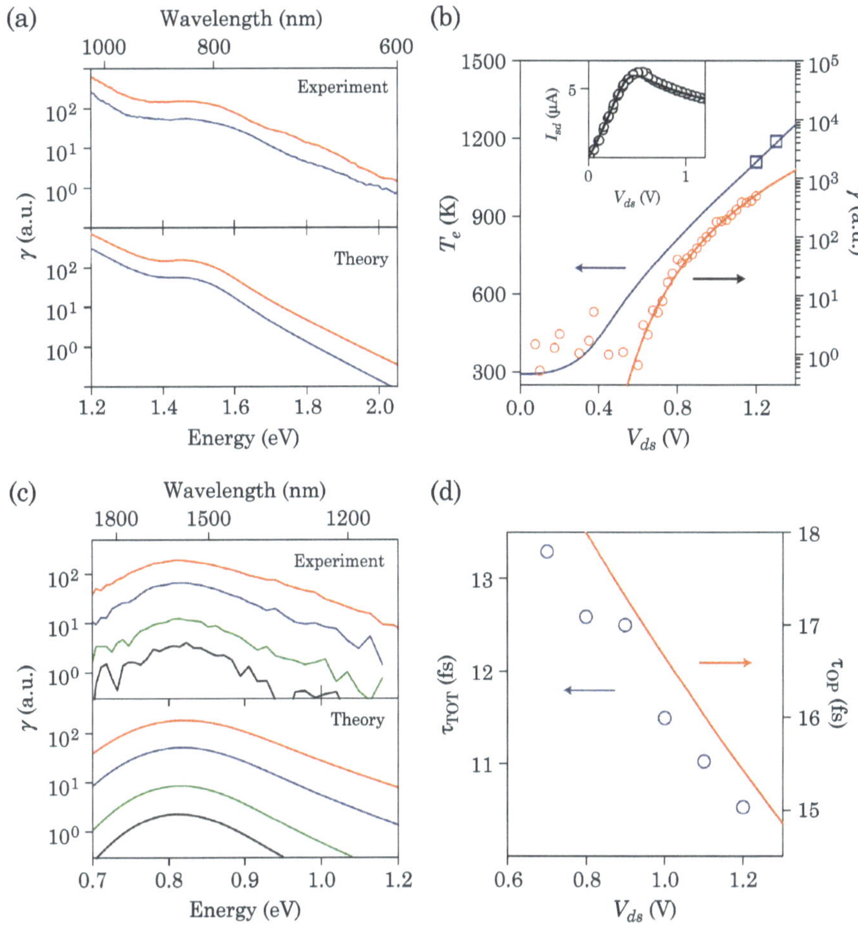

Figure 7.41 Comparison of measured and calculated properties of electrically driven thermal light emission from an individual suspended carbon nanotube. The top figure in (a) is the measured optical emission intensity in the visible as a function of photon energy for source–drain biases of 1.2 V (top line) and 1.3 V (bottom line). The bottom figure is the computed intensity from Eq. (7.48) in the text. (b) shows the measured (data points) and calculated (solid lines) electronic temperature and emission intensity as a function of source–drain bias. (c) compares the measured and calculated emission intensity in the infrared. (d) shows the extracted hot electron lifetime (symbols) compared with the calculated electron–phonon scattering time (solid line). Figure from Ref. [28].

represents a significant portion of the electricity consumed for lighting [30]. The reason for the continued presence of this old technology is the simplicity and low cost of implementing thermal radiation from electrical heating. Incandescent lighting is inherently a materials problem, where breakdown of the heated filament ultimately determines the lifetime of the light bulb. Despite the fact that the light bulb technology has been around for decades, little evolution has occurred in the materials that make the heated filaments. It is quite

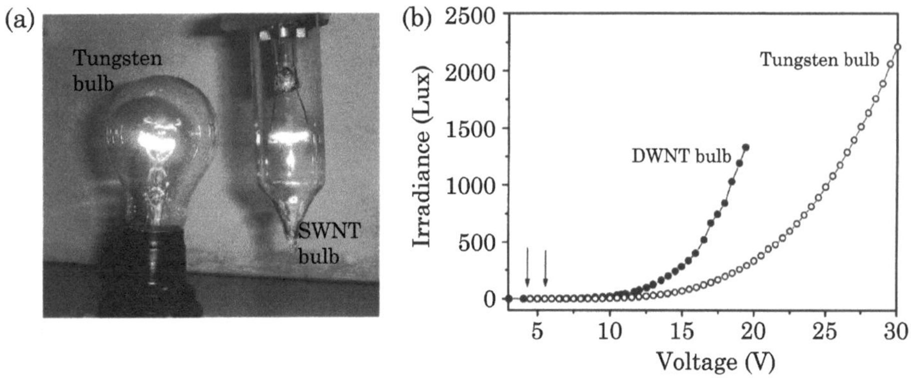

Figure 7.42 (a) Carbon nanotube light bulb next to a tungsten light bulb. (b) The irradiance of the nanotube bulb shows a lower onset of light emission and higher irradiance compared with the tungsten bulb. Figure from Ref. [31].

intriguing therefore that carbon nanotubes have been shown to make excellent filaments for conventional light bulbs [31]. To demonstrate the superiority of carbon nanotubes, strands of single- and double-walled carbon nanotubes were assembled into long filaments, and replaced tungsten filaments in household light bulbs (Fig. 7.42 (a)), with the glass bulb sealed under high vacuum. The measured irradiance of the nanotube light bulb is found to be higher than that of a tungsten bulb (Fig. 7.42 (b)), including a much lowered voltage turn-on. Furthermore, the nanotube filaments are found to be stable over long periods of operation, over many cycles: nanotube light bulbs worked well after more than 5000 cycles, and were found to be stable for up to 360 hours at an operation voltage of 25 V, where the temperature is about 1400 K.

7.5 Optical Detection with Functionalized Nanotubes

As we have shown in the previous sections, it is possible to generate photocurrents in carbon nanotubes using monochromatic illumination at laser intensities. However, for optical detection and switching, much lower light intensities need to be converted into electrical signals; one approach to achieve this goal is to functionalize carbon nanotubes with molecules that are optically active, and use the response of the molecules to modulate the current in a carbon nanotube field-effect transistor [32–34]. This approach relies on three mechanisms in optically-active molecules: (1) chromophores that undergo a change of dipole moment when exposed to light [32], (2) molecules that transfer charge when exposed to light [34], and (3) molecules that cause an increase in scattering in the nanotube when irradiated [33]. These approaches can be used to modulate the current in a field-effect transistor either by effectively changing

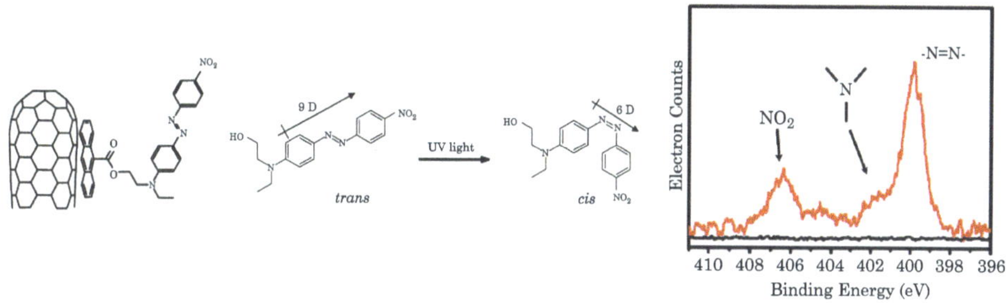

Figure 7.43 Sketch of a single-wall carbon nanotube functionalized with Disperse Red 1 (DR1), through an anthracene linker. Under UV light, DR1 isomerizes from the equilibrium *trans* conformation to the metastable *cis* conformation. In doing so, the molecular dipole moment changes from 9 to 6 D. Right: $N(1s)$ core level photoelectron spectrum of the physisorbed anthracene-DR1 functionality. Figure from Ref. [32].

the gate voltage felt by the nanotube, by effectively changing the nanotube doping or by causing a reduction of the current due to increased scattering.

7.5.1 Modulation of Molecular Dipole Moment

Fig. 7.43 shows a sketch of a carbon nanotube functionalized with Disperse Red 1 (DR1), an azobenzene molecule that is used to impart red color to acrylics. This molecule is known to isomerize under UV illumination, accompanied by a significant change in the dipole moment. Experiments have shown that at equilibrium, the DR1 chromophore is in the *trans* conformational state (Fig. 7.43) in which there is significant orbital overlap between the phenyl rings. Because of this conjugation and the presence of the strongly electronegative nitro end group, the *trans* conformation has a considerable, 9 D dipole moment. Under UV light, the chromophore isomerizes to the *cis* conformation (Fig. 7.43) where the orbital overlap is significantly reduced, leading to a smaller, 6 D dipole moment. If left in ambient conditions, the *cis* isomer will relax to the more stable and less sterically hindered *trans* conformation.

The nanotubes are functionalized using noncovalent attachment of the DR1 chromophore via an anthracene tether. Though such chromophores can be attached to the nanotube sidewalls either covalently or noncovalently, noncovalent attachment is advantageous because it only weakly perturbs the nanotube electronic states. Polycyclic aromatic hydrocarbons such as pyrene and anthracene have been shown to physisorb onto nanotubes by forming π–π bonds to the nanotube sidewall with little charge transfer. Since most π–π bound molecules can be easily removed using common solvents, noncovalent attachment also offers the ability to reversibly functionalize nanotubes. The anthracene modified DR1 is synthesized from 9 anthracenecarboxylic acid and DR1 using a dicyclohexylcarbodiimide esterification reaction. The crude product is purified by silica gel chromatography and the structure is confirmed using ^{1}H-NMR. After purification, the anthracene-DR1 is dissolved in dimethylformamide for application to the nanotubes. Pure anthracenecarboxylic acid is also dissolved for use as a nonfunctional control.

The individual nanotube transistors are fabricated on highly doped silicon wafers with a 500 nm thermal oxide using chemical vapor deposition (CVD) and subsequent lithography. The nanotubes are grown using iron and molybdenum nanoparticle catalysts at 900 °C using methane feedstock with hydrogen coflow. Under these CVD conditions, the average nanotube diameter is about 1.6 nm. After growth, electron and atomic force microscopy are used to locate individual nanotubes, and electron beam lithography, metal deposition, and 400 °C forming gas anneal (4:1 Ar/H2) are used to form contacts. After annealing, a drop of the anthracene-DR1 (or anthracenecarboxylic acid) solution is placed onto the chip and the sample is washed to remove the nonspecifically bound chromophores. X-ray photoelectron spectroscopy (XPS, Fig. 7.43) measurements confirm that the chromophore remains after washing and yield a coverage of 1–2 molecules per 100 nanotube carbon atoms, calculated from the $N(1s)$ to $C(1s)$ intensity ratio. I–V_g characteristics of individual single-wall carbon nanotube transistors before and after adsorption of the chromophore show minimal changes in the drain current, indicating that the anthracene-DR1 molecules cause minimal scattering of the charge carriers, and no degradation in device performance. Devices can exhibit threshold voltage shifts due to small chirality-dependent charge transfer from anthracene. Gate voltage scans to high positive voltages show no indication of the n-channel opening up, indicating that the transistors are unipolar. UV-induced switching is performed using a handheld UV lamp with 254 and 365 nm lines and a low intensity of 100 μW/cm^2. Electrical measurements are performed in a nitrogen purged cell to eliminate the potential for ozone oxidation from UV excitation of the atmospheric oxygen.

While measuring the drain current through a nanotube transistor, the gate voltage is varied to acquire a series of I–V_g characteristics (Fig. 7.44). Before illumination with UV light, the transistor shows p-type behavior with a threshold voltage of about 1 V. When the chromophore is isomerized to the *cis* conformation, the threshold voltage is shifted to the right, in this case by 0.7 V, and does not depend on the wavelength of UV light used. The threshold voltage shift under illumination was measured for several nanotube transistors; all show positive threshold shifts with a range of 0.6–1.2 V. This shift could indicate a charge-transfer mechanism [34] or a change in the local electrostatic environment [17,35]. Charge transfer from the chromophore to the nanotube is inhibited by the alkane spacer and anthracene tether separating the chromophore and the nanotube and control experiments show no threshold voltage shift, indicating that there is no photoinduced charge transfer from the anthracene tether. An estimate of the amount of charge transfer needed for a 1 V threshold voltage shift gives 0.07 e/molecule, so charge transfer cannot be entirely ruled out. However, since the UV photoisomerization and concurrent dipole moment change of DR1 is well established, it is proposed that the dipole moment change acts as a small local negative gate voltage. The large

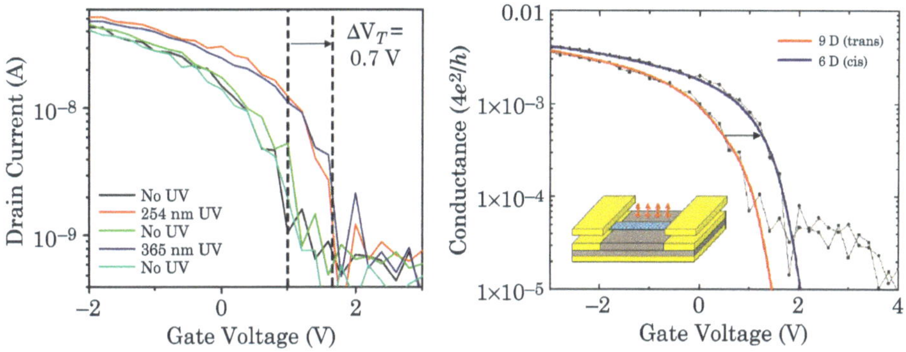

Figure 7.44 Left: transistor characteristics showing threshold voltage shifts under UV light. The threshold voltage is shifted to the right by 0.7 V for both 254 and 365 nm light and is fully reversible. Right: device simulations of the nanotube conductivity for both chromophore isomers (solid lines) compared with the experimental data (gray symbols). The transistor characteristic for the 9 D *trans* isomer (red curve, left) shifts toward positive gate voltages when the chromophore switches to the 6 D *cis* isomer (blue curve, right). As seen in the experiments, the threshold voltage is shifted to the right by 700 mV. Inset: sketch of nanotube transistor used in the simulations, with the red arrows indicating the molecular dipoles. Figures from Ref. [32].

transistor threshold voltage shift is due to the relatively short spacer group used to separate the chromophore from the nanotube (1 nm for anthracene-DR1).

The impact of the molecular dipoles on the conductance of the carbon nanotube can be understood by considering the change in electrostatic potential on the nanotube created by dipoles oriented perpendicular to the nanotube axis, as the dipole moment is changed from 9 to 6 D. For a single dipole, the potential change is

$$\Delta V_{\text{dip}}(z) = \frac{\Delta\mu}{4\pi\varepsilon_0}\left(\frac{1}{d\sqrt{h^2+z^2}} - \frac{1}{d\sqrt{(h+d)^2+z^2}}\right) \tag{7.50}$$

where $\Delta\mu$ is the change in dipole moment, and h $(h+d)$ is the distance between the positive (negative) charge of the dipole and the nanotube. For a single dipole this creates a localized potential perturbation, which causes weak scattering of the electrons. However, for a collection of dipoles separated by distance ξ on the order of h the potential of each individual dipole adds up to give a uniform potential on the tube

$$\begin{aligned}\Delta V_{\text{dip}}(z) &= \frac{\Delta\mu}{4\pi\varepsilon_0 d}\sum_{n=-\infty}^{\infty}\left(\frac{1}{\sqrt{h^2+(z-n\xi)^2}} - \frac{1}{\sqrt{(h+d)^2+(z-n\xi)^2}}\right)\\ &\approx \frac{\Delta\mu h}{4\pi\varepsilon_0\xi^3}\sum_{n=-\infty}^{\infty}\frac{1}{\left[(h/\xi)^2+n^2\right]^{3/2}}.\end{aligned} \tag{7.51}$$

The magnitude of the electrostatic potential change can be estimated using the values $h = 1$ nm, $\xi = 0.35$ nm and $\Delta\mu = 3$ D, to give $\Delta V_{\text{dip}} \approx 0.5$ V.

To further test this mechanism, quantum transport calculations of the nanotube transistor characteristics in the presence of dipoles were performed [32]. In brief, the nonequilibrium Green's function formalism implemented in a tight-binding formalism in conjunction with a Poisson solver is used to self-consistently calculate the charge and the electrostatic potential in the device, and the conductance [36,37]. A schematic of the simulated device is shown in Fig. 7.44; it consists of a (17,0) nanotube (bandgap 0.55 eV, radius 0.66 nm) sitting on SiO_2. At the ends of the computational cell, the nanotube is sandwiched between two metallic plates, forming the source and drain contacts. The bottom and top metallic plates have thicknesses of 2.25 and 2 nm, respectively, and are separated from the nanotube by 0.3 nm. The metal work function is 1 eV below the nanotube midgap, and after self-consistency, sits slightly below the valence band edge, creating an ohmic contact. Because of the computational demands of the calculations, the actual experimental device cannot be directly simulated; instead the calculations are done for a smaller device of 35 nm channel length and with a gate oxide thickness of 4.5 nm. The results of the computations are then related to the experimental device by scaling the gate voltage by the ratio of the capacitances of the experimental and simulated device, using the expression for the capacitance $2\pi\varepsilon/\ln(2t/R)$, where t is the gate oxide thickness and R is the nanotube radius. Because of the high density of molecules on the nanotube surface, the dipole moment is assumed to be perpendicular to the nanotube surface, and is modeled by a positive (negative) point charge of magnitude equal to the electron charge and at distance h $(h + d)$ above the nanotube. From the molecule configuration we have $h = 1$ nm, and the charge is distributed using a three-dimensional Gaussian distribution. The dipoles are evenly distributed on the nanotube using the range of experimentally measured chromophore densities. The *trans* to *cis* isomerization is studied by changing the value of d from 0.1 to 0.067 nm, reproducing the molecular dipole moments of 9 and 6 D. The calculated conductance as a function of gate voltage is shown in Fig. 7.44 for the *trans* and *cis* cases, overlaid on the experimental data. A dipole spacing of 0.35 nm is used, corresponding to the midpoint of the range of chromophore densities measured by XPS. (The calculated conductance was rescaled by a factor 1/140 to match the experimental conductance in the ON state of the nanotube transistor. The lower conductance of the experimental devices may be due to scattering in the nanotube or the presence of a small Schottky barrier at the contacts. The numerical results were also shifted by 0.65 V to match the threshold voltage of the *trans* experimental data. The need for this shift may be due to doping of the nanotube. The rescaling of the conductance and shift of the gate voltage are only necessary to reproduce the conductance curve prior to illumination. Once this is obtained, no further adjustment is made, and the shifted conductance curve is obtained by changing only the dipole moment.) A threshold voltage shift of 700 mV is found, in excellent agreement with the experiments and supporting

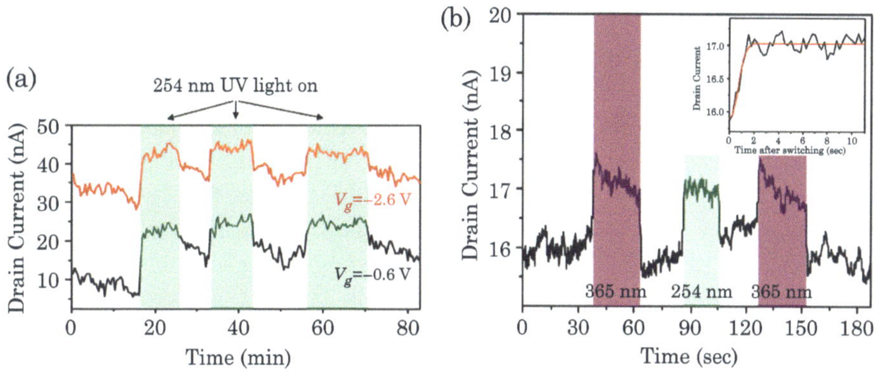

Figure 7.45 Time traces at fixed gate voltage for two different chromophore-functionalized nanotube transistors showing multiple repeatable switching events under both 254 and 365 nm light. Right inset: close up view of the transition under 254 nm light from the right graph. The smooth red line is a fit to the data using the Avrami equation (see text) with a time constant of ~2 s. Figure from Ref. [32].

the dipole change mechanism as the reason for the modulation of the nanotube conductance. Calculations were performed for several dipole densities. From these results, the uncertainty in the measured dipole density yields a range in the predicted threshold shift of approximately 0.45–0.95 V, correlating well with the 0.6–1.2 V shifts measured for other devices.

We turn next to the kinetics of the switching event, shown in Fig. 7.45. In panel (a) of this figure, *trans* to *cis* isomerization under 254 nm UV light leads to a uniform increase in the drain current. In addition, the magnitude of the current change is dependent on the gate voltage, with the largest signal arising from the subthreshold region of the I–V_g characteristic near $V_g = 0$ (see Fig. 7.45). In both cases shown, the sample was illuminated 3 times, each of which led to an appreciable increase in the measured drain current. All fabricated transistors show increased drain current under UV illumination, with abrupt and repeatable transitions. The transistors have shown repeatable switching for more than 100 cycles and over several months. DR1 has a finite absorption window in the UV region such that most UV wavelengths will isomerize the chromophore. Specifically, DR1 is sensitive to both 254 and 365 nm UV light, with an approximate 5:1 absorbance ratio (254 nm:365 nm). Indeed, we see in Fig. 7.45 (b) that 365 nm UV light also leads to modulation of the nanotube conduction, though the magnitude of change is the same for both wavelengths. Given the 5:1 absorbance contrast, the equal modulation for both UV wavelengths indicates that the isomerization has saturated under the low intensity UV lamp.

The inset to Fig. 7.45 shows a close-up view of the transition region under 254 nm UV light. The line through the data is a fit using the Avrami equation, $\Delta I = 1 - \exp(-Kt^n)$, where $n = 2$ [38]. Using the time it takes for the current to change from 10% to 90% of the full signal, we can extract a rise time of 2 s. The off-time is approximately the same, as are the switching times for 365 nm light. This rise time seems surprisingly long for a molecular switching event, but is consistent with solid state implementations of DR1 in polymer matrices where molecule–molecule interactions can inhibit the kinetics of the isomerization [39]. Because the chromophores form a dense layer on the nanotube side walls, steric interactions between chromophore units could reduce the overall switching rates. Indeed, the $n = 2$ Avrami exponent is characteristic of a 2D growth mechanism from randomly dispersed nuclei, suggesting that the isomerization proceeds from a few *cis* conformers in the DR1 layer. The measured switching time is also as fast or faster than other chromophore–nanotube systems. Experiments using other chromophore systems show switching times on the order of tens to thousands of seconds [33,34,40]. Assuming that sterics dominate the transition rate, diluting the chromophore concentration on the nanotube side wall could accelerate the switching.

It is important to point out that the low (100 $\mu W/cm^2$) intensities necessary to optically modulate the transistor are in stark contrast to measurements of intrinsic nanotube photoconductivity which typically require 1 kW/cm^2 intensities as discussed in the previous sections. In addition, synthetic control of the nanotube-hybrid systems may provide the ability to tune the absorption window and magnitude of dipole switching.

7.5.2 Charge Transfer

Because field-effect transistors are essentially devices that control the charge in the channel, mechanisms that lead to charge transfer to the channel can be used to impact the channel conductance. This effect can be particularly strong in carbon nanotubes because of the nanometer size of the channel. This effect has been exploited to modulate the conductance of carbon nanotube field-effect transistors with light by optically inducing charge transfer between the nanotube and molecules on the nanotube surface [34].

One particular molecule that is well-known to undergo photoinduced charge transfer is porphyrin, which plays an important role in photosynthesis. Experimental studies have shown that charge separated states exist in porphyrin-functionalized carbon nanotubes [41] with relatively long lifetimes. To explore the possibility of using such photoinduced charge transfer for optical detection and switching, carbon nanotube field-effect transistors were fabricated by growing single wall carbon nanotubes by chemical vapor deposition on a silicon wafer with 500 nm of gate oxide. This growth process yielded a network of nanotubes with a density of 1.6 nanotubes/μm^2; Pd/Cr electrodes were patterned on the nanotube network to produce field-effect transistors with channel lengths of 500 μm and widths of 1000 μm. The channel length is much longer than the nanotube length, so that transport through the nanotube network is through percolation; and since only 1/3 of the nanotubes are metallic, there are very few direct metallic paths across the channel, ensuring

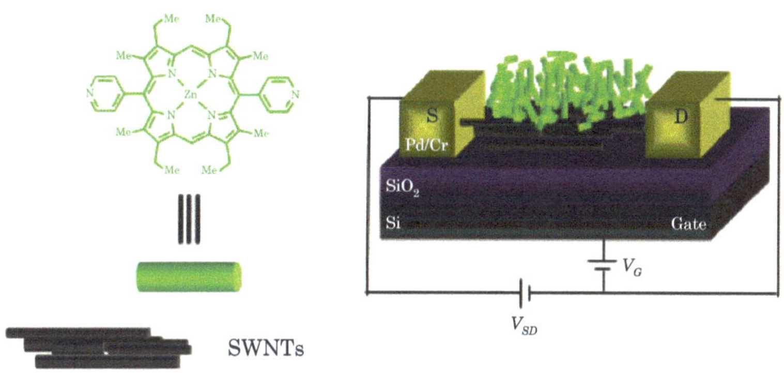

Figure 7.46 Sketch of the zinc(II) metalloporphyrin used to functionalize a carbon nanotube field-effect transistor, as shown on the right. The transistor channel is several microns long and the channel is made of a network of carbon nanotubes. Figure from Ref. [34].

a semiconducting behavior and thus transistor action. By making the channel length several hundred microns long the device resistance is dominated by the channel and not by the contacts. The nanotube network was functionalized with zinc(II) metalloporphyrin (see sketch of molecule and device in Fig. 7.46) with only the middle portion of the device exposed to the porphyrin molecules. The combination of the long channel length and the absence of functionalization near the contacts implies that any change to the device resistance upon exposure to light is not due to the contacts.

Fig. 7.47 shows the measured transfer characteristics of the carbon nanotube field-effect transistor before and after deposition of the porphyrin. The bare nanotube device shows relatively weak gate dependence, although this might be due to a threshold voltage beyond the measured values. After porphyrin functionalization, the device shows clear ON and OFF states with a threshold voltage of about 4 V. However, the magnitude of the current is decreased by more than an order of magnitude upon attachment of the porphyrin. This is in contrast with the DR1 functionalized nanotubes described in the previous section, where the noncovalent bonding with the anthracene linker did not affect the magnitude of the current. The origin of the current reduction can be due to increased electron scattering because of the molecules on the nanotube surface and/or changes in the conductivity at the nanotube–nanotube crossing points. The presence of a threshold voltage in the gate voltage window suggests that there is a decrease in the hole concentration in the nanotubes—this could arise from electron transfer from the porphyrin or from the displacement of oxygen from the surface of the nanotubes. The transformation of the transfer characteristics from the bare to the porphyrin-covered behavior is found to depend on the density of molecules on the surface of the nanotubes, with the curve of Fig. 7.47 corresponding to the saturated behavior.

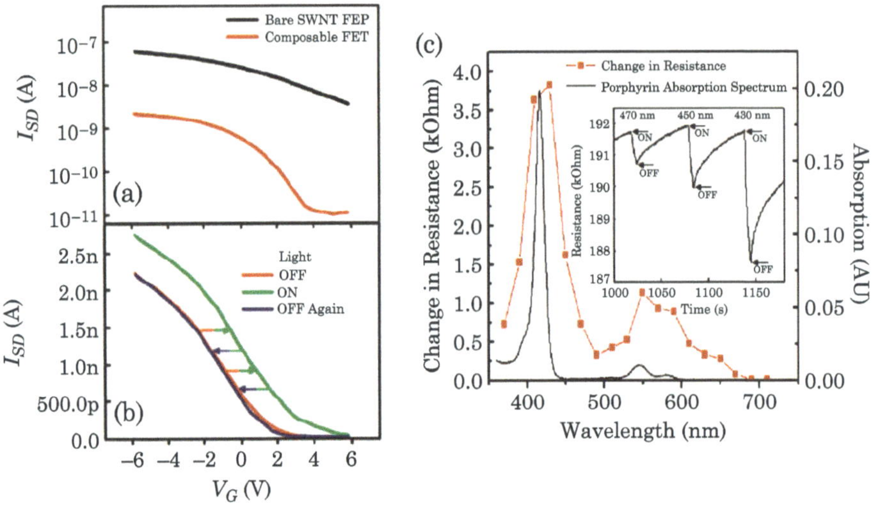

Figure 7.47 (a) Transfer characteristics of a carbon nanotube field-effect transistor before and after functionalization. (b) Transfer characteristics upon illumination with 420 nm light of intensity 10 mW/cm^2. (c) Change in resistance of the nanotube field-effect transistor as a function of the illumination wavelength compared with the absorption spectrum of the porphyrin molecule. Figures from Ref. [34].

The photoresponse of the porphyrin-functionalized carbon nanotube field-effect transistor is first measured by illuminating the device with light of 420 nm wavelength, which corresponds to the strongest absorption peak of porphyrin. The light intensity is 10 mW/cm^2. As the data of Fig. 7.47 indicates, the transfer characteristics are uniformly shifted to the right upon illumination, a behavior that is entirely reversible. This behavior is consistent with electron transfer from the nanotube to the porphyrin. To verify that the prophyrin molecule is indeed responsible for the photoswitching, the optical wavelength is varied from 370 nm to 710 nm in steps of 20 nm. Fig. 7.47 shows a comparison of the measured change in resistance with the absorption spectrum of porphyrin in solution: the two curves agree remarkably well.

The amount of charge transferred per prophyrin molecule can be estimated by relating the threshold voltage shift with the amount of charge in the nanotube

$$\Delta V_{\text{th}} = C^{-1}\Delta Q = enC^{-1}\frac{A}{a} \tag{7.52}$$

where A is the area of the channel covered by nanotubes, $a = 1$ nm^2 is the area that a molecule occupies on the surface of the nanotube, and n is the number of electrons transferred per molecule. For the particular device of Fig. 7.47, there are 1.6 nanotubes/μm^2; assuming that the area occupied by each nanotube is $L \times d$ where $L = 3$ μm is the nanotube length and $d = 1.5$ nm the diameter, we obtain $A/a = 3.6 \times 10^9$. The capacitance of the nanotube network is more

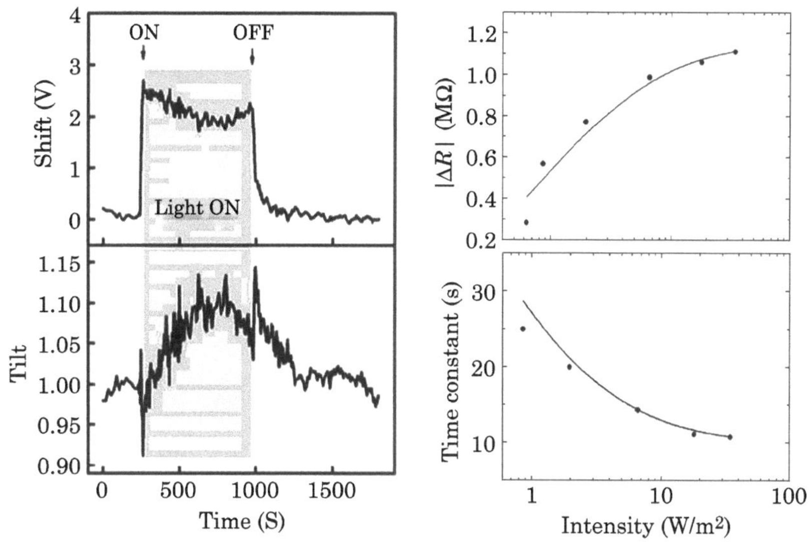

Figure 7.48 Left: time dependence of the threshold voltage shift and the inverse of the subthreshold swing as a function of time for the porphyrin-functionalized carbon nanotube field-effect transistor. The graph on the right shows the dependence of the change in device resistance and the time constant on the light intensity, with the solid lines best fits using Eqs. (7.57) and (7.59). Left figures from Ref. [34].

difficult to calculate precisely. However, one can consider the capacitance of an array of parallel nanotubes embedded in the oxide and separated from the gate by distance h. In that case, the capacitance is given by

$$C = \frac{2\pi\varepsilon LN}{\left[\ln(4h/d) + 2\sum_{n=1}^{N/2} \ln\left(\sqrt{h^2 + n^2 l^2}/nl\right)\right]} \tag{7.53}$$

where N is the total number of nanotubes in the array and l is the separation between nanotubes. Assuming that the nanotubes are aligned end-to-end to bridge the channel length, we have $N = 4800$ and $l = 0.208$ μm giving

$$C = \frac{2\pi\varepsilon LN}{\ln(4h/d) + 4.83}; \tag{7.54}$$

the capacitance is that of nanotubes in parallel, but with a correction in the denominator. For $h = 500$ nm and $d = 1.5$ nm the contribution for the single tube $\ln(4h/d) = 7.2$ is comparable to that of the other nanowires in the array. Thus, because the average separation between nanotubes is comparable to the gate thickness, there is a correction to the capacitance, reducing it by a factor of ~2. The numerical value $C = 43$ pF for the channel length of $L = 500$ μm is obtained. Therefore, using Eq. (7.52) and the measured threshold voltage shift of 1.7 V gives the number of electrons transferred per molecule $n = 0.13$. (The particular value depends on details of the capacitance, number of molecules on the nanotubes, etc. and was estimated to be 0.36 in Ref. [34].)

The dynamics of the charge transfer between the porphyrin and the nanotubes can be analyzed by monitoring the time-dependence of the threshold voltage shift when the light is turned ON and OFF (Fig. 7.48). The turn-on behavior occurs over a timescale of less than a microsecond, with an initial threshold voltage shift of almost 3 V. This voltage shift seems to fluctuate in a range between 2 and 3 V when the light is ON. The device recovers its initial behavior when the light is turned off, and this occurs over a somewhat longer timescale than the turn-on. A simultaneous monitoring of the tilt of the transfer characteristics near the threshold voltage (essentially the inverse of the subthreshold swing) shows that there is little change in this quantity, indicating that the mobility is not changing substantially during irradiation.

If we consider a dynamical system consisting of N_{ex} molecules in the excited state and N_g molecules in the ground state (with $N_{\text{tot}} = N_g + N_{\text{ex}}$) then we have

$$\frac{dN_{\text{ex}}}{dt} = \Gamma_1 N_g - \Gamma_2 N_{\text{ex}} \tag{7.55}$$

where Γ_1 is the transition rate for a molecule between the ground state and the excited state and Γ_2 is the transition rate between the excited state and the ground state. In the steady-state we have $dN_{\text{ex}}/dt = 0$ giving

$$\frac{N_{\text{ex}}}{N_{\text{tot}}} = \frac{\Gamma_1}{\Gamma_1 + \Gamma_2}. \tag{7.56}$$

Under the assumption that the transition rate between the ground state and the excited state is proportional to the light intensity, $\Gamma_1 = \alpha I_{\hbar\omega}$, and that the change in device resistance is proportional to the fraction of molecules in the excited state, we obtain

$$|\Delta R| = \beta \frac{\alpha I_{\hbar\omega}}{\alpha I_{\hbar\omega} + \Gamma_2}. \tag{7.57}$$

The experimental values of the change in resistance as a function of light intensity were extracted from the the data in Ref. [34] and are plotted in the right panel of Fig. 7.48. The change in resistance clearly increases with increase in light intensity, and is reasonably well described by Eq. (7.57). Further evidence that the time dependence of the response is governed by a dynamic equilibrium between excitations and relaxations is obtained by looking at the time-dependent resistance. By solving for the time dependence of the excited

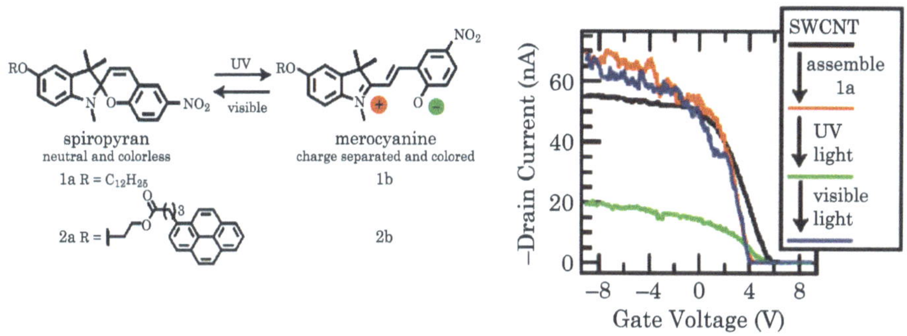

Figure 7.49 Left: sketch of the spiropyran-merocyanine transformation upon irradiation. Irradiation with UV light transforms spiropyran to merocyanine while visible light reverses the transformation. 1a and 2a show the photoactive molecule coupled with a linker for attachment to carbon nanotubes. Right: measured transfer characteristics of a carbon nanotube field-effect transistor functionalized with spiropyran. Figures from Ref. [33].

state population,

$$N_{\rm ex}(t) = \frac{\Gamma_1 N_{\rm tot}}{\Gamma_1 + \Gamma_2}\left[1 - e^{-(\alpha I_{\hbar\omega}+\Gamma_2)t}\right], \tag{7.58}$$

we obtain the time constant

$$\tau = \frac{1}{\alpha I_{\hbar\omega} + \Gamma_2}, \tag{7.59}$$

which decreases with increasing light intensity. Fig. 7.48 shows the time constant extracted from the experimental results of Ref. [34] compared with the expected functional form. The general behavior is generally good provided that a saturation time constant be added to the expression; this may indicate that the transition rate also depends on the number of molecules in the excited state.

7.5.3 Scattering

Photoswitching of carbon nanotube field-effect transistors has also been observed with funtionalization of nanotubes with spiropyran, a molecule that undergoes a transformation to a charge-separated state upon exposure to UV light (Fig. 7.49). These molecules were attached to the surface of single wall carbon nanotubes using pyrene and alkane groups, as these are thought to bind noncovalently. In contrast to the threshold voltage shift described above for dipole changes and charge transfer, current–voltage measurements indicate that the threshold voltage is not shifted when the devices are exposed to light; instead, the magnitude of the current is found to decrease upon exposure to

UV light. Further irradiation with visible light allows for a complete recovery of the initial transfer characteristics.

The most probable explanation for the reduction in current is increased scattering in the nanotube due to the molecules in the merocyanine form. While there may be several explanations for this behavior, they seem to rely on the belief that the molecules are not tightly packed on the nanotube surface [33]. For example, it is possible that the molecules are farther apart than the decay length of the electrostatic potential created by dipoles present in the merocyanine; in this case, localized potential perturbations are expected rather than an overall shift of the electrostatic potential as discussed in Section 7.5.1. A similar situation could arise if the charge transfer between the molecules and the nanotube is localized.

References

1. M.F. Islam, D.E. Milkie, C.L. Kane, A.G. Yodh and J.M. Kikkawa, "Direct measurement of the polarized absorption cross section of single-wall carbon nanotubes", *Phys. Rev. Lett.*, Vol. 93, p. 037404, 2004.
2. A. Ruini, M.J. Caldas, G. Bussi and E. Molinari, "Solid state effects on exciton states and optical properties of PPV", *Phys. Rev. Lett.*, Vol. 88, p. 206403, 2002.
3. F. Léonard and J. Tersoff, "Dielectric response of semiconducting carbon nanotubes", *Appl. Phys. Lett.*, Vol. 81, p. 4835, 2002.
4. C.D. Spataru, S. Ismail-Beigi, L.X. Benedict and S.G. Louie, "Excitonic effects and optical spectra of single-walled carbon nanotubes", *Phys. Rev. Lett.*, Vol. 92, p. 077402, 2004.
5. E. Chang, G. Bussi, A. Ruini and E. Molinari, "Excitons in carbon nanotubes: an ab initio symmetry-based approach", *Phys. Rev. Lett.*, Vol. 92, p. 196401, 2004.
6. C.D. Spataru, S. Ismail-Beigi, L.X. Benedict and S.G. Louie, "Excitonic effects and optical spectra of single-walled carbon nanotubes", *AIP Conf. Proc.*, Vol. 772, p. 1061, 2005.
7. H. Zhao and S. Mazumdar, "Electron–electron interaction effects on the optical excitations of semiconducting single-walled carbon nanotubes", *Phys. Rev. Lett.*, Vol. 93, p. 157402, 2004.
8. V. Perebeinos, J. Tersoff and Ph. Avouris, "Scaling of excitons in carbon nanotubes", *Phys. Rev. Lett.*, Vol. 92, p. 257402, 2004.
9. S.M. Bachilo, M.S. Strano, C. Kittrell, R.H. Hauge, R.E. Smaley and R.B. Weisman, "Structure-assigned optical spectra of single-walled carbon nanotubes", *Science*, Vol. 298, p. 2361, 2002.
10. F. Wang, G. Dukovic, L.E. Brus and T.F. Heinz, "The optical resonances in carbon nanotubes arise from excitons", *Science*, Vol. 308, p. 838, 2005.
11. R. Saito, G. Dresselhaus and M.S. Dresselhaus, "Trigonal warping effect of carbon nanotubes", *Phys. Rev.* B, Vol. 61, p. 2981, 2000.
12. V. Perebeinos and Ph. Avouris, "Exciton ionization, Franz-Keldysh and Stark effects in carbon nanotubes", *Nano Lett.*, Vol. 7, p. 609, 2007.
13. D.A. Stewart and F. Léonard, "Photocurrents in nanotube junctions", *Phys. Rev. Lett.*, Vol. 93, p. 107401, 2004.
14. D.A. Stewart and F. Léonard, "Energy conversion efficiency in nanotube optoelectronics", *Nano Lett.*, Vol. 5, p. 219, 2005.
15. M. Freitag, Y. Martin, J.A. Misewich, R. Martel and Ph. Avouris, "Photoconductivity of single carbon nanotubes", *Nano Lett.*, Vol. 3, p. 1067, 2003.

16. S.Y. Lin, R.Y. Tsai and S.C. Lee, "High-performance InAs/GaAs quantum-dot infrared photodetectors with a single-sided $Al_{0.3}Ga_{0.7}As$ blocking layer", *Appl. Phys. Lett.*, Vol. 78, p. 2784, 2001.
17. M.S. Marcus, J.M. Simmons, O.M. Castellini, R.J. Hamers and M.A. Eriksson, "Photogating carbon nanotube transistors", *J. Appl. Phys.*, Vol. 100, p. 084306, 2006.
18. J.U. Lee, "Photovoltaic effect in ideal carbon nanotube diodes", *Appl. Phys. Lett.*, Vol. 87, p. 073101, 2005.
19. J.U. Lee, "Band-gap renormalization in carbon nanotubes: Origin of the ideal diode behavior in carbon nanotube p–n structures", *Appl. Phys. Lett.*, Vol. 87, p. 073101, 2005.
20. J.U. Lee, P.J. Codella and M. Pietrzykowski, "Direct probe of excitonic and continuum transitions in the photocurrent spectroscopy of individual carbon nanotube $p-n$ diodes", *Appl. Phys. Lett.*, Vol. 90, p. 053103, 2007.
21. M.E. Itkis, F. Borondics, A. Yu and R.C. Haddon, "Bolometric infrared photoresponse of suspended single-walled carbon nanotube films", *Science*, Vol. 312, p. 413, 2006.
22. J.A. Misewich, R. Martel, Ph. Avouris, J.C. Tsang, S. Heinze and J. Tersoff, "Electrically induced optical emission from a carbon nanotube FET", *Science*, Vol. 300, p. 783, 2003.
23. M. Freitag, J. Chen, J. Tersoff, J.C. Tsang, Q. Fu, J. Liu and Ph. Avouris, "Mobile ambipolar domain in carbon-nanotube infrared emitters", *Phys. Rev. Lett.*, Vol. 93, p. 076803, 2004.
24. M. Freitag, J. Tersoff, J. Chen, J.C. Tsang and Ph. Avouris, "Electroluminescence in carbon nanotubes", *AIP Conf. Proc.*, Vol. 786, p. 477, 2005.
25. J. Tersoff, M. Freitag, J.C. Tsang and Ph. Avouris, "Device modeling of long-channel nanotube electro-optical emitter", *Appl. Phys. Lett.*, Vol. 86, p. 263108, 2005.
26. J. Chen, V. Perebeinos, M. Freitag, J. Tsang, Q. Fu, J. Liu and Ph. Avouris, "Bright infrared emission from electrically induced excitons in carbon nanotubes", *Science*, Vol. 310, p. 1171, 2005.
27. L. Marty, E. Adams, L. Albert, R. Doyon, D. Ménard and R. Martel, "Exciton formation and annihilation during 1D impact excitation of carbon nanotubes", *Phys. Rev. Lett.*, Vol. 96, p. 136803, 2006.
28. D. Mann, Y.K. Kato, A. Kinkhabwala, E. Pop, J. Cao, X. Wang, L. Zhang, Q. Wang, J. Guo and H. Dai, "Electrically driven thermal light emission from individual single-walled carbon nanotubes", *Nature Nanotech.*, Vol. 2, p. 33, 2007.
29. E. Pop, D. Mann, J. Cao, Q. Wang, K. Goodsen and H. Dai, "Negative differential conductance and hot phonons in suspended nanotube molecular wires", *Phys. Rev. Lett.*, Vol. 95, p. 155505, 2005.
30. Basic research needs for solid state lighting, Report of the Basic Energy Sciences Workshop on Solid State Lighting, Office of Science, United States Department of Energy (2006).
31. J. Wei, H. Zhou, D. Wu and B. Wei, "Carbon nanotube filaments in household light bulbs", *Appl. Phys. Lett.*, Vol. 84, p. 4869, 2004.
32. J. Simmons, I. In, V.E. Campbell, T.J. Mark, F. Léonard, P. Gopalan and M.A. Eriksson, "Optically modulated conduction in chromophore-functionalized single wall carbon nanotubes", *Phys. Rev. Lett.*, Vol. 98, p. 086802, 2007.
33. X. Guo, L. Huang, S. O'Brien, P. Kim and C. Nuckolls, "Directing and sensing changes in molecular conformation on individual carbon nanotube field effect transistors", *J. Am. Chem. Soc.*, Vol. 127, p. 15045, 2005.
34. D.S. Hecht, R.A. Ramirez, E. Artukovic, M. Briman, K. Chichak, J.F. Stoddart and G. Grüner, "Bioinspired detection of light using a porphyrin-sensitized single-wall nanotube field effect transistor", *Nano Lett.*, Vol. 6, p. 2031, 2006.
35. L. Larrimore, S. Nad, X. Zhou, H. Abruna and P.M. McEuen, "Probing electrostatic potentials in solution with carbon nanotube transistors", *Nano Lett.*, Vol. 6, p. 1329, 2006.
36. F. Léonard and D.A. Stewart, "Properties of short channel ballistic carbon nanotube transistors with ohmic contacts", *Nanotechnology*, Vol. 17, p. 4699, 2006.

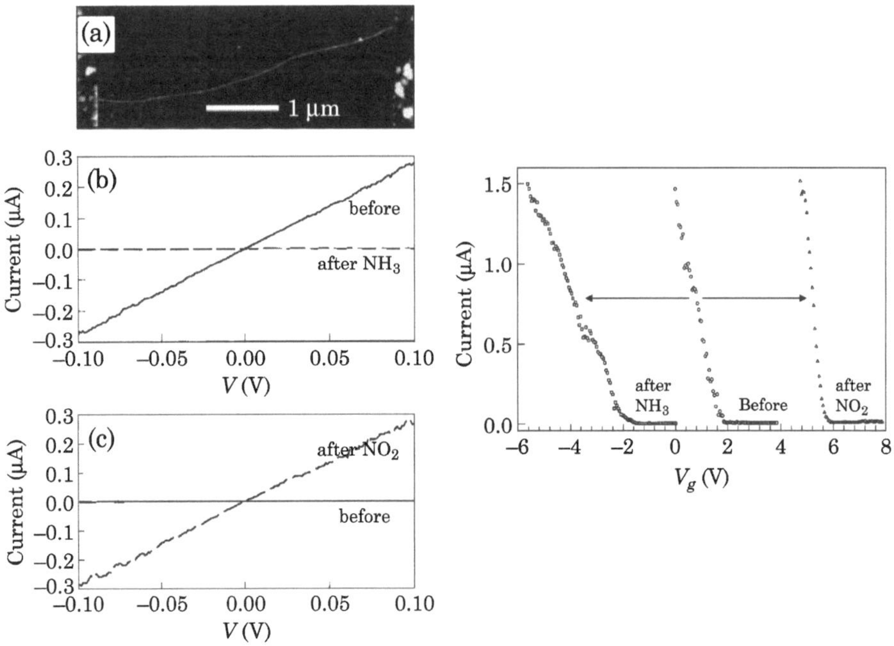

Figure 8.1 Left: (a) atomic force microscope image of a single carbon nanotube in a field-effect device. (b) Current versus voltage measured before and after exposure to NH_3. (c) Current versus voltage measured before and after exposure to NO_2. Right: transfer characteristics of the carbon nanotube field-effect transistor before and after exposure to NH_3 and NO_2. Exposure of the device to these analytes leads to a shift of the threshold voltage in the positive or negative direction. Figures from Ref. [1].

conductance at a gate voltage of 0 V is considerably decreased. For nitrogen dioxide, the small initial conductance at $V_G = +4$ V is considerably increased after exposure. These initial experiments clearly demonstrated the potential of carbon nanotubes for chemical sensing, and have spurred fundamental work to identify the mechanisms that lead to such a large electrical response in carbon nanotubes. We note that the reader will have encountered related mechanisms at the end of Chapter 7 on optoelectronic devices, in the context of optical detection with functionalized carbon nanotubes.

8.1 Sensing Mechanisms

8.1.1 Charge Transfer

Near the threshold voltage V_{th}, the charge in the channel of a carbon nanotube field-effect transistor depends linearly on the gate voltage, but the dependence of the conductance is exponential. Thus, the conductance is extremely sensitive to the charge in the channel, and this is the basis for utilizing carbon nanotube

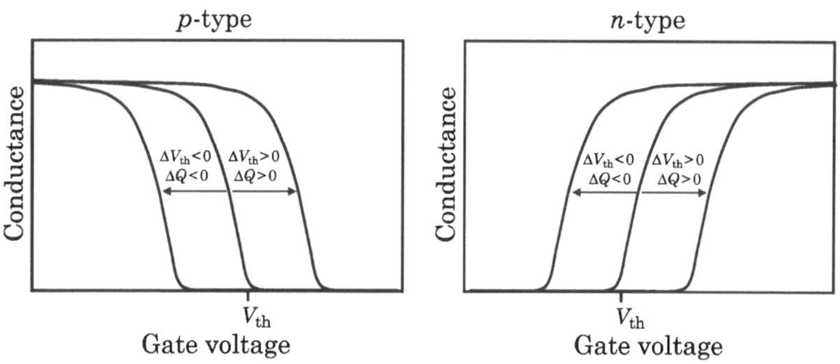

Figure 8.2 Sketch of the impact of charge transfer on the characteristics of carbon nanotube field-effect transistors, for p-type and n-type devices. ΔV_{th} is the shift of the threshold voltage upon transfer of charge ΔQ to the carbon nanotube.

field-effect transistors as ultrasensitive sensors. The concept is illustrated in Fig. 8.2. Whether the field-effect transistor is p-type or n-type, addition of negative charge in the channel leads to a shift of the threshold voltage to more negative values, while addition of positive charge leads to a positive threshold voltage shift. The transfer of positive charge to the carbon nanotube is detected with a p-type device by working in the OFF state, with the added positive charge causing a turn-on of the device. Similarly, negative charge is detected when a p-type device initially in the ON state turns off upon addition of negative charge to the nanotube. (These are simply reversed for a n-type device.) This general mechanism applies regardless of whether the device consists of a single nanotube or a network of nanotubes, as long as it has a clear gating effect.

Near the threshold voltage, the charge in the channel is proportional to the gate voltage according to

$$en = C\left(V_g - V_{\text{th}}\right) \tag{8.1}$$

where C is the capacitance per unit length between the gate and the nanotube and n is the number of electrons per unit length. In the presence of analytes on the nanotube surface, this equation is modified to

$$en + e\alpha\theta a^{-1}d = C\left(V_g - V_{\text{th}}\right) \tag{8.2}$$

where α is the number of electrons transferred per molecule, a is the area that a molecule occupies on the nanotube surface, d is the nanotube diameter, and θ is the surface coverage. By writing $V_{\text{th}} = V_{\text{th}}^0 + \Delta V_{\text{th}}$ we obtain

$$\Delta V_{\text{th}} = -e\alpha\theta C^{-1}a^{-1}d. \tag{8.3}$$

The largest change in the nanotube field-effect transistor conductance will occur when the device is operated near the threshold voltage. In this regime, the relative change in conductance is

$$\frac{\Delta G}{G} = \mp S^{-1} \Delta V_{\text{th}} = \pm S^{-1} e \alpha \theta C^{-1} a^{-1} d \tag{8.4}$$

where S is the subthreshold swing as discussed in Chapter 4. The top and bottom signs in this equation refer to n-type and p-type transistors respectively. Using the expression for the capacitance per unit length $C = \frac{2\pi\varepsilon}{\ln(4h/d)}$ we can obtain an expression for the maximum relative change in conductance

$$\left(\frac{\Delta G}{G}\right)_{\max} = \frac{e^2 d a^{-1} \ln(4h/d)}{2\pi\varepsilon kT \ln 10} \text{ decades.} \tag{8.5}$$

This expression is valid for complete coverage of the nanotube with analytes, in a device with the minimum possible subthreshold swing. As an example, for a nanotube of 1 nm diameter with a SiO_2 gate oxide thickness of 100 nm and an analyte occupying a surface area of 1 nm^2, we obtain a maximum ratio of about 75 decades. Of course, this value is unphysically high because the device conductance saturates, but the estimate drives the point that the change in nanotube conductance can be extremely large. At the other end of the spectrum, one can consider the impact of a single analyte on the nanotube conductance. Under the assumption that the transferred charge is delocalized over the entire channel length, we can estimate the relative change in conductance to be

$$\left(\frac{\Delta G}{G}\right)_{\min} = \frac{e^2 \alpha \ln(4h/d)}{2\pi\varepsilon kT \ln 10} \frac{1}{L} \text{ decades.} \tag{8.6}$$

The appearance of the channel length is made explicit in this expression. With the same parameters as above, we estimate a value for the ratio of α (75 nm$/L$) decades; for an analyte that transfers a full electron, a change of a factor of two in the conductance requires a channel length of less than 125 nm.

For detection of analytes of concentration c in a gas or liquid phase, it is useful to relate the surface coverage θ to the analyte concentration [2,3]. This can be accomplished by considering equilibrium surface coverage with analyte binding energy E_b and analyte chemical potential in the gas or liquid μ. The partition function is then given by

$$Z = 1 + z_{\text{vib}} e^{-(\mu - E_b)/kT} \tag{8.7}$$

where z_{vib} is the vibrational contribution. The chemical potential can be written as

$$\mu = \mu_0 + kT \ln x \tag{8.8}$$

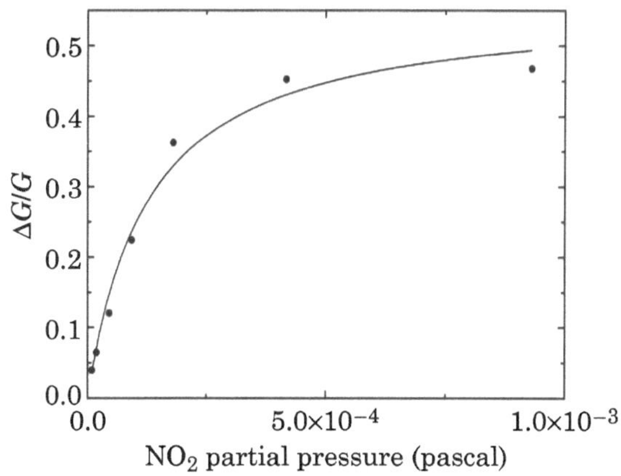

Figure 8.3 Relative conductance change of carbon nanotube transistors when exposed to NO_2. The symbols were extracted from Ref. [3] and the solid line is the best fit using Eq. (8.11).

where x is the mole fraction of the analyte, and μ_0 is the chemical potential of the pure substance. For small mole fractions, the nanotube surface coverage is calculated to be [4]

$$\theta = \frac{c}{c + c_0} \tag{8.9}$$

with

$$c_0 = \bar{c} z_{\text{vib}}^{-1} e^{-(\mu_0 - E_b)/kT}. \tag{8.10}$$

The expression for the concentration dependence on the coverage can be combined with that for the threshold voltage shift, Eq. (8.3), to obtain

$$\frac{\Delta G}{G} = \pm e\alpha\, dS^{-1}C^{-1}a^{-1}\frac{c}{c + c_0}. \tag{8.11}$$

From this equation and measurements of the threshold voltage shift as a function of the analyte concentration, the amount of charge transfer can be calculated. An example of the application of this approach is shown in Fig. 8.3. There, experimental data [3] for the relative conductance change of carbon nanotube network transistors when exposed to NO_2 is plotted as a function of the partial pressure of NO_2. Assuming that the concentration is proportional to the partial pressure, Eq. (8.11) can be used to obtain a relatively good description of the experimental data, as the solid line in the figure indicates.

The best fit gives a value for the prefactor $e\alpha S^{-1}C^{-1}a^{-1}d$ equal to 0.56. For this device, the channel is 4 microns long, the oxide is 500 nm thick, and there are 20 to 30 nanotubes in the channel separated by ~4 microns. Using Eq. (7.54)

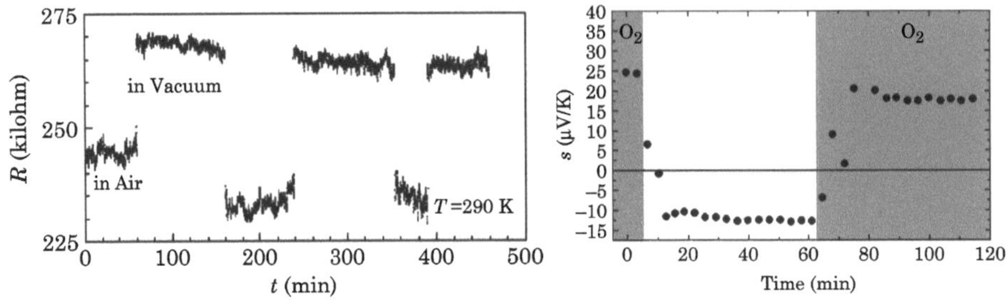

Figure 8.4 Measured resistance (left, from Ref. [5]) and Seebeck coefficient (right, after Ref. [5]) of carbon nanotube films in vacuum and when exposed to air.

with $N = 25$ and assuming a nanotube diameter of 2 nm, we obtain a device capacitance $C = 7.5 \times 10^{-10}$ F/m. From the device transfer characteristics, a subthreshold swing of $S = 14.2$ V/decade can be extracted, and for a molecular size $a = 10^{-19}$ m^2 a charge transfer of 0.07 electrons per molecule is obtained.

One of the early experiments that supported the charge transfer model consisted in measuring the resistance of networks of carbon nanotubes in vacuum and in oxygen-rich environments. The time-dependence of the resistance during such an exposure to air is shown in Fig. 8.4, where one can see that the resistance of the nanotube network decreases in air, with a full recovery observed as the device is returned to ultrahigh vacuum conditions. Based on the above discussion, the reduction in the resistance could arise from charge transfer if the nanotubes are n-type in vacuum and positive charge is transferred to them, or if the nanotubes are p-type and negative charge is transferred. While the simplest way to distinguish between these two possibilities is to sweep the gate voltage and obtain the transfer characteristics, a complementary method is to measure the Seebeck coefficient s. In thermoelectric materials, the Seebeck coefficient, defined as $s = dV/dT$, is a measure of the voltage generated across the material when a temperature difference is applied across it. The sign of s depends on the type of charge carrier: for a p-type (n-type) material s is positive (negative). Thus, concomitant measurements of the resistance and the Seebeck coefficient can provide evidence for the type of doping in the carbon nanotubes. Fig. 8.4 shows the measured Seebeck coefficient for a nanotube network device, indicating that in vacuum, the nanotubes are n-type, but that exposure to oxygen gives p-type doping. The implication is that oxygen transfers positive charge to the carbon nanotubes.

First principles calculations have been performed to study the properties of single carbon nanotubes with oxygen adsorbates[6,7]. The central result of these calculations is that oxygen molecules bind weekly to pristine zigzag and armchair nanotubes, whether they are semiconducting or metallic. The calculated binding energies for these situations are on the order of 0.05 eV, indicating physisorption on the nanotube wall. Furthermore, it is found that charge transfer from the nanotube to the oxygen molecule is weak, on the order

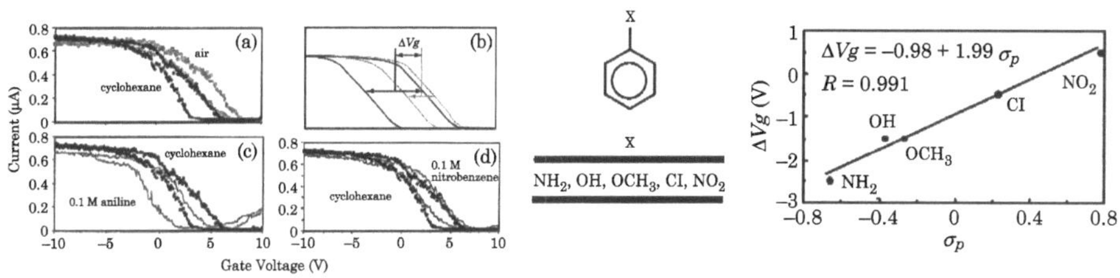

Figure 8.5 Left: impact of aromatic compounds on the transfer characteristics of carbon nanotube network transistors. (b) shows the definition of the threshold voltage shift taking into account hysteresis. Middle: the monosubstituted aromatic compounds studied. Right: measured threshold voltage shift of carbon nanotube field-effect transistors when exposed to aromatic compounds as a function of their Hammet parameter σ_p. Figures from Ref. [9].

of 0.01 electrons. The same calculations have found, however, that defects in the carbon nanotube lattice can have a profound effect on oxygen absorption. For example [6], the binding energy of an oxygen molecule to a so-called 7-5-5-7 defect is on the order of 0.3 eV, with 0.4 electrons transferred from the nanotube to the oxygen molecule. As will be discussed in a later section, experiments have also suggested that oxygen may have an impact on the properties of the nanotube/metal contacts [8], which may dominate the sensor response.

The shift in the threshold voltage of carbon nanotube transistors when exposed to analytes has been observed in a number of experiments. Support for the charge transfer model can be obtained by studying analytes with differing electron donating properties [9]. For example, monosubstituted benzenes such as aniline, phenol, toluene, chlorobenze, and nitrobenzene are believed to bind noncovalently to carbon nanotubes, but possess much different electron donating properties. A measure of this property is the Hammett parameter σ_p, which was introduced by Hammett [10] to describe the relationship between reaction rates and equilibrium properties of organic reactions. This parameter is related to the electron donating or withdrawing properties of the substituents on the benzene ring. The parameter is defined as zero for benzene, is positive for electron withdrawing species, and negative for electron donating species.

Fig. 8.5 shows the transfer characteristics of carbon nanotube network transistors when exposed to solutions of cyclohexane, and solutions of cyclohexane with 0.1 M aniline and nitrobenzene. (The conductivity of these liquids is low, and the device conductance is dominated by the carbon nanotube.) The general behavior is for the transfer characteristics to be shifted to the left or right, with the conductance in the ON state unaffected by the presence of the aromatic compounds. This behavior suggests that the analytes do not cause additional scattering in the carbon nanotube, while possibly causing a charge transfer that shifts the transfer characteristics. The devices show hysteresis; an average threshold voltage shift can be calculated using the procedure depicted in Fig. 8.5. From this procedure, the threshold voltage

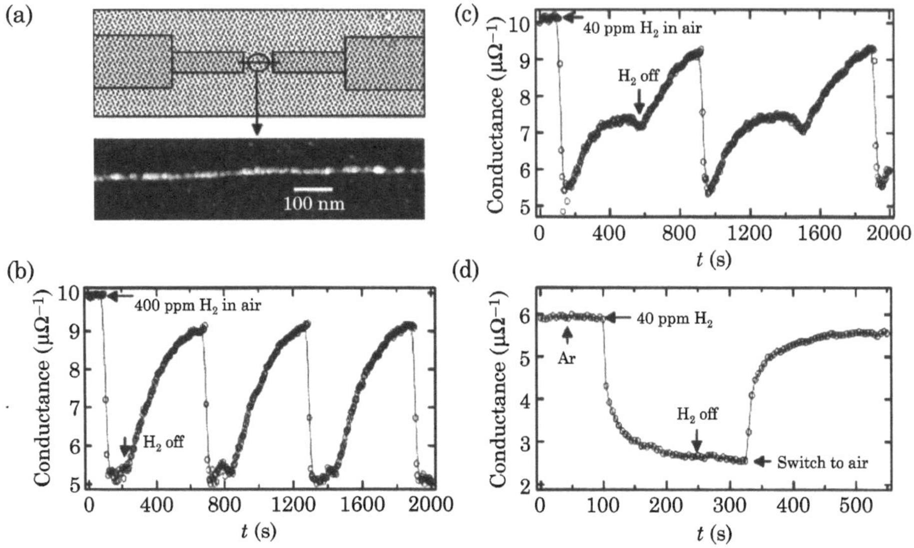

Figure 8.6 (a), top: schematic of the nanotube device with Pd nanoparticles coating the nanotube; bottom: atomic force microscopy image of an individual carbon nanotube decorated with Pd nanoparticles of diameters ranging from 2 nm to 3.5 nm. Conductance of nanotube device when exposed to (b) 400 ppm of hydrogen in air, (c) 40 ppm of hydrogen in air, and (d) 40 ppm of hydrogen in Ar, and then switched to air. Figure from Ref. [11].

shift is found to depend linearly on the Hammett parameter of the different substituents, as shown in Fig. 8.5.

Charge transfer effects can also arise when the work function of metal clusters on carbon nanotubes is modified by analytes [11]. For example, it is well-known that hydrogen diffuses readily into palladium, which can lead to a change in its work function. Indeed, this mechanism forms the basis of hydrogen gas sensor technologies. It is possible to take advantage of this mechanism in carbon nanotube devices by decorating the nanotube surface with nanoparticles of Pd. This is accomplished by first assembling a carbon nanotube electronic device, followed by electron beam evaporation of Pd on the whole device, leading to Pd nanoparticles decorating the nanotube sidewalls (Fig. 8.6). (The nanoparticle layer is not continuous, and the electronic transport is through the nanotube.)

The conductance of the Pd-decorated nanotube device decreases substantially when it is exposed to 400 parts per million of hydrogen in an air flow (Fig. 8.6 (b)), with a response time on the order of 5 to 10 s. Device recovery upon turn-off of the hydrogen gas is complete and occurs on a timescale of 400 s. It is believed [11] that the sensing mechanism is electron transfer from the Pd nanoparticles to the carbon nanotube due to a lowering of the Pd work function, which reduces the hole carrier concentration and decreases the conductance. It is interesting to note that competing hydrogen reactions can lead to an overshoot of the conductance reduction at low hydrogen concentrations. For example, Fig. 8.6 (c) shows that for 40 parts per million of hydrogen in air, the initial

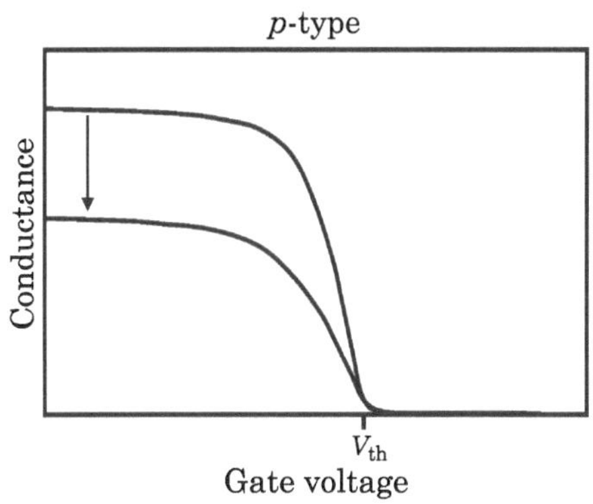

Figure 8.7 Illustration of the impact of increased scattering on the transfer characteristics of a *p*-type carbon nanotube field-effect transistors. The threshold voltage is unchanged, but the conductance in the ON state and the subthreshold swing decrease.

large drop of the conductance on the 5 to 10 s timescale is followed by a partial recovery of the conductance while the hydrogen is still flowing in the chamber. Experiments with the same concentration of hydrogen in Ar show that this behavior is not present, but that switching to air gives a full device recovery (Fig. 8.6 (d)). A likely scenario is that the hydrogen dissolved in the Pd reacts with oxygen in air, causing the hydrogen to leave the Pd in the form of water.

8.1.2 Scattering

The impact of analytes on the conductance of carbon nanotubes can also occur through an increased scattering of electrons in the channel. In this case, and in the absence of charge transfer and contact effects, the transfer characteristics of carbon nanotube transistors are expected to be modified according to Fig. 8.7. The figure shows that the threshold voltage is unaffected by the analytes, but that the ON state conductance, and hence the subthreshold swing, is reduced because of increased scattering.

The simplest model to describe the impact of analytes on the nanotube conductance is one where each analyte is treated as a point scatterer for coherent carrier scattering. In Eq. (4.102) and Fig. 4.47, we considered the transmission probability for an electron in the presence of two scatterers in series. Generalization to N scatterers in series each with transmission probability T_0 gives the expression for the total transmission probability

$$T = \frac{T_0}{T_0 + N(1 - T_0)}. \tag{8.12}$$

For a single scatterer ($N = 1$) the total transmission $T = T_0$, while in the limit of large N such that $N \gg \frac{T_0}{1-T_0}$ we have $T \sim \frac{1}{N}\frac{T_0}{1-T_0}$. This last expression

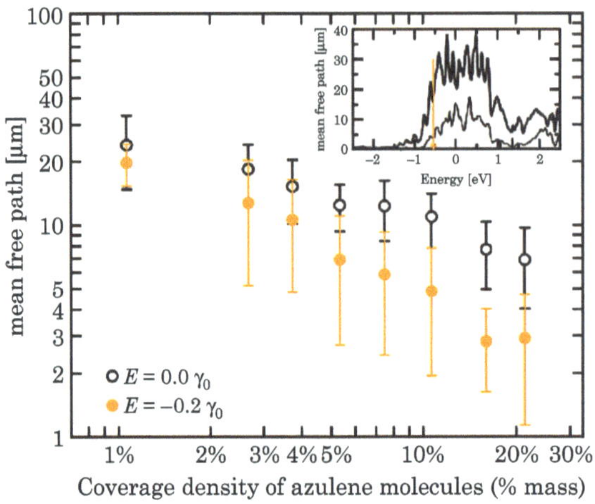

Figure 8.8 Calculated mean free path for electron scattering with physisorbed azulene molecules. At the Fermi level (open circles), the mean free path is longer than 7 microns, even at the highest coverages. Figure from [12].

is also valid at any N (including $N = 1$) if the transmission probability of an individual scatterer is small.

The time-dependence of the conductivity can be obtained by considering the number of analytes on the surface of the nanotube as a function of time. For analytes that attach to the surface of the nanotube with probability λ, the number of analytes depends on time as

$$N(t) = \frac{Ld}{a}\left(1 - e^{-\lambda \Phi a t}\right) \tag{8.13}$$

where Φ is the number of analytes impinging on the nanotube per unit time per unit area, a is the area that the analyte occupies on the surface of the nanotube, L is the length of the nanotube, and d its diameter. The ratio Ld/a, representing the maximum number of analytes that can attach on the surface, is somewhat approximate, but reflects the fact that nanotubes typically sit on a substrate, so only a portion of the nanotube surface is available for attachment. Combining Eqs. (8.12) and (8.13) with the Landauer expression for the conductance$G$ (Eq. (2.11) at zero temperature for a single, degenerate transmission channel), we obtain the time-dependent conductance as

$$G(t) = \frac{4e^2}{h} \frac{T_0}{T_0 + \frac{Ld}{a}\left(1 - T_0\right)\left(1 - e^{-\lambda \Phi a t}\right)}. \tag{8.14}$$

By fitting to experimental data, this formula can be utilized to extract the transmission coefficient of a single scatterer T_0 as well as the sticking probability λ.

More detailed theoretical work has been performed to understand the role of analytes on nanotube conductance[12]. For example, the impact of benzene (C_6H_6) and azulene ($C_{10}H_8$) on the nanotube electronic structure and electronic transport has been studied by calculating the distortions of the nanotube electronic structure in the presence of analytes, mapping this distortion into a tight-binding parametrization, and calculating the scattering mean free path. It is found that these simple molecules do not disturb the nanotube electronic structure significantly, especially around the Fermi level. Thus the scattering mean free paths are found to be larger than 7 microns for azulene, and larger than 100 microns for benzene even at coverages up to 20% by mass (Fig. 8.8). This indicates that surface functionalization of carbon nanotubes with molecules through π-stacking interactions will not appreciably distort the electronic structure or the transport properties. It also indicates that sensing through monitoring of the ON state conductance is more effective with molecules that chemisorb on the nanotube surface rather than physisorbed species like aromatic hydrocarbons (sensing of physisorbed analytes can also be accomplished if charge transfer takes place using the approach described in the previous section).

8.1.3 Contacts

As we have seen in previous chapters, contacts play an important role in carbon nanotube devices. In particular, the band alignment at nanotube/metal contacts depends strongly on the metal work function. Thus, analytes can have a serious impact on the conductance of nanotube devices if they modify the metal work function at the contact. Experiments using Kelvin probes have shown that the band alignment at nanotube/Au interfaces can be changed by as much as 0.1 eV upon exposure to oxygen [13].

Initial evidence for the importance of contacts in carbon nanotube sensors was provided by the study of single-nanotube field-effect transistors with Au contacts first annealed in vacuum and then exposed to various doses of oxygen [8]. Contrary to as-prepared devices, annealing the devices in vacuum leads to n-type field-effect transistors, as we have discussed in the context of logic circuits in Chapter 4. Exposure of these devices to oxygen leads to a recovery of the p-type behavior, as illustrated in Fig. 8.9. Importantly, it is found that the threshold voltage of the nanotube transistor is essentially unaffected by the presence of oxygen; instead, a gradual reduction of the conductance for positive voltages and a gradual increase of the conduction for negative gate voltages is observed. This behavior indicates modification of the Schottky barriers at the contacts. As the sketches of Fig. 8.9 illustrate, a device in vacuum initially has a metal Fermi level aligned close to the conduction band edge, leading to n-type behavior. As the oxygen dosage is increased, the Au Fermi level moves deeper into the bandgap, leading to ambipolar behavior. Finally, at the highest oxygen dosage, the Schottky barrier for holes is smaller than that for electrons, and the device behaves as a p-type transistor.

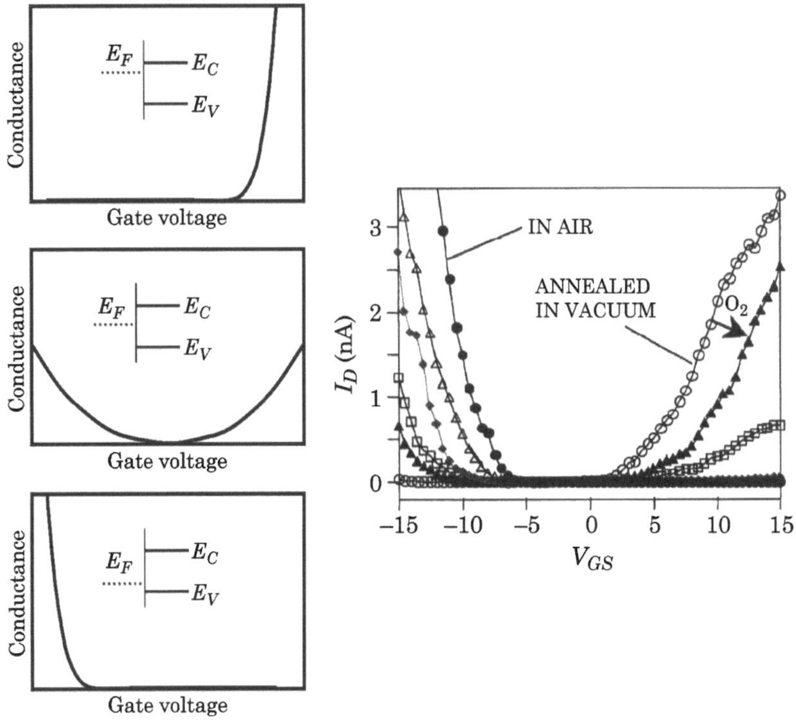

Figure 8.9 Left: sketch of the impact of a change in metal work function due to analytes on the transfer characteristics of carbon nanotube transistors. Right: measured transfer characteristics in vacuum and at different doses of oxygen, indicating a gradual change from n-type to p-type behavior. Right figure from Ref. [8].

To further explore the role of contacts, recent experiments have looked at the sensing properties of carbon nanotube transistors with contacts protected by polymeric layers [14] or self-assembled monolayers [15]. Fig. 8.10 shows the transfer characteristics of a carbon nanotube transistor with unprotected top Pd contacts of 75 nm thickness upon exposure to NO_2. In this experiment, the concentration of NO_2 in the chamber depends on time according to $n(t) = n_0\left(1 - e^{-t/t_0}\right)$ where $t_0 = 23$ min. Measurement of the transfer characteristics of the nanotube transistor shows that the current in the ON state increases with exposure time, and can be as much as a factor of three larger than for the device before exposure. A comparison of the time dependence of the conductance shows that it correlates with the time dependence of the NO_2 concentration in the chamber (Fig. 8.10). Similar experiments with a SU-8/PMMA polymeric protective layer of 2 micron thickness over the contacts (but leaving the nanotube channel unprotected) gives a different result: it is found that the time dependence of the conductance no longer correlates with the time dependence of the NO_2 concentration in the chamber, but instead varies much more slowly with time.

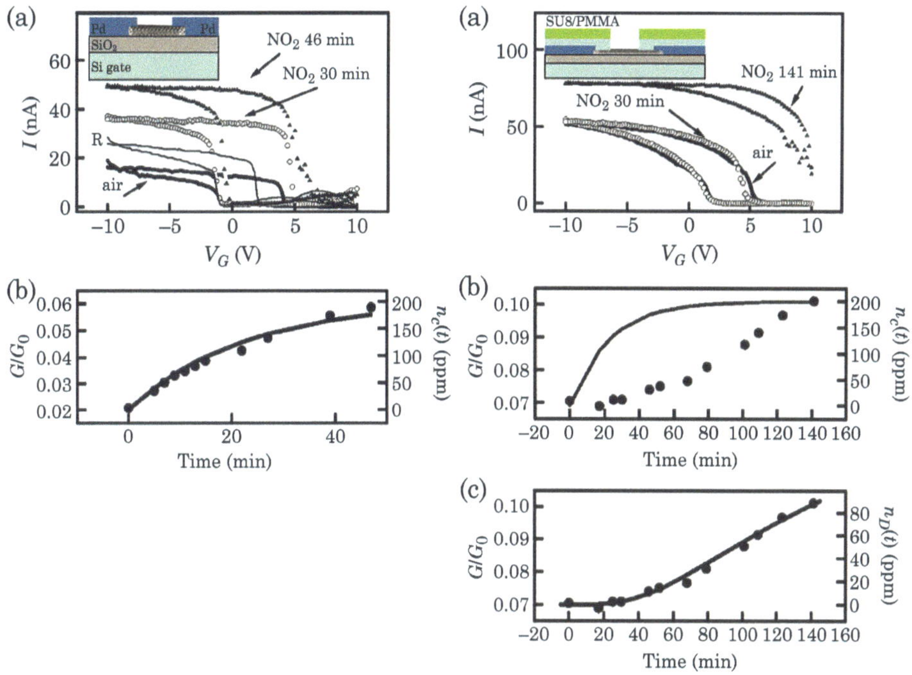

Figure 8.10 Transfer characteristics of a carbon nanotube field-effect transistor upon exposure to NO_2. The left and right panels show the results without and with a SU-8/PMMA protective layer on the contacts. The panels labeled (b) in both figures compare the measured time-dependence of the conductance with the expected time dependence of the concentration inside the chamber (solid line). The bottom right panel shows a comparison of the measured conductance with the calculated concentration of NO_2 from diffusion across the SU-8/PMMA layer. Figures from Ref. [14].

This observation can be explained by considering the concentration of NO_2 that diffuses through the protective layer and reaches the metal/nanotube interface. The situation is illustrated in Fig. 8.11. There, a protective layer of thickness L sits on top of a carbon nanotube. At the outer surface, exposed to the chamber, there is a time-dependent concentration of NO_2 given by that in the chamber. The time dependence of the concentration throughout the layer is obtained by solving the diffusion equation for the concentration

$$\frac{\partial c(x,t)}{\partial t} = D\frac{\partial^2 c(x,t)}{\partial x^2} \tag{8.15}$$

with the boundary condition at the surface

$$c(0,t) = c_0\left(1 - e^{-t/t_0}\right) \tag{8.16}$$

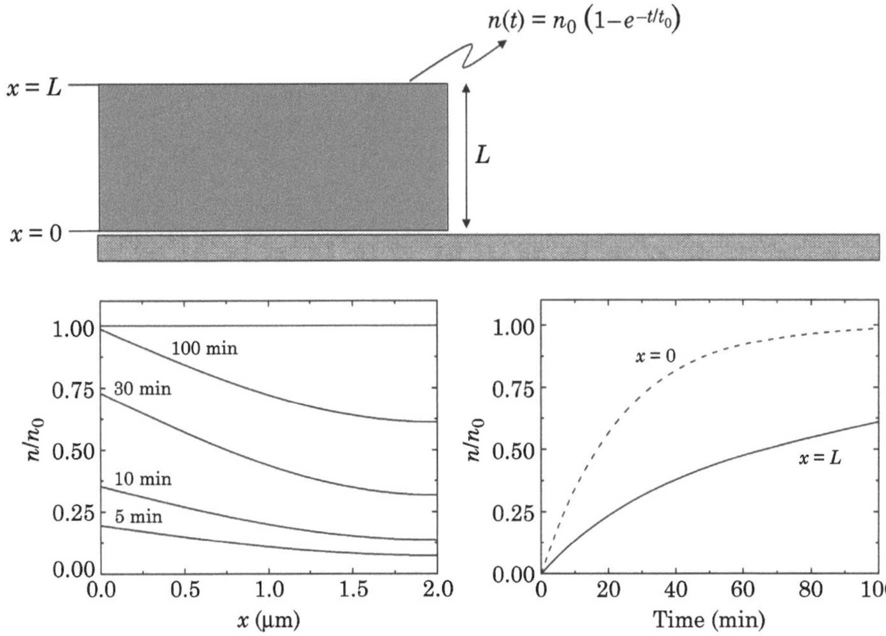

Figure 8.11 Top: schematic of the protective layer sitting on top of a carbon nanotube. Bottom left: calculated concentration profile for diffusion through a 2 micron thick layer, with a time-dependent boundary condition at the surface exposed to the chamber. Because of the finite diffusion coefficient, the concentration at the unexposed end of the layer ($x = L$) lags behind the concentration at the surface. Bottom right: calculated time dependence of the concentration at the nanotube surface (solid line) compared to that at the exposed surface (dashed line).

which corresponds to the concentration in the chamber. We assume that there is no flux of NO_2 through the bottom surface of the polymer layer, so that the boundary condition there is

$$\left.\frac{\partial c}{\partial x}\right|_{x=L} = 0. \tag{8.17}$$

The initial condition is $n(x, 0) = 0$. The diffusion equation in the presence of time-dependent boundary conditions can be solved with the change of variables

$$u(x,t) = c(x,t) - c_0\left(1 - e^{-t/t_0}\right) \tag{8.18}$$

leading to the nonhomogeneous differential equation

$$\frac{\partial u(x,t)}{\partial t} - D\frac{\partial^2 u(x,t)}{\partial x^2} = \frac{c_0}{t_0}e^{-t/t_0} \tag{8.19}$$

with the boundary conditions $u(0,t) = 0$ and $\partial_x u(L,t) = 0$. The original differential equation has been transformed to a nonhomogeneous differential equation but with homogeneous boundary conditions. The solution of this equation can be obtained by following standard procedures [16], leading to the full solution for the concentration:

$$c(x,t) = c_0 \left(1 - e^{-t/t_0}\right) - c_0 \times \sum_{m=1,3,5,\ldots}^{\infty} \frac{2}{\pi m} \frac{e^{-t/t_0} - e^{-\frac{D\pi^2 m^2}{4L^2}t}}{\frac{D\pi^2 m^2}{4L^2} t_0 - 1} \sin\left(\frac{\pi m}{2L} x\right). \tag{8.20}$$

Fig. 8.11 shows the time evolution of the concentration profile for a film thickness of 2 microns, a diffusion constant $D = 1.8 \times 10^{-16}$ m$^2/s$, and $t_0 =$ 23 min. Because the diffusion constant leads to a characteristic diffusion time $\sqrt{L/D} \approx 1500$ min much longer than the time necessary to achieve the steady-state concentration in the chamber, the concentration at the nanotube/metal interface lags behind that in the chamber, and is typically about 50% less for the parameters used here. The calculated concentration at the nanotube/metal interface correlates well with the time dependence of the conductance as shown in Fig. 8.10 (c). Furthermore, experiments where the whole device was covered in SU-8/PMMA showed the same behavior as the devices where only the contacts were protected, providing further evidence that for NO_2 the sensing mechanism is due to changes in the contact properties.

The change in the transfer characteristics due to diffusion of NO_2 to the nanotube/metal interface is most likely due to a decrease of the Schottky barrier. This is consistent with the increase of the ON state conductance with increased exposure. An alternative mechanism that could also apply to sensors with metallic carbon nanotubes arises if there is a tunneling barrier at the nanotube/metal interface. In this case, the presence of analytes can change the height of the tunneling barrier, causing an increase or decrease of the conductance. To explore this possibility, we consider tunneling across a thin vacuum layer at the interface between the metallic carbon nanotube and the metal contact, forming a metal/insulator/metal device, as illustrated in Fig. 8.12.

The tunneling current in such a system is given by [17]

$$I = I_0 \left[(\phi - V/2) \exp\left(-\sqrt{\phi - V/2}/V^*\right) - (\phi + V/2) \exp\left(-\sqrt{\phi + V/2}/V^*\right)\right] \tag{8.21}$$

where ϕ is the height of the tunnel barrier, and

$$V^* = \frac{\hbar}{eb}\sqrt{\frac{\phi}{m*}}. \tag{8.22}$$

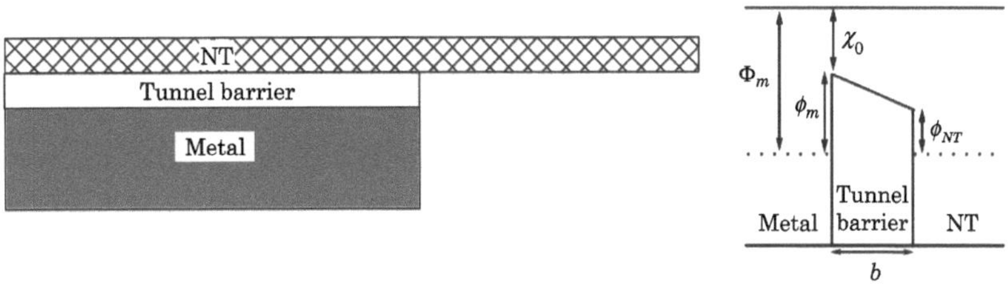

Figure 8.12 Illustration of a metal/metallic–nanotube interface with a thin tunneling barrier, and the band alignment defining the thickness of the tunnel layer b, the tunnel barrier heights on the metal ϕ_m and nanotube sides ϕ_{NT}, the metal work function Φ_m and tunnel barrier electron affinity χ_0.

In this last equation $m*$ is the effective mass of the tunneling electrons, and b is the tunneling distance. For simplicity we assume that $\phi = (\phi_{NT} + \phi_m)/2$ where ϕ_{NT} is the height of the tunnel barrier on the nanotube side, and ϕ_m is the height of the tunnel barrier on the metal side. For small bias voltages, the current depends on the barrier height as

$$I = \tilde{I} \exp\left(-\frac{2b\sqrt{2m*}\sqrt{\phi}}{\hbar}\right) \tag{8.23}$$

where the voltage dependence has been included in the prefactor $\tilde{I}$. We assume that the height of the tunnel barrier depends linearly on the concentration of analyte at the contact (for example, through a change in the metal work function). For an unprotected contact, or for fast diffusion of the analyte through the protective layer, we have

$$\phi(t) = \phi_0 + \Delta\phi\left(1 - e^{-\lambda \Phi L\, dt}\right) \tag{8.24}$$

where $\Delta\phi$ is the change in barrier height with concentration, λ is the sticking coefficient, Ld is the nanotube area available to analytes, and Φ is the flux of analytes on the nanotube surface. This sensing mechanism therefore gives a unique signature in the time-dependence of the conductance which allows to distinguish it from other mechanisms.

8.1.4 Capacitance

While this chapter has focused mainly on changes of the nanotube conductance due to analytes, an alternative approach to detect the presence of analytes is through the measurement of the capacitance. This approach has so far been demonstrated using networks of carbon nanotubes, and applied to sensing of various chemical species and agents [18,19]. Fig. 8.13

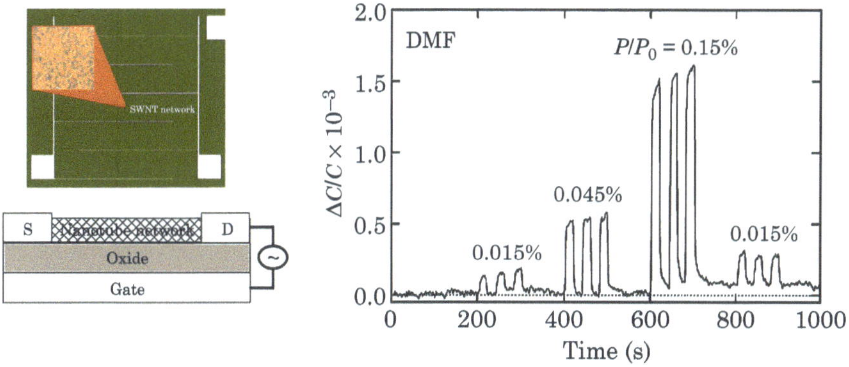

Figure 8.13 Left: optical micrograph and cross-sectional sketch of a carbon nanotube network capacitor used for sensing experiments. The capacitance is measured by applying an AC voltage between the nanotube film and the gate. Right: relative change in capacitance when the device is exposed to dimethyl formamide (DMF). Figures from Ref. [18].

illustrates a device utilized to perform these measurements. It consists of a network of carbon nanotubes, with a mixture of semiconducting and metallic nanotubes. An interdigitated array of Pd electrodes is patterned on top of the nanotube network, which sits on SiO_2. An AC voltage is applied between the nanotube network (through the Pd electrodes) and a backgate, and the capacitance is measured by detecting the out-of-phase AC current with a lock-in amplifier. Fig. 8.13 shows the results of such capacitance measurements when the nanotube network is exposed to dimethyl formamide of varying vapor concentrations. The figure clearly indicates a noticeable change in the capacitance on a short timescale, both for the turn-on and the recovery (the response time is less than 4 s, limited by the vapor-delivery system). A broad range of analytes have been shown to give a capacitance response, and notably, a minimum detectable level of 50 parts per billion has been achieved for dimethylmethylphosphonate (DMMP), a simulant for the nerve agent sarin [18]. Comparisons with commercial chemicapacitors are quite favorable. For example, commercial chemicapacitors can detect acetone with a minimum detection limit of 2 parts per million and detection time of 228 s. This can be compared with the values of 0.5 parts per million and detection time of less than 4 s for the nanotube network capacitor. Similar comparisons for the detection of DMMP are equally favorable.

It has been proposed that the change in capacitance is due to the dipole moments of the analytes, which can change the effective dielectric constant of the capacitor [18]. While some polar molecules have shown increased capacitance sensitivity compared to nonpolar molecules, measurements across a broad range of molecular dipole moments do not show a strong correlation. This has lead to controlled experiments [19] to further explore the sensing mechanism in these types of measurements, as we now discuss.

The proposed concept is that the nanotube network device (channel width w and channel thickness t) behaves like a transmission line with resistance per length r, back-gate electrostatic capacitance $c_c = \varepsilon w/t$, inductance $l = \mu_0 t$, and oxide conductance$g$. The impedance of such a transmission line is given by

$$Z = \sqrt{\frac{r + i\omega l}{g + i\omega c_c}}. \tag{8.25}$$

Assuming that the oxide conductance and nanotube network inductance are negligible, (at frequencies $\omega \ll r/l$) the nanotube network device can be modeled as a transmission line with a characteristic length

$$l_0 = \sqrt{\frac{2}{r\omega c_c}}. \tag{8.26}$$

For channel lengths much larger than this characteristic length scale, the impedance is

$$Z = \frac{(1-i)}{\sqrt{2}}\sqrt{\frac{r}{\omega c_c}}. \tag{8.27}$$

This can be compared with the impedance of a RC circuit

$$Z = R_{NT} - \frac{i}{\omega C_{NT}}, \tag{8.28}$$

to obtain the effective capacitance of the nanotube network as

$$C_{NT} = \sqrt{\frac{2c_c}{\omega r}} = c_c l_0. \tag{8.29}$$

Note that this equation indicates that there is an intimate connection between the resistivity of the nanotube network and its capacitance, $C_{NT} \sim r^{-1/2}$. Because of this relationship, the relative sensitivities of the capacitance and the resistance to analytes follows a relation

$$\frac{\Delta C}{C} = -\frac{1}{2}\frac{\Delta r}{r}. \tag{8.30}$$

This equation indicates that the relative change in capacitance can be entirely due to a relative change in the resistivity of the nanotube network. This relationship is confirmed by measurements comparing the capacitance and resistivity changes of carbon nanotube networks in ultra-high vacuum and upon exposure to acetone, water, and argon. As shown in Fig. 8.14, in all of these

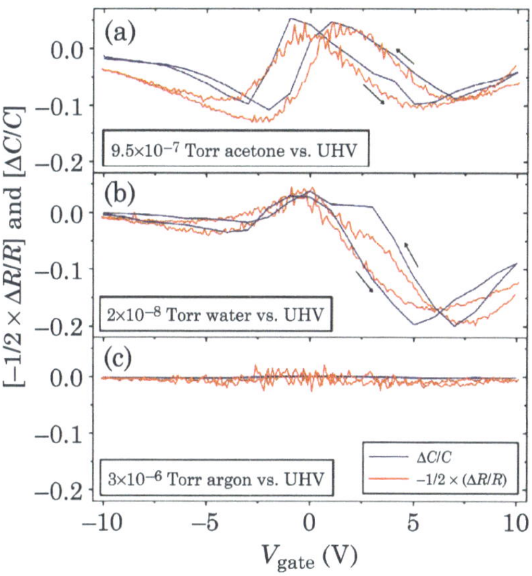

Figure 8.14 Comparison of the measured relative changes in capacitance and resistance for carbon nanotube networks exposed to acetone, water and argon. Figure from Ref. [19].

cases the relationship (8.30) is satisfied, indicating that resistivity changes are the most likely mechanism for sensing of these analytes. While the ratio of $-1/2$ has also been seen by other researchers [20], deviations from this relation also occur [20].

8.2 Liquid Gating

Liquid gating is a technique to achieve strong field-effect behavior in carbon nanotubes when the nanotube is immersed in solution. This is important because many sensing applications require the ability to detect analytes or to monitor reactions occurring in the liquid phase, including the ability to detect biochemical reactions in, on, and around cells. In this section, we derive an expression for the capacitance of an electrolyte-gated carbon nanotube by adapting the Gouy–Chapman theory of the double-layer capacitance.

The system under consideration is illustrated in Fig. 8.15. There, a carbon nanotube of radius R with linear charge density λ is immersed in a charge-neutral electrolyte, and a reference electrode sets the potential at V_{lg} where the subscript "lg" stands for "liquid gate". For simplicity, we will assume that the reference potential is set infinitely far from the nanotube surface. In the electrolyte, Poisson's equation is

$$\frac{\partial^2 V}{\partial r^2} + \frac{1}{r}\frac{\partial V}{\partial r} = -\frac{\rho(r)}{\varepsilon} \tag{8.31}$$

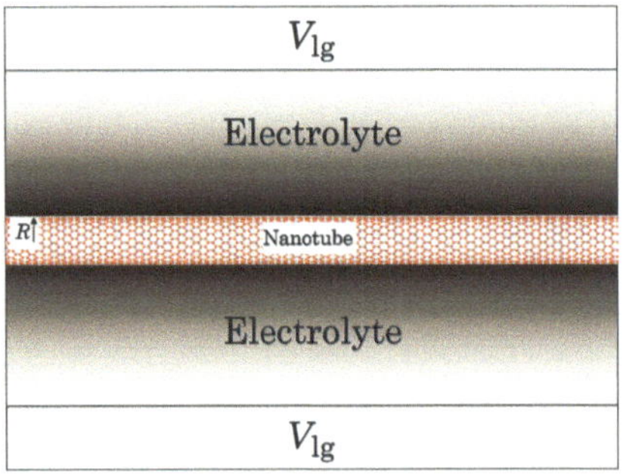

Figure 8.15 Cross-section of the system for calculation of the capacitance of a liquid gate. A carbon nanotube is immersed in an electrolyte, and a reference potential V_{lg} is set far from the nanotube. The gray shading represents the distribution of charge in the electrolyte that screens the charge on the nanotube.

where $\rho(r)$ is the charge distribution in the electrolyte, and ε is the dielectric constant of the electrolyte.

In the Gouy–Chapman theory, it is assumed that the excess concentration $c(r)$ of ions in the solution is given by a Boltzmann distribution

$$c(r) = c_0 - c_0 \exp\left[-\frac{E(r)}{kT}\right] \tag{8.32}$$

where c_0 is the concentration in the neutral solution and $E(r)$ is the energy required to bring an ion from the reference potential to position r. In our case that energy for an ion of valence z is simply $E(r) = ze\delta V(r)$ and we have

$$c(r) = c_0 - c_0 \exp\left[-\frac{ze\delta V(r)}{kT}\right] \approx \frac{ze\delta V(r)}{kT} \tag{8.33}$$

where the last approximation assumes that the potential is small. Using the relation $\rho(r) = -ezn_0 c(r)$ with n_0 the number of ions per unit volume, we obtain Poisson's equation as

$$\frac{\partial^2 \delta V}{\partial r^2} + \frac{1}{r}\frac{\partial \delta V}{\partial r} = \frac{z^2 e^2 n_0}{\varepsilon kT}\delta V(r). \tag{8.34}$$

The solution of this equation with the boundary conditions that $V = V_{lg}$ as $r \to \infty$ and $V = 0$ at $r = R$ is

$$V(r) = V_{lg}\left[1 - \frac{K_0(r/l)}{K_0(R/l)}\right] \tag{8.35}$$

where $K_0(x)$ is the modified Bessel function of order 0. The parameter

$$l = \sqrt{\frac{\varepsilon kT}{z^2 e^2 n_0}} \tag{8.36}$$

has units of length and represents the screening length in the electrolyte. Indeed, for $r \gg l$ the potential can be approximated as

$$V(r) \approx V_{\mathrm{lg}} \left[1 - \frac{1}{K_0(R/l)} \sqrt{\frac{\pi l}{2r}} e^{-r/l} \right] \tag{8.37}$$

so the potential decays over a distance l.

To obtain the capacitance we calculate the charge induced on the nanotube from

$$Q_{\mathrm{tot}} = \frac{2\pi z^2 e^2 n_0 V_{\mathrm{lg}} L}{kT} \int_R^\infty \frac{K_0(r/l)}{K_0(R/l)} r dr \approx \frac{2\pi\varepsilon V_{\mathrm{lg}} L}{K_0(R/l)} \tag{8.38}$$

where L is the nanotube length. The capacitance per unit length is then

$$\frac{C_{\mathrm{lg}}}{L} = \frac{2\pi\varepsilon V_{\mathrm{lg}}}{K_0(R/l)} \approx \frac{2\pi\varepsilon}{\ln(2l/R\gamma)}. \tag{8.39}$$

In this last equation γ is Euler's constant and we assumed that $R \ll l$. This expression for the capacitance can be compared with that of a backgate

$$\frac{C_{bg}}{L} = \frac{2\pi\varepsilon}{\ln(2h/R)}. \tag{8.40}$$

While both have the same functional form, there are two crucial differences. First, the dielectric constant of electrolytes is typically much larger than that of gate insulators (water has a dielectric constant of 80 for example). Second, the gate-oxide thickness is replaced with the length scale $2l/\gamma$ and for water this is on the order of 1 nm. The combination of these two factors leads to a liquid-gate capacitance per unit length on the order of 10 aF/nm, two orders of magnitude larger than a typical backgate capacitance. An important implication of this result is that the intrinsic capacitance of the nanotube, which is usually neglected because it is much larger than the backgate capacitance, becomes the dominant capacitance. Liquid gating has been utilized to detect protein binding [21] and, as will be discussed in Section 8.3.1, enzymatic reactions [22]. We note that the liquid gate can be utilized in two distinct ways: first, it can serve as a very effective gate to probe the conductance of the carbon nanotube, and therefore explore the role of analytes in terms of charge transfer, scattering and contacts. Second, the binding of analytes, as well as chemical and biological reactions, can lead to a disruption of the double-layer near the carbon nanotube, and thus a change in the device properties. Which mechanism operates in practice depends on the particular event being monitored.

8.3 Functionalized Nanotubes

From the discussion in the previous sections, it is clear that carbon nanotubes are quite sensitive to their environment. While this can be advantageous for sensing applications, the extreme sensitivity to the environment also implies that analyte specificity is more difficult to achieve. To this end, functionalization of the nanotube surface is necessary to target specific chemical and biological agents. The functionalization schemes fall into two classes: selective schemes where the functionalization reacts only with a limited range of analytes, and blocking schemes where a surface layer allows only a few analytes to reach the carbon nanotube.

8.3.1 DNA Functionalization

Single-stranded DNA (ssDNA) was initially utilized in the carbon nanotube arena as a surfactant to isolate individual nanotubes in solution [23]. Electrical measurements subsequently showed that the ssDNA does not alter the conductance of individual nanotubes appreciably [24]. Because of this preservation of the high nanotube conductance and the unique recognition capability of ssDNA, sensors with ssDNA-functionalized carbon nanotubes have been explored [25–27].

In one example [25], a carbon nanotube field-effect transistor made of a single carbon nanotube is fabricated, with ssDNA applied to the carbon nanotube by deposition of a drop of distilled water with diluted ssDNA. Two different sequences of ssDNA were applied to the carbon nanotube: a 21-mer sequence (5' GAG TCT GTG GAG GAG GTA GTC 3', sequence 1) and a 24-mer sequence (5' CTT CTG TCT TGA TGT TTG TCA AAC 3', sequence 2). These sequences are chosen because previous experiments in the context of artificial noses demonstrated their sensitivity to small, vapor-phase molecules. Atomic force microscopy of the same nanotube before and after application of the ssDNA solution indicates an increase in the apparent height from 5.4 nm to 7.2 nm, indicating the presence of a 1–2 nm thick layer of ssDNA on the surface of the nanotube. Support for the formation of ssDNA/carbon nanotube hybrids has also been provided by other experiments where the nanotubes were functionalized with ssDNA in solution, with subsequent heating of the solution above 80 °C leading to precipitation of the carbon nanotubes, indicating dissociation of ssDNA from nanotube surface [24].

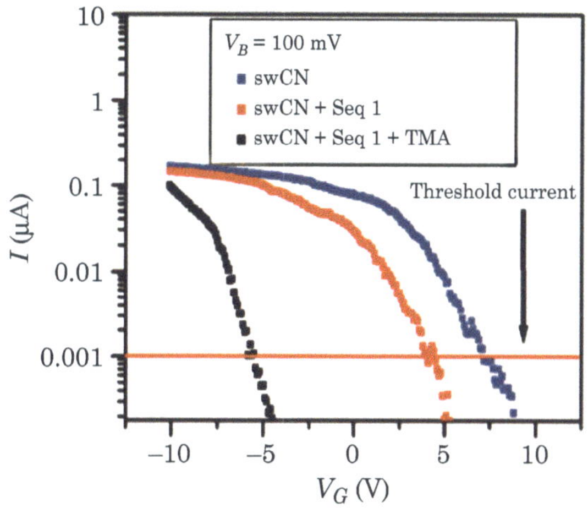

Figure 8.16 Transfer characteristics of the bare nanotube device, the device upon functionalization with ssDNA, and the functionalized device exposed to trimethylamine (TMA). Figure from Ref. [25].

The mild effects of ssDNA sequence 1 on the conductance of carbon nanotubes is illustrated in Fig. 8.16, which plots the measured transfer characteristics of the nanotube device before and after functionalization. A small reduction in the ON state conductance is observed, accompanied by a shift of the threshold voltage from 10 V to 5 V. The threshold voltage shift may be an indication that ssDNA of this sequence transfers electrons to the carbon nanotube. Exposure of the ssDNA/nanotube hybrid device to trimethylamine (TMA) by flowing a mixture of air/analyte over the device shows a very strong shift of the threshold voltage by about 10 V. We note however that the bare carbon nanotube also shows a response to TMA, although as discussed further below, the response is not as strong. Since TMA has a large pK value of 9.8, it is proposed that TMA is protonated by residual water. This would be consistent with the presence of residual water on bare nanotubes, and is also expected after exposure to the ssDNA diluted in distilled water. The presence of the ssDNA may enhance the protonation of TMA. Thus, the change in nanotube conductivity, through the shift of the threshold voltage, most likely arises due to a charge transfer effect.

The enhanced response in the presence of ssDNA is not unique to TMA. Indeed, an enhanced response has been seen for several chemicals. In the case of methanol, the bare nanotube device shows essentially no response, but a strong decrease of the current is observed when the nanotube is functionalized with sequence 2; in addition, the response is reversible and repeatable. Similar experiments with TMA show that the presence of sequence 2 also enhances the response. The magnitude and sign of the current change is specific to each analyte: a nanotube functionalized with sequence 1 shows a current reduction for methanol, and a current increase for propionic acid (Fig. 8.17).

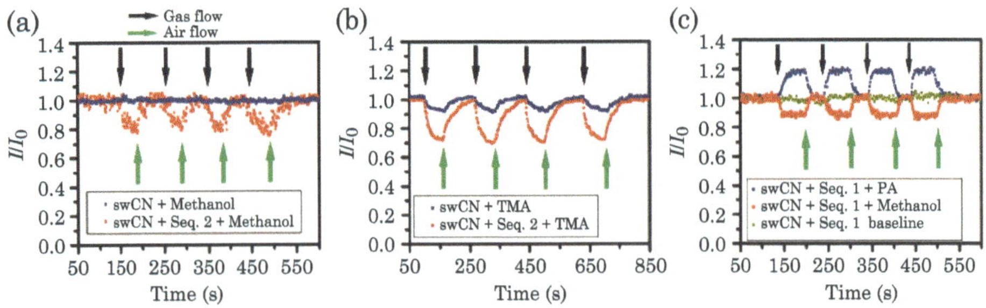

Figure 8.17 Current change of carbon nanotube field-effect transistor when exposed to methanol, trimethyamine (TMA), and propionic acid (PA), with and without DNA functionalization of the nanotube. Figure from Ref. [25].

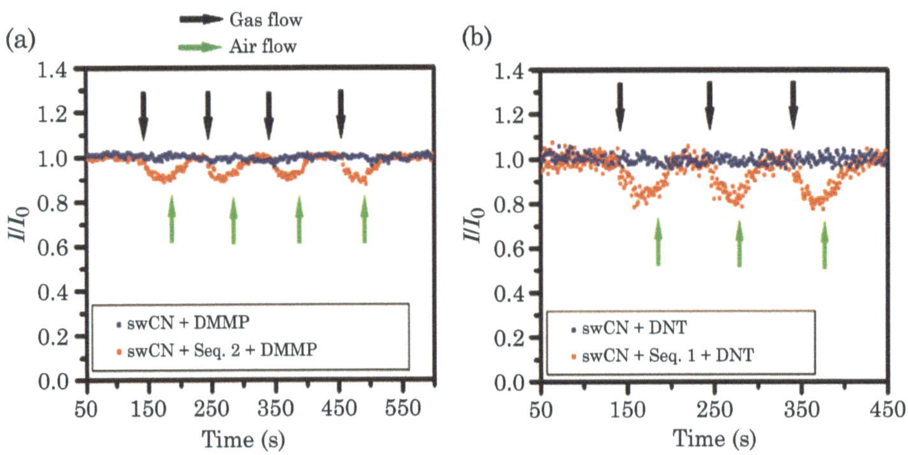

Figure 8.18 Change in current through a carbon nanotube field-effect transistor when exposed to dimethyl methylphosphonate (DMMP) and dinitrotoluene (DNT). The bare nanotube device shows no response to either agent, but functionalization with different ssDNA sequences gives a response for each gas. Figure from Ref. [25].

The nanotube/ssDNA hybrids are also useful for the detection of chemicals used as simulants of explosives and nerve gas. (Detecting explosives is difficult because their low vapor pressure requires highly sensitive sensors.) Fig. 8.18 shows the response of the nanotube device to dimethyl methylphosphonate (DMMP) and dinitrotoluene (DNT), simulants for explosives and nerve gases, respectively. As can be seen in the figure, the functionalized nanotubes show a strong response to each of these agents, while the bare nanotubes show no response. Moreover, the response is sequence-specific: at a concentration of 25 parts per million, DMMP gives a 7% reduction of the current with sequence 1, and a 14% reduction with sequence 2. Similar results are obtained with DNT.

Other experiments with DNA-functionalized carbon nanotubes have focused on carbon nanotube network transistors [26], with the aim of detecting single-nucleotide mismatches in target DNA sequences. To demonstrate this ability, a combination of fluorescent imaging and electronic transport measurements were first performed to demonstrate the functionalization of the carbon nanotubes and the preferential attachment of a matched DNA sequence. Fig. 8.19 shows a series of fluorescent microscopy images taken after functionalization with the capture DNA and after subsequent exposure to the target DNA. First, a DNA sequence of 5'-CCT AAT AAC AAT-3' labeled with a small fluorescent molecule was applied to the carbon nanotube network followed by a thorough washing to remove unbound DNA. The image in Fig. 8.19 (a) indicates that the DNA sequence attached predominantly to the carbon nanotubes (including over the electrodes), but not on the silicon dioxide substrate. The unlabelled capture sequence does not show any fluorescence (Fig. 8.19 (b)). Exposure of the carbon nanotube network functionalized with the unlabelled sequence to the fluorescent-labelled complementary DNA sequence shows that the target DNA binds primarily to the carbon nanotube network (Fig. 8.19 (c)). When a different capture probe with little homology with the target probe is used, little fluorescence is observed from the carbon nanotube network (Fig. 8.19 (e)). These results indicate the ability to functionalize the nanotube network, and the strong recognition capabilities of the ssDNA on the nanotube surface.

The binding of ssDNA to its complementary strand can be observed by monitoring the changes in the transfer characteristics of the carbon nanotube network transistor. When functionalized with the capture probe, it is found that the ON state conductance is somewhat decreased from its value for the bare network, and that the threshold voltage is reduced by 2–5 V, consistent with the results presented earlier for the single nanotube devices (Fig. 8.20 (a)). While exposure of the device to the noncomplementary ssDNA makes little difference in the transfer characteristics (Fig. 8.20 (b)), exposure to the complementary strand gives a further shift of the threshold voltage by 1–2 V. These experiments demonstrate the ability to discriminate between two different DNA sequences using a rapid, label-free technique. Such approaches are currently being explored to enable the rapid diagnostic of diseases. As an example of the applicability of the carbon nanotube sensor for this purpose, it was utilized to detect single-nucleotide mutation in the HFE gene, which is responsible for hemochromatosis [26]. In these experiments, the nanotubes were functionalized with 17-mer sequences of wild-type and mutant ssDNA which differ only in a single nucleotide. Hybridization with a 51-mer sequence containing the target sequence complementary to the wild-type was conducted on devices with only the wild-type or the mutant capture probes. Measurement of the transfer characteristics (Fig. 8.21) and fluorescence microscopy after washing show that little wild-type hybridization occurred on the mutant-functionalized nanotubes, while significant hybridization occurred on the wild-type-functionalized nanotubes, leading to a reduction of the ON state conductance and a reduction of the threshold voltage.

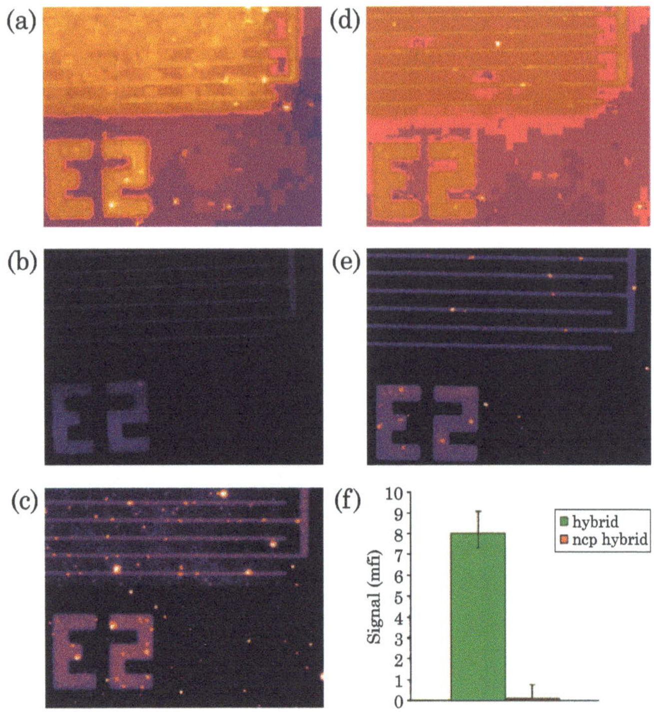

Figure 8.19 Fluorescence microscopy images of networks of carbon nanotubes between interdigitated electrodes after DNA incubation for one hour and removal of the unbound DNA oligomers. Image (a) was taken after incubation with 12-mer capture probes that were labeled with a fluorescent dye. Image (b) is for the same functionalization but without the dye. (c) Image taken after functionalization with the unlabelled capture sequence and exposure to the target sequence labeled with a fluorescent dye. (d) Image of the nanotube network functionalized with a different fluorescent-labelled DNA sequence. (e) A device functionalized with the nonlabelled sequence of (d) has very low binding affinity to a mismatched fluorescent-labelled sequence. The graph in (f) shows the difference in the fluorescence intensity for capture of the complementary and noncomplementary strands. Figure from Ref. [26].

Many aspects of sensing with DNA-functionalized carbon nanotubes remain to be explored. The properties of the nanotube/ssDNA hybrid itself require further study to better understand the structure of the ssDNA, its impact on the nanotube electronic structure, the role of buffers, the importance of salts in the solution [26], the differences between solution and on-chip functionalization, etc. Modeling work has been performed to address some of these issues [28,29]. Recent experiments [27] have proposed that DNA hydridization at the contacts is the dominant sensing mechanism, with the changes in the transfer characteristics due to an increase of the Schottky barrier at the contact. In these experiments, Au was utilized as the contact material, and the increase in the Schottky barrier is believed to originate from a reduction of the Au work function upon hybridization on the Au surface. This conclusion is supported by

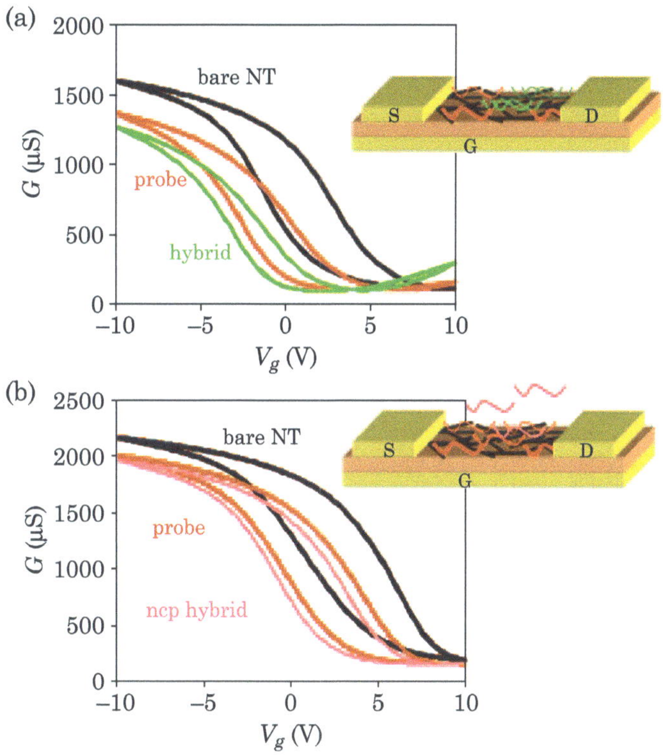

Figure 8.20 Transfer characteristics of a carbon nanotube network field-effect transistor when functionalized with ssDNA. (a) shows the response to the complementary strand, while (b) is the response to a mismatched strand. Figure from Ref. [26].

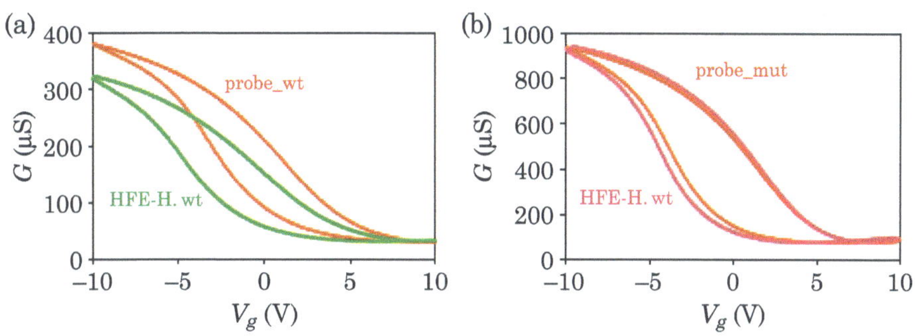

Figure 8.21 Response of ssDNA-functionalized carbon nanotube network transistor showing single-nucleotide discrimination between the wild-type (wt) and mutated (mut) forms of a target DNA sequence. Figure from Ref. [26].

quartz crystal microbalance and x-ray photoelectron spectroscopy on ssDNA-functionalized nanotubes dispersed on a substrate, which showed little binding of the complementary strand.

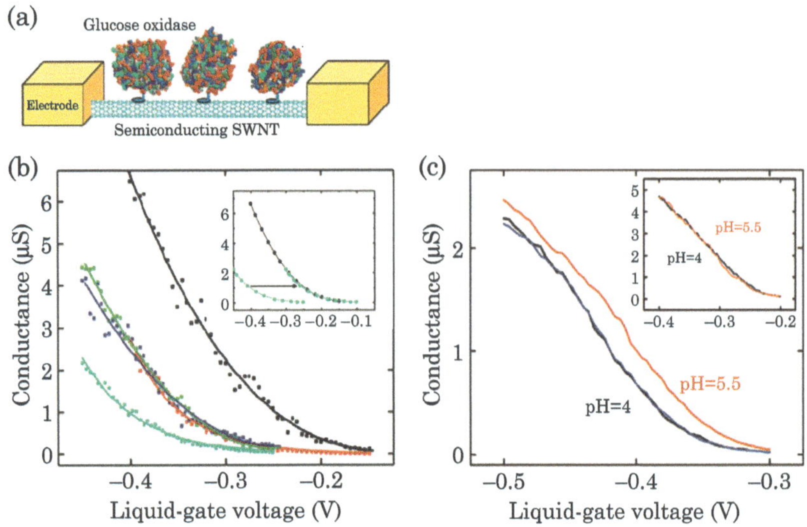

Figure 8.22 (a) Sketch of a carbon nanotube functionalized with the enzyme glucose oxidase. (b) The sequences of data from top to bottom correspond to the bare nanotube, the nanotube after soaking in DMF solution for 2 and 4 hours, after 2 hours in DMF with the linking molecule, and finally, after immobilization of glucose oxidase. The inset shows the collapse of the data for the bare nanotube and the enzyme-coated nanotube when the gate voltage is shifted. (c) Sensitivity of the enzyme-coated device to pH. Inset shows the response of the nanotube before immobilization of glucose oxidase. Figures adapted from Ref. [22].

8.3.2 Enzyme Coatings

Earlier in this chapter, we discussed liquid-gating of carbon nanotube transistors. By taking advantage of this approach, research has shown that enzyme-coated carbon nanotubes can serve as pH sensors, as well as to detect enzymatic activity [22]. As shown in Fig. 8.22, enzymes can be immobilized on carbon nanotubes through a linker molecule that binds noncovalently to the surface of carbon nanotubes. A particular example is the enzyme glucose oxidase which catalyses the oxidation of glucose. Attachment of this enzyme to the surface of carbon nanotubes results in a strong decrease of the conductance of the nanotube device (Fig. 8.22 (b)) when the conductance is measured in deionized water. Thus, this device is a good sensor for glucose oxidase in liquid. It was originally proposed that the sensing mechanism is the disruption of the double-layer near the nanotube and the decrease of the capacitance of the liquid gate [22]. However, rescaling of the gate voltage to represent this effect does not make the current–voltage curves overlap. Instead, a simple shift of the gate voltage gives excellent data overlap (inset in Fig. 8.22 (b)). Thus, it appears that charge transfer, as discussed in Section 8.1.1 is a likely mechanism to explain the experimental data.

The enzyme-functionalized carbon nanotube field-effect transistors can be utilized to perform sensing in the liquid environment. For example, these devices

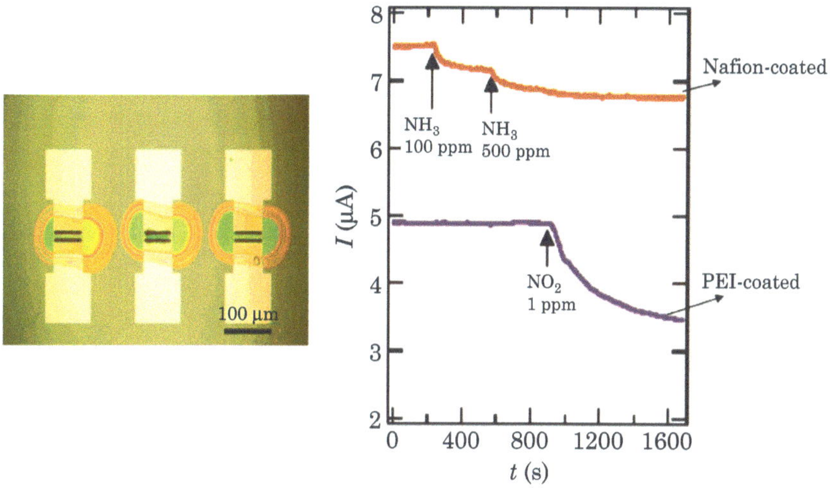

Figure 8.23 Multiplex detection of ammonia and nitrogen dioxide with carbon nanotubes coated with polymeric blocking layers. The optical image on the left shows three nanotube devices after each device is microspotted with droplets of different polymer solutions. The three devices are then simultaneously exposed to ammonia and nitrogen dioxide vapors. The right graph shows that the Nafion-coated nanotube device responds to ammonia but not to nitrogen dioxide, with the reverse response for the PEI-coated devices. Figures from Ref. [3].

are sensitive to the solution pH, with a decrease of the pH leading to a decrease of the conductance (Fig. 8.22 (c)). The impact of decreasing the pH on the transfer characteristics is a shift of the threshold voltage to more negative values, and it is proposed that charged groups on the glucose oxidase become less negatively charged in decreasing pH [22]. It is also interesting to note that addition of glucose to the solution produced an increase of the device conductance, allowing the real-time detection of enzymatic activity.

8.3.3 Polymer Coatings

One strategy to impart specificity to carbon nanotube sensors is to cover the nanotube device with a polymeric layer that blocks most analytes except the targeted ones [3,30]. This strategy has been employed to achieve multiplex sensing of ammonia and nitrogen dioxide on a substrate containing multiple polymer-coated devices [3]. An optical image of one of these devices is shown in Fig. 8.23. There, three different carbon nanotube field-effect transistors have been microspotted with two different polymers: polyethyleneimine (PEI) and Nafion (a polymeric perfluorinated sulfonic acid ionomer). Ammonia has a low affinity to PEI because of the high density of amines. Likewise, Nafion blocks species that do not contain –OH groups, such as nitrogen dioxide, but is permeable to molecules such as NH_3 which forms NH_4OH. These attributes

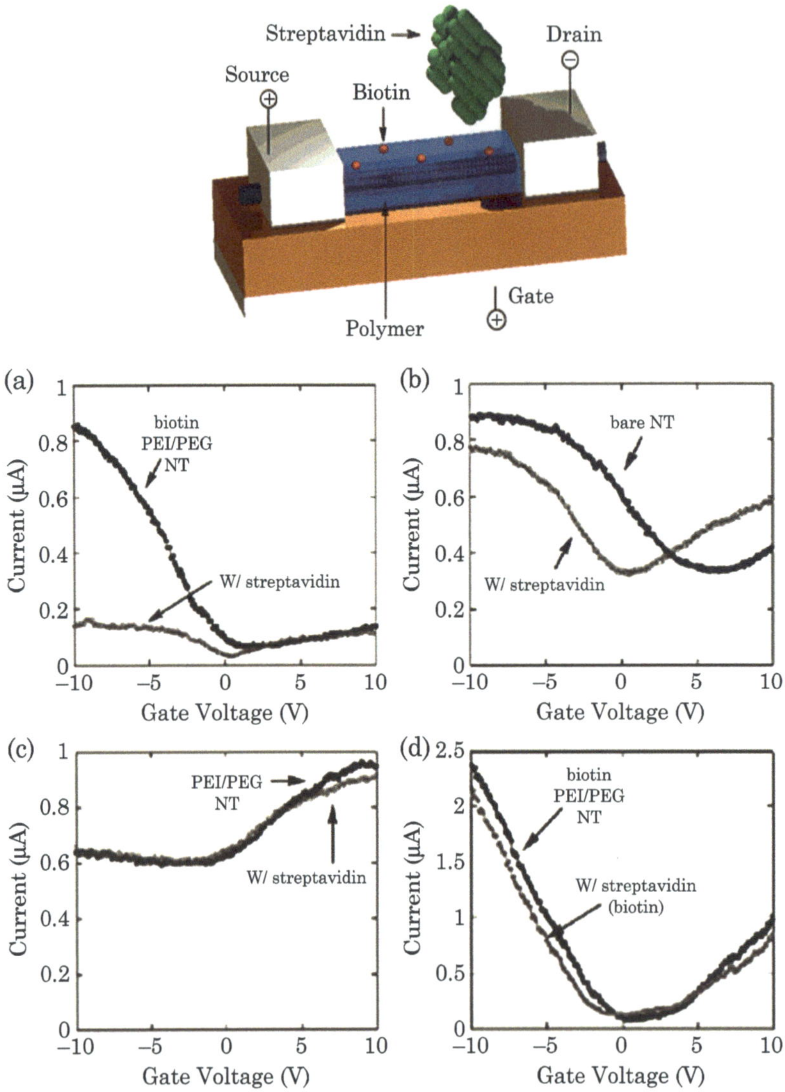

Figure 8.24 Schematic of a carbon nanotube field-effect transistor with a functionalized polymeric blocking layer. The bottom panels show the measured transfer characteristics of several nanotube devices with different functionalization before and after exposure to streptavidin. Figures from Ref. [30].

were combined to selectively detect ammonia and nitrogen dioxide on separate devices. Fig. 8.23 shows a time trace of the current flowing through two of the devices, one coated with PEI and the other coated with Nafion. Simultaneous exposure of both devices to NH_3 and NO_2 indicates that the PEI-coated device responds to NO_2 but not to NH_3; the reverse situation is observed for the Nafion-coated device.

This strategy can be taken a step further by adding functionalization to the blocking layer [21,30]. In these cases, experiments to detect protein binding on carbon nanotubes indicated nonspecific binding to the carbon nanotubes. To prevent this nonspecific binding [30], a coating of polyethyleneglycol (PEG) and PEI was deposited on the nanotubes by submerging them in solution. The nanotube devices exhibited marked changes in their transfer characteristics after functionalization with these polymers. However, the initial *p*-type behavior can be recovered by adding a biotin functionalization to the polymeric layer. Biotin is a receptor molecule for the protein streptavidin, and the biotin–streptavidin system is often used as a model system for studies of protein binding. Functionalizing the polymeric layer with biotin allows streptavidin to bind covalently to the polymeric layer without disturbing the nanotube electronic properties. Demonstration of the viability of this approach is shown in Fig. 8.24. In these figures, the transfer characteristics of the carbon nanotube network transistor is shown for different functionalizations upon exposure to streptavidin. Panel (b) in this figure indicates that the bare nanotube device shows a response to streptavidin consisting of a reduction of the ON current and a shift of the threshold voltage. This type of behavior has been discussed above and is not specific to streptavidin. Coating the nanotube device with PEI/PEG completely blocks streptavidin (panel (c)) and the device shows no response. However, when the PEI/PEG is functionalized with biotin, a large decrease of the current is observed, as indicated in Fig. 8.24 (a). The actual mechanism that causes this strong sensing response is unclear. Increased scattering could certainly cause such an overall decrease of the current. However, the impact of the contacts has not been fully explored, and the modulation of Schottky barriers at the contacts could also play a role.

References

1. J. Kong, N.R. Franklin, C. Zhou, M.G. Chapline, S. Peng, K. Cho and H. Dai, "Nanotube molecular wires as chemical sensors", *Science*, Vol. 287, p. 622, 2000.
2. A.A. Talin, L.L. Hunter, F. Léonard and B. Rokad, "Large area, dense silicon nanowire array chemical sensors", *Appl. Phys. Lett.*, Vol. 89, p. 153102, 2006.
3. P. Qi, O. Vermesh, M. Grecu, A. Javey, Q. Wang and H. Dai, "Toward large arrays of multiplex functionalized carbon nanotube sensors for highly selective and selective molecular detection", *Nano Lett.*, Vol. 3, p. 347, 2003.
4. J.M. Blakeley and J.C. Shelton, J.M. Blakeley, Ed., *Surface Physics of Materials,* Vol. 9, Academic, New York, p. 208, 1975.
5. P.G. Collins, K. Bradley, M. Ishigami and A. Zettl, "Extreme oxygen sensitivity of electronic properties of carbon nanotubes", *Science*, Vol. 287, p. 1801, 2000.
6. M. Grujicic, G. Cao and R. Singh, "The effect of topological defects and oxygen adsorption on the electronic transport properties of single-walled carbon-nanotubes", *Appl. Surf. Sci.*, Vol. 211, p. 166, 2003.
7. A. Tchernatinsky, S. Desai, G.U. Sumanasekera, C.S. Jayanthi, S.Y. Wu, B. Nagabhirava and B. Alphenaar, "Adsorption of oxygen molecules on individual single-wall carbon nanotubes", *J. Appl. Phys.*, Vol. 99, p. 034306, 2006.

8. V. Derycke, R. Martel, J. Appenzeller and Ph. Avouris, "Controlling doping and carrier injection in carbon nanotube transistors", *Appl. Phys. Lett.*, Vol. 80, p. 2773, 2002.
9. A. Star, T.-R. Han, J.-C.P. Gabriel, K. Bradley and G. Grüner, "Interaction of aromatic compounds with carbon nanotubes: correlations to the Hammett parameter of the substituent and measured carbon nanotube FET response", *Nano Lett.*, Vol. 3, p. 1421, 2003.
10. L.P. Hammett, "The effect of structure upon the reactions of organic compounds. Benzene derivatives", *J. Am. Chem. Soc.*, Vol. 59, p. 96, 1937.
11. J. Kong, M.G. Chapline and H. Dai, "Functionalized carbon nanotubes for molecular hydrogen sensors", *Adv. Mater.*, Vol. 13, p. 1384, 2001.
12. S. Latil, S. Roche and J.-C. Charlier, "Electronic transport in carbon nanotubes with random coverage of physisorbed molecules", *Nano Lett.*, Vol. 5, p. 2216, 2005.
13. X. Cui, M. Freitag, R. Martel, L. Brus and Ph. Avouris, "Controlling energy-level alignments at carbon nanotube/Au contacts", *Nano Lett.*, Vol. 3, p. 783, 2003.
14. J. Zhang, A. Boyd, A. Tselev, M. Paranjape and P. Barbara, "Mechanism of NO_2 detection in carbon nanotube field effect transistor chemical sensors", *Appl. Phys. Lett.*, Vol. 88, p. 123112, 2006.
15. R.J. Chen, H.C. Choi, S. Bangsaruntip, E. Yenilmez, X. Tang, Q. Wang, Y.-L. Chang and H. Dai, "Investigation of the mechanisms of electronic sensing of protein adsorption on carbon nanotube devices", *J. Am. Chem. Soc.*, Vol. 126, p. 1563, 2004.
16. W.E. Boyce and R.C. DiPrima, *Elementary differential equations and boundary value problems*, John Wiley and Sons, New York, 1986.
17. S.M. Sze, *Physics of semiconductor devices*, John Wiley & Sons, New York, 1981.
18. E.S. Snow, F.K. Perkins, E.J. Houser, S.C. Badescu and T.L. Reinecke, "Chemical detection with a single-walled carbon nanotube capacitor", *Science*, Vol. 307, p. 1942, 2005.
19. G. Esen, M.S. Fuhrer, M. Ishigami and E.D. Williams, "Transmission line impedance of carbon nanotube thin films for chemical sensing", *Appl. Phys. Lett.*, Vol. 90, p. 123510, 2007.
20. E.S. Snow and F.K. Perkins, "Capacitance and conductance of single-walled carbon nanotubes in the presence of chemical vapors", *Nano Lett.*, Vol. 5, p. 2414, 2005.
21. R.J. Chen, S. Bangsaruntip, K.A. Drouvalakis, N. Wong Shi Kam, M. Shim, Y. Li, W. Kim, P.J. Utz and H. Dai, "Noncovalent functionalization of carbon nanotubes for highly specific electronic biosensors", *Proc. Nat. Acad. Sci.*, Vol. 100, p. 4984, 2003.
22. K. Besteman, J-O. Lee, F.G.M. Wiertz, H.A. Heering and C. Dekker, "Enzyme-coated carbon nanotubes as single-molecule biosensors", *Nano Lett.*, Vol. 3, p. 727, 2003.
23. M. Zheng, A. Jagota, M.S. Strano, A.P. Santos, P. Barone, S.G. Chou, B.A. Diner, M.S. Dresselhaus, R.S. McLean, G.B. Onoa, G.G. Samsonidze, E.D. Semke, M. Usrey and D.J. Walls, "Structure-based carbon nanotube sorting by sequence-dependent DNA assembly", *Science*, Vol. 302, p. 1545, 2003.
24. A.A. Talin, P.M. Dentinger, F.E. Jones, S. Pathak, L. Hunter, F. Léonard and A.M. Morales, "Assembly and electrical characterization of DNA-wrapped carbon nanotube devices", *J. Vac. Sci. Technol.* B, Vol. 22, p. 3107, 2004.
25. C. Staii, A.T. Johnson, M. Chen and A. Gelperin, "DNA-decorated carbon nanotubes for chemical sensing", *Nano Lett.*, Vol. 5, p. 1774, 2005.
26. A. Star, E. Tu, J. Niemann, J.-C.P. Gabriel, C.S. Joiner and C. Valcke, "Label-free detection of DNA hybridization using carbon nanotube network field-effect transistors", *P. Nat. Acad. Sci.*, Vol. 103, p. 921, 2006.
27. X. Tang, S. Bansaruntip, N. Nakayama, E. Yenilmez, Y.-I. Chang and Q. Wang, "Carbon nanotube DNA sensor and sensing mechanism", *Nano Lett.*, Vol. 6, p. 1632, 2006.
28. G. Lu, P. Maragakis and E. Kaxiras, "Carbon nanotube interaction with DNA", *Nano Lett.*, Vol. 5, p. 897, 2005.

29. A.N. Enyashin, S. Gremming and G. Seifert, "DNA-wrapped carbon nanotubes", *Nanotechnology*, Vol. 18, p. 245702, 2007.
30. A. Star, J.-C.P. Gabriel, K.l Bradley and G. Grüner, "Electronic detection of specific protein binding using nanotube FET devices", *Nano Lett.*, Vol. 3, p. 459, 2003.

Index

MICRO & NANO TECHNOLOGIES

PUBLISHED 2008

Microdrops and Digital Microfluidics · Jean Berthier · 978-0-8155-1544-9

Micromixers: Fundamentals, Design and Fabrication · Nam-Trung Nguyen · 978-0-8155-1543-2

Fabrication and Design of Resonant Microdevices · Behraad Bahreyni · 978-0-8155-1577-7

Physics of Carbon Nanotube Devices · François Léonard · 978-0-8155-1573-9

FORTHCOMING 2008-2009

Hot Embossing: Theory and Technology of Microreplication · Matthias Worgull · 978-0-8155-1579-1

Introduction to Quantum Information Processing (QIP) · Timothy P. Spiller and William J. Munro · 978-0-8155-1575-3

Nanotechnology Applications for Clean Water · Edited by Mamadou Diallo, Jeremiah Duncan, Nora Savage, Anita Street, Richard Sustich · 978-0-8155-1578-4

Micro-machining Using Electrochemical Discharge Phenomenon: Fundamentals and Application of Spark Assisted Chemical Engraving · Rolf Wüthrich · 978-0-8155-1587-6

Industrial Micro and Nano Fabrication · J.G.E. Gardeniers and R. Luttge · 978-0-8155-1582-1

Small Scale Mechanics: Principles and Applications · David Mendels · 978-0-8155-1590-6

Emerging Nanotechnologies for Manufacturing · Edited by Waqar Ahmed and M.J. Jackson · 978-0-8155-1583-8

Risk Governance of Nanotechnology: Environmental, Health and Safety Concerns About Nanotechnology and Their Implication for the Nanotechnology Industry · Edited by Steffi Friedrichs · 978-0-8155-1586-9

Handbook of MEMS Materials and Technologies · Edited by Veikko Lindroos, Markku Tilli, Ari Lehto, and Teruaki Motooka · 978-0-8155-1594-4

Micromanufacturing Engineering and Technology · Edited by Yi Qin · 978-0-8155-1545-6

OF RELATATED INTEREST

MEMS: A Practical Guide to Design, Analysis and Applications · Edited by Jan Korvink and Oliver Paul · 978-0-8155-1497-8 · 2006

Nanostructured Materials: Processing, Properties and Applications, 2nd Edition · Edited by Carl C. Koch · 978-0-8155-1534-0 · 2007

Ultrananocrystalline Diamond: Synthesis, Properties, and Applications · Edited by Olga A. Shenderova and Dieter M. Gruen · 978-0-8155-1524-1 · 2006

For the latest information on related titles visit **www.williamandrew.com/MNT/**

FSC
www.fsc.org
MIX
Papier | Fördert
gute Waldnutzung
FSC® C083411

Kolibri 360 GmbH
Ferdinand-Jühlke-Straße 7
99095 Erfurt, Deutschland
produktsicherheit@kolibri360.de